RADIO FREQUENCY

How to Find It and Fix It

Editors
Ed Hare, KA1CV
Robert Schetgen, KU7G

Production
Dianna Roy
David Pingree
Sue Fagan
Michelle Bloom
Joe Shea

Cover photos

Clockwise from upper left—

A transparent telephone shows how most phones "look" to RFI/EMI.

Computers and Amateur Radio transmitting equipment can work together. This book shows how.

Harmonic interference or fundamental overload distorts a TV image of astronaut Ron Parise, WA4SIR, from ARRL's *Ham Radio in Space* videotape. Techniques in the Televisions chapter can eliminate this type of interference.

(photos by Kirk Kleinschmidt, NTØZ)

Published by
The American Radio Relay League
Newington, CT

Foreword

This book is a major step forward in RFI/EMI literature for the layman. Most Amateur Radio operators are familiar with radio-frequency interference (RFI), but what's EMI? Electromagnetic interference (EMI) is a term that includes the RFI we have all fought for years, but it recognizes that radiation is not the only mechanism of interference. The term is broad, it encompasses more than RFI, it embodies a better understanding of reality and parallels another term: electromagnetic compatibility (EMC).

The term EMC implies recognition that electronic equipment can interact. In some cases, the interaction is so severe that the normal operation of the equipment is affected.

Solving interference problems requires cooperation. Manufacturers and amateurs are working together to establish EMI-susceptibility design guidelines, so problems are solved before the equipment leaves the factory. Other problems can be solved in the field. In nearly all EMI cases, interaction can be reduced to acceptable levels.

At the time of publication, only about 2% of electronic engineers are actively involved in EMC work, but the number is growing. As EMC technology grows the world will become a better place for all spectrum users.

In this book, we have brought together the knowledge of many EMI experts, both professional and amateur. There has been a lot of input from ARRL Field Organization volunteers who deal with EMI at the local, hands-on, level. This book presents reliable information about some previously thorny problems such as touch-controlled lamps and interference to VCRs.

This book will never be complete. You, the reader, can help with the next edition. Please share with our editors the benefit of your own experiences with RFI/EMI. Together, we can make the world a more electromagnetically compatible place—a more pleasant place for ourselves and our neighbors.

David Sumner, K1ZZ
Executive Vice President

Newington, CT
December 1991

Preface

Welcome to the world of RFI/EMI. This new approach to an old Amateur Radio problem should be very helpful to you. The chapters were each written by different authors, so you get the benefit of many viewpoints and a wide range of expertise. This also means that there is some duplication of coverage in the various chapters. In laying out a complete picture for their chapters, several of the authors include information that pertains to other RFI/EMI topics. Thus, when treating a specific RFI/EMI problem, it is a good idea to look at other chapters as well.

All RFI/EMI problems are fundamentally similar. Some device is a signal source (intentional or otherwise), some medium carries the signal to some receptor (affected device). Interference occurs because the isolation between the source and receptor is insufficient. The source may produce energy in the receptor's design bandwidth. The receptor may accept energy outside its design bandwidth. Some third device may take energy from the source and distribute it in the receptor's bandwidth.

We can eliminate interference by increasing the isolation between the interacting devices. For example, a low-pass filter on an HF transmitter can suppress harmonics on TV channels. Or, a high-pass filter on a TV protects it from HF fundamental overload. Often we can alter the medium between the source and receptor to reduce signal strength at the receptor (thereby reducing the interference). That is, if the HF signal enters the TV via the shield of a coaxial feed line, a common-mode choke can reduce the HF signal reaching the TV. In many cases a combination of approaches is needed.

Because cases are fundamentally similar, we see this scenario played over and over throughout the chapters. Realize that seemingly unrelated chapters may hold a suggestion that will help with a given problem. While the chapters are specialized, every chapter is useful; you should read them all before an RFI/EMI problem develops. That way you can benefit from the experience of every author, and you stand a better chance of quickly understanding any problem you may encounter.

If you need information immediately, read First Steps (it tells how to deal with irate complainants and get help from the ARRL Field Organization), then read EMI Fundamentals and proceed to the chapter most suited to the affected device.

Acknowledgments

Many people have contributed knowledge, encouragement and enthusiasm during the production of this book. It truly represents the collective total of amateur and professional knowledge about interference problems. This total could not have been assembled without energetic support from the Amateur Radio and Electromagnetic Compatibility (EMC) communities. It is important to recognize that support.

The names of the chapter authors appear at the beginnings of their respective chapters. In addition, several other persons have performed similar work without actually authoring a chapter. In alphabetical order, they are:

Dana Craig, KC6USX, Senior EMC Engineer
Hartley Gardner, W1OQ, ARRL Official Observer
Howard Liebman, W2QUV, ARRL NYC/LI Section Technical Coordinator
Al Marquardt, W5PXH, ARRL Technical Advisor
Jerry Meyerhoff, WA9FIY, Automotive EMC Engineer
John Norback, W6KFV, ARRL SB Section Technical Coordinator

In addition, many others have made important contributions to sidebars and additions to the chapter authors' work. In alphanumeric order, they are:

AA1AA	WM1S	N3HLU	W4TAH	N6ARE	WB7CYO	KF8NH	WØEO
N1ACB	KA1SNA	KB3JA	N4TMI	KN6B	WA7CYP	WD8OYG	KØGRM
N1BAQ	KA1YBC	W3JW	N4UA	W6BF	W7FF	WN8P	NØHFD
WB1CEI	WA1YKL	AE3T	KB4UET	N6BWX	N7HMV	W8TR	WDØHWL
KB1CU	NP2B	WA3TQJ	KC4WN	WU6D	K7ICW	W8YZ	NØIHG
KA1CWM	N2BK	K3TX	AA4XE	K6DQ	W7JWJ	WA8ZNH	WAØITU
WJ1D	NN2C	KA3VGD	AA5AN	KH6GI	AL7MP	WA9ACI	NØKAW
N1DVJ	K2CIB	KQ3W	AA5BC	KA6GMA	WB7NXH	KB9AMM	WAØKUH
N1EXA	W2CQH	NG3Z	N5BNW	N6GZI	NW7O	K9CGD	NØLL
KY1F	WB2CUW	WD4AFY	N5DDT	WN6I	WB7OML	K9CQE	WAØNJF
W1FXQ	WA2DCI	W4BJT	WA5EUJ	K6JEY	KY7P	W9EPT	WØOGS
WA1GPO	KC2FU	K4CAV	N5GFX	WA6KLA	W7PMD	WA9FIY	WØPEA
W1HJT	W2ILP	K4CHS	W5KKD	KD6MR	K7RNZ	K9GWT	WAØQBC
N1HYC	KB2MB	K4ERO	WA5KZA	AA6NM	K7RXV	N9HHG	WØQGN
N1IEP	W2QUV	KA4FKU	N5MPN	W6OFF	WC7S	AE9I	KØWOP
K1JW	W2VDX	AA4FQ	W5PXH	W6OIZ	WS7S	WB9IVR	DL3BAA
W1JY	W2WVC	WB4GHU	W5QMJ	W6ONT	WA7SSO	K9KMR	GØCNR
NA1L	WA2YVL	WA4ITY	W5QMU	WA6QYR	K7UU	WB9MRI	GØMEG
KA1LPW	WN3A	KJ4KB	W5QX	WK6R	W7ZOI	WB9NQQ	VE3FOZ
KB1MJ	W3ABC	WA4KER	W5RIY	W6RGG	N8AGS	NK9Ø	VE3SUN
KQ1N	K3BRS	WD4KSH	WX5T	N6RJ	W8ARH	W9VE	
KA1NH	KC3BU	W4MFD	W5URI	KC6USX	K8CFU	W9WFV	
KA1OTN	N3EA	N4MOK	K5WGQ	KG6VI	WD8EBH	WØBJH	
KV1P	N3ELM	W4NTO	KD5YC	KE6XI	N8FNC	WBØBZP	
NK1P	W3FM	WD4PKZ	K5ZC	W6XM	K8GK	KØCX	
KB1RP	W3GRG	W4PSC	NK6A	N7BBW	WA8KZN	KØDAS	
K1RQG	W3HII	N4SS	W6ANQ	AA7BJ	K8LMN	NØEBH	

Each and every contributor (including any who were inadvertently omitted from these lists) deserves our gratitude. Therefore, on behalf of the ARRL and the Amateur Radio Service, the editors say "Thanks."

Table of Contents

Chapter 1

First Steps

By Ed Hare, KA1CV
Senior ARRL Laboratory Engineer

and

Robert Schetgen, KU7G
ARRL Assistant Technical Editor

Every electromagnetic interference (EMI) case involves people, and sometimes people compatibility is far more difficult to achieve than electromagnetic compatibility (EMC). You may have already been accosted by an angry neighbor with an EMI complaint. If so, take a few deep breaths, relax and read this chapter.

The first part of this chapter explains the realities of the EMI world. Next, a discussion of personal diplomacy tells how to interact successfully with the complainant. In many cases, a third party can defuse the situation. An ARRL Field Organization volunteer may be that needed third party. The chapter closes with a description of the Field Organization as it relates to EMI cases.

Effective EMI solutions require cooperation, investigation and action. "First Steps" tells how to establish the needed cooperation. The rest of the book covers the investigation and action requirements. Rest assured that this book contains the information you need to resolve the vast majority of all EMI cases satisfactorily.

Imagine the following scenario: Joe Hamm has just passed his FCC examination and awaits the arrival of his "ticket" so he can have his first amateur contact. The big day finally arrives and Joe rushes into his newly built station to make use of the privileges he has worked so hard to earn. After about an hour, he finally gets over his initial nervousness and establishes contact with another ham—in Ireland! After about 15 minutes of excited conversation, Joe hears a loud knock on his door.

When Joe opens the door he is confronted by Sam Neighbor, who tells him that he is causing interference because Sam can hear Joe on his telephone. If Joe is lucky, Sam is pleasant and understanding. If Joe is not so lucky, Sam might be a bit more forceful, spouting demands and threats of FCC (or personal) intervention. Joe, like most hams first confronted with this situation, doesn't know what to do. Is it Joe's fault? Should Joe get off the air? Who can he turn to for help? Joe is not alone. Thousands of hams and consumers have faced this situation since radio was first developed.

Hams Can Help

In this case, Sam Neighbor is actually fortunate. At least hams have some technical ability (as demonstrated by an FCC Amateur Radio license). They can help determine the causes of EMI and suggest cures.

Interference is caused by all radio services and many other unintentional sources of radio noise. If the problem were caused by a CB operator or the local fire-department transmitter, there might not be an easy source of technical help.

Even brand-new hams, like Joe, have access to the expertise of other hams, through the American Radio Relay League (ARRL) Field Organization. The ARRL Section Technical Coordinator, Technical Specialists and local RFI Committees (Radio Frequency Interference is a category of EMI) have all volunteered to help with EMI trouble. Other hams in the area may be willing to help as well.

THE EMI WORLD
Two Erroneous Beliefs

What has brought Joe and Sam to this unfortunate situation? Sometimes consumers believe that amateur stations spew out-of-band signals across the radio spectrum, and ruin reception for all around them. On the other hand, hams sometimes think consumer-equipment manufacturers deliberately leave out filters and shields needed to eliminate interference problems. Such *blame-oriented beliefs hinder the EMI-resolution process. In truth, manufacturers and hams both work diligently to prevent EMI problems.*

Present Engineering Resources

There are an estimated 300,000 electrical engineers in the United States. They are competent and dedicated people designing systems and products that must compete in a difficult marketplace. Unfortunately, most colleges don't offer courses in electromagnetic compatibility (EMC).

These engineers are asked to design systems that are compatible with other systems (including radio transmitters). The lack of EMC training and experience, coupled with the ever-increasing complexity and scope of electronic systems, can result in unexpected problems.

Of these 300,000 engineers, perhaps 6,000 of them are actively engaged in EMC-related engineering. With so much engineering to be done, and such a relatively small percentage of engineers actually familiar with EMC issues, the potential for incompatibility is fairly large. This is a worldwide problem, as demonstrated by a QSL card from Germany (in Fig 1). EMI or TVI (television interference) obviously affects amateurs all over the world.

Increased Electronic Population

In addition, the electronics revolution has resulted in the proliferation of all sorts of home electronic devices. Twenty years ago, there were only TVs, telephones and various types of audio amplifiers. The opportunity for incompatibility

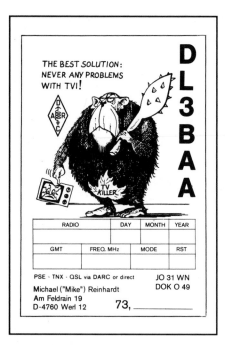

Fig 1—Apparently, German amateurs can call on some *special* help in EMI cases.

was small, and the circuitry was unsophisticated (by today's standards).

The 1980s and '90s brought many new electronic technologies and associated new equipment. For example, one of the ARRL Lab Engineers recently saw an advertisement for a computer-controlled ac-power outlet strip. He jokingly noted that now a station might cause interference to a neighbor's extension cord. It was a joke, but it could also be reality! The use of microprocessors in such devices as toasters and set-back thermostats has multiplied the interference potential many times.

Existing Standards

Amateur operators are self-trained to understand, and comply with, FCC requirements for spectral purity. The FCC limits how much energy a transmitter can radiate outside its intended operating band (on frequencies allocated to other users). This energy is called "spurious emissions" (spurious means unwanted), which hams sometimes call "spurs."

The manufacturers of amateur equipment are aware of the FCC requirements. Commercially built amateur equipment must meet the FCC criteria ("type acceptance") before it may be sold. All modern amateur equipment meets these federal standards. The ARRL regularly tests sample amateur transmitters as part of

their Product Review process and advertising-acceptance policies. This helps ensure that modern amateur transmitters do not generate interference.

Consumer electronics manufacturers have EMI-rejection programs and standards in place. They voluntarily design equipment to reject out-of-band signals. Groups like the Electronic Industries Association (EIA) and Institute of Electrical and Electronic Engineers (IEEE) are constantly working to improve designs and technology to increase the performance of consumer electronic equipment. EMI problems are not usually caused by malice on the part of the ham or the equipment manufacturer.

Manufacturers Who Don't Comply with EMI Design Standards

What might have gone wrong? Not all models from all manufacturers meet modern EMI design standards. Some (very few) manufacturers do not comply. (This is becoming less of a problem than it was in years past.)

FCC Clear on Audio Devices

The FCC places responsibility for interference to telephones and other audio devices. The FCC *Interference Handbook* (1990 edition) is clear: telephones or other audio devices that pick up radio signals *improperly function* as radio receivers. Contact the manufacturer or seller with any problems.

Who Is Responsible?

The problems that are part of the design and manufacture of consumer elec-

tronics equipment can also affect Amateur Radio equipment. Design problems or component failure can cause even modern amateur transmitters to generate spurious signals. Older amateur equipment (manufactured prior to 1977) did not have the stringent spectral purity requirements that are imposed on new amateur equipment. Home-constructed amateur equipment can also have problems.

Interference to other radio services such as television or radio broadcast can be caused by proximity, improper design or component failure in either the affected consumer radio (television) equipment or the transmitter. *Only a thorough technical investigation can determine the true cause of an EMI problem.*

Where To Turn

So where should Joe Hamm and Sam Neighbor turn for help? What can be done to ensure a happy ending to this situation? *First accept that most EMI problems can be cured!* The application of education, personal diplomacy, manufacturer contact and technical solutions (such as external filters and shields) can cure most EMI cases.

EDUCATION

Joe Hamm took the first step toward solving this EMI problem before he ever got on the air. The material and books that Joe studied to get his license discussed the concepts of interference and some of the basic remedies.

Some EMI cases call for more information than is presented in license manuals. Amateur Radio magazines have covered the subject since the 1920s. This book explains in detail how basic EMI problems occur and what can be done to cure them. Even so, EMI can present complex problems, and not all of these problems have standard solutions. Learn the fundamentals of EMI. Study troubleshooting techniques and the application of technical solutions. Learn all that you can, and you should understand nearly every situation that you encounter.

PERSONAL DIPLOMACY

Many ARRL Field Organization volunteers and EMI professionals provided input for this book. Many agree that personal diplomacy is an important factor in the solution of EMI problems. This applies to *all* affected parties. In some of the less-successful cases (in the ARRL HQ files), poor (or hostile) communication often resulted in unsatisfactory resolutions. Both hams and their neighbors have been guilty of bad attitudes, hostile conversations, and sometimes just plain bad manners.

Don't Get Off On The Wrong Foot!

The first contact between a ham and a neighbor with an EMI complaint can set the tone of all future relations between the parties. If a neighbor approaches with threats, demands and hostility, human nature urges the ham to respond in kind. On the other hand, if the ham reacts to a complaint by spouting off about "right to operate" and doesn't offer either sympathy or assistance, the neighbor may refuse to cooperate in any future investigation that may be required. In such cases the FCC often serves as a mediator rather than a regulatory agency. Although the FCC Field Offices have solved a lot of EMI cases, mediation is not their purpose. The process is often slow, and the FCC solution may please neither the ham nor the neighbor.

When The Complainant Is Angry—

By the time the complainant is angry, he has probably decided that *you* are the cause of his problem. He may also feel that your actions are imposing on him, and that you simply don't care. To sum up, he may imagine you as a selfish, uncaring, obnoxious troublemaker.

If things have already gotten out of hand, all parties may have contributed

some of the ill will that led to the impasse. If both parties want a satisfactory resolution (and they usually do), they must put aside resentment, pride and bad feelings and get back to the business of locating a solution. Apologies never hurt anyone!

Why should you help? Consider for a moment that national governments extend to amateurs the *privilege* to operate in valuable radio spectrum. In a metropolitan area, amateur frequencies can have a commercial value of millions of dollars per MHz. The governments of the world

have extended these privileges to amateurs because the world benefits from our existence. In addition to the emergency communications services often provided by amateurs, the world gains a reservoir of self-trained radio operators, skilled in operating practice and electronics technology. The amateur station may or may not be at fault in any particular case of interference (only a technical investigation can determine that for certain), but the amateur is *involved* in the problem. There is no better place to apply your technical skill, which has been gained

---

Table 1
The Facts of EMI

1) All licensed transmitters must meet specifications set by the FCC. Those specifications are set in order to prevent interference to other communications services.
2) Transmitter owners are responsible for the proper operation of their transmitters and conformance to the FCC regulations.
3) Transmitter owners must reduce spurious (unwanted) emissions to whatever level is necessary to prevent interference to other *communications* services. (This provision does *not* apply to fringe-area TV reception.)
4) Transmitter (including Amateur Radio) operators are obligated to conform to FCC regulations (items 1 through 3). Transmitter operators are not obligated to help consumers with EMI complaints that do not involve their transmissions (although they may *elect* to do so).
5) FCC considers audio equipment (telephones, turntables, alarm systems and so on) that receives EMI to be *improperly functioning* as a radio receiver. Such improper function is a design inadequacy that should be corrected by the manufacturer. The FCC recommends that owners return such equipment (to the seller or manufacturer) as defective.
6) The EMI susceptibility of consumer electronic equipment is limited only by the manufacturers' voluntary compliance with committee developed standards.
    The voluntary standards (and therefore manufacturers' designs of consumer equipment) do not provide for equipment to operate in close proximity to strong communications transmitters. Transmitter operators are not responsible for EMI in such situations.
7) In general, all equipment owners are responsible for the proper operation of their equipment. *If the equipment is operating as intended and EMI persists, it is the responsibility of the equipment owner to modify the equipment as necessary in order to achieve the owner's operating goals.*
    Example 1: EMI occurs as a result of the proper and legal operation of a licensed transmitter. The owner of the equipment *receiving* the EMI is responsible for corrective measures, which may include (but are not limited to) added filters, shields and equipment modifications by qualified service personnel.
    Example 2: Consumer equipment, such as a scanner, accepts RF energy from a nearby transmitter and produces EMI. The equipment owner is required to eliminate the EMI or remove the equipment from service.
    Example 3: Amateur Radio transmissions prevent reception of "fringe area" TV stations. Since the TV receiver is in a fringe area (not in the intended TV-coverage range), the FCC does not protect the user. The user may either accept the interference or modify the TV system as needed to receive the desired channel (if possible).

INTERFERENCE CLASSIFICATION

From the information you've read so far in this book, you might correctly assume that a number of ARRL volunteers are at your beck and call for many types of technical difficulties that you might encounter. For them to best serve you, do some homework before calling. For example, are you certain that the interference is the sort best handled by a technical volunteer? There are different kinds of interference, and different ARRL Leadership Officials to deal with them. Let's talk about some of these.

Amateur-To-Amateur Interference

Let's assume that you're already clear on amateur-to-nonamateur interference; that is handled by your Technical Coordinator (TC). What about the other kinds of interference? There are two general areas of "other" interference. They are *amateur-to-amateur* and *nonamateur-to-amateur*. Does that sound simple enough? Often it is not. Suppose, for example, that you're operating on your favorite 40-meter frequency when you are suddenly plagued by a carrier. Without the appropriate skills and equipment, it is not simple to determine whether it is an amateur or a nonamateur source.

Eventually, let's say that you determine the source of the interference to be amateur. Now you must define it even more closely:

Inadvertent Interference

We're probably all guilty of such interference at one time or another. Because of inattention or some other reason, the offender is unaware that he is creating interference. He may not be able to hear the affected station. With repeaters, a user may simultaneously key more than one. (This is very common in large metropolitan areas.) The inadvertent or unintentional offender may be a new amateur who has not yet learned good amateur operating procedures. In any case, inadvertent or unintentional interference does not represent a major problem to the Amateur Radio Service. Education seems to be the best and most logical solution.

Careless Interference

The second kind of amateur-to-amateur interference is careless interference. Here, "careless" means intentional but not premeditated or recurring. While this type of interference is deliberate, it is usually short-lived and caused by temporary eruptions of temper. A person might get "hot under the collar" while operating and tell someone off—or refuse to relinquish a frequency when it is the proper thing to do. When the individual cools off and thinks about his actions, he is likely to be a bit ashamed of himself. The person who commits this interference does not intend to repeat it. Although the occurrence is unfortunate, and the perpetrator might be prone to serious FCC enforcement actions, the Amateur Radio Service generally does not experience repeated interference from such individuals.

Harassment

The third kind of interference is classified as harassment. This is a more serious form of intentional interference. A harassing offender is likely to do anything (short of the destruction of life and property) to make life miserable for his victims. While inadvertent and careless interference is usually isolated and of short duration, harassment is a long-term program. It ranges from a "kerchunker" on a repeater to a jammer who uses foul language or any other means to disrupt the operation of others.

Malicious Interference

Although we frequently think of anything that upsets us from license study (and perhaps years of experimentation).

At this point, it is very important to build a foundation of cooperation so that the EMI problem can be resolved. Since the complainant is already angry, it is doubly important that you keep calm. If you cannot do so, contact one of the volunteers listed later in this chapter. Sometimes a third party can build a bridge between warring neighbors.

If you are willing to build that bridge, begin by expressing sympathy, understanding and concern. You have an equal desire for a satisfactory solution. The simple truth is that you both own equipment that was purchased, and is operated, with the best of intentions. Some of that equipment interacts with some other equipment in an undesired fashion. It is possible to reduce that interaction, but teamwork is needed. *Together*, you can determine how the EMI is happening and select the most practical remedy.

Communicate that as a transmitter owner, you bear limited responsibility and are willing to make any needed corrections to your station. *The presence of EMI, however, does not necessarily indicate equipment malfunction*. If *any* equipment is malfunctioning, the owner of that equipment is responsible for its repair. (Table 1 may help with this.) It may well be that your transmitter plays no role in the EMI. The complainant's equipment may not interact with your transmitter, but with some other source—possibly in his own house. Unfortunately, once he perceives your station as the problem, you may be forced to locate the actual source in order to convince him you are innocent.

Good EMI-problem diplomacy removes the need to determine who is at fault. Positive conflict resolution requires that both parties relinquish adversarial roles and realize that they share a common problem. Working *together* resolves the conflict to the benefit of both parties.

Neighbors usually cooperate if you offer assurance that solutions are possible and outline how you can help. If you are willing to locate solutions to the problem, the neighbor should be willing to put those solutions into place.

as being "malicious," only a tiny amount of interference on the amateur bands is truly malicious. A maliciously interfering offender *intends to damage* people or things. Malicious interference includes harassment and more.

Intentional, harassing and malicious offenders are usually obvious. They often change frequency with their victim. In fact, amateurs reporting malicious interference should gather documentation that illustrates that very behavior.

So, What Now?

If you have proof of intentional, harassing, or malicious amateur-to-amateur interference, keep your chin up; there *is* help. This assistance is yet another free ARRL service: The Amateur Auxiliary to the FCC Field Operations Bureau. This program is the result of Public Law 97-259. It allows the ARRL Amateur Auxiliary to formally liaise with and assist the FCC with data gathering for some of the heretofore-mentioned interference problems. Contact your ARRL Section Manager for information about the Amateur Auxiliary in your Section.

The Local Interference Committee

Much of the information presented above deals with nontechnical matters. They are often best resolved by amateur-to-amateur peer pressure and groups such as the Amateur Auxiliary to the FCC Field Operations Bureau. Clearly, however, a growing number of difficulties can be solved by individuals who have a clear understanding of the technical reasons behind certain kinds of interference. It's no secret that as more poorly designed consumer devices enter the marketplace, the potential for EMI problems to (and from!) Amateur Radio equipment increases dramatically.

Fortunately, one or more of your local ARRL affiliated clubs probably has an RFI committee. Sometimes such committees are dubbed "Local Interference Committees," yet (to make matters worse) some LICs are designed for a much different purpose. The "RFI or TVI Committee" concept was developed long before the idea of Local Interference Committees, and their purposes are often quite different. While a TVI or RFI committee will likely assist with EMI or RFI, Local Interference Committees may deal only with amateur-to-amateur operational difficulties. LICs can often provide direction finding equipment and expertise to locate local caused-by-amateur problems.

Although you need to do some homework before locating assistance for your interference problem, help is available. Once you have determined whether the problem is best solved by either the technical wizards (TCs) or the Amateur Auxiliary folks, the assistance you need requires only a phone call to your ARRL Section Manager. The SM can supply the name and address of either the Technical Coordinator or Affiliated Club Coordinator. Each should be able to guide you to the local people best able to help

Nonamateur-To-Amateur Interference

Finally, there's the matter of nonamateur interference to our operations in the Amateur Radio Service. At present, the ARRL Monitoring System routinely reports about 350 instances of nonamateur intrusion on amateur frequencies each month. This information is relayed to the FCC Treaty Branch in Washington, DC, for appropriate governmental action.

Clearly, ARRL programs help with each kind of interference you're likely to experience. Each Amateur Radio operator is encouraged to do his fair share of the job by determining the kind of interference before making contact with the appropriate volunteer.—*Luck Hurder, KY1T, Deputy Manager, ARRL Field Services Department*

Local Help

By Luck Hurder, KY1T
   Deputy Manager,
   ARRL Field Services Department

What's A TC?

One of the best-kept secrets of the extensive ARRL Field Organization is that an elusive appointee called the Technical Coordinator (TC) lurks in just about every ARRL Section in the nation. Even murkier is the fact that he has often appointed for himself yet another tier of technical expertise in the form of Technical Specialists (TS). These two important ARRL field volunteers provide technical support and guidance to amateurs in their Section. They should be contacted at the beginning of every EMI problem. Most TCs prefer to be called *before* tempers have flared and a situation has become difficult.

Did you say you've never even heard of TCs or TSs? And, you didn't realize that the ARRL promoted yet another free service for the benefit of Amateur Radio licensees and their neighbors? This service even helps those who don't know one end of a soldering iron from the other.

Well, the secret's out now, folks. I'm about to lay bare the until-now well-concealed facts of life regarding TCs: how to find them, and what they do. Y'all listen up!

How to Locate Your Section ARRL TC

Grab a *QST*; make it a recent one. Flip to page 8 and find the Section Manager's (SM) name for your ARRL Section. Ask the SM for the name and address of your

Section TC. The TC may be listed in the Section News (for your Section) at the back of the *QST*.

Did you say you haven't got a *QST* handy? (For shame!) Well, all isn't lost. Give ARRL HQ a call; ask for the Field Services Department, and inquire about the TC for your Section.

So, What Do They Do?

Technical Coordinators are volunteers. As such, they each have certain capabilities and areas of understanding about Amateur Radio technology. Here are some general guidelines for TCs:

The Technical Coordinator

1) Supervises and coordinates the work of the Section Technical Specialists.

2) Encourages amateurs in the Section to share their technical achievements with others through the pages of *QST*, at club meetings, hamfests and conventions.

3) Promotes technical advances and experimentation at VHF/UHF and with specialized modes, and works closely with enthusiasts in these fields within the Section.

4) Serves as an advisor to radio clubs that sponsor training programs for obtaining amateur licenses or upgraded licenses in cooperation with the ARRL Affiliated Club Coordinator.

5) In times of emergency or disaster, functions as the coordinator for establishing an array of equipment

for communications use and is available to supply technical expertise to government and relief agencies to set up emergency communications networks, in cooperation with the ARRL Section Emergency Coordinator.

6) Refers amateurs in the Section who need technical advice to appropriate local TSs.

7) Encourages TSs to serve on RFI and TVI committees in the Section for the purpose of rendering technical assistance as needed, in cooperation with the ARRL OO Coordinator.

8) Is available to assist local technical program committees in arranging suitable programs for ARRL hamfests and conventions.

9) Conveys the views of Section amateurs and TSs about the technical contents of *QST* and ARRL books to ARRL HQ. (Suggestions for improvements should also be called to the attention of the ARRL HQ technical staff.)

10) Works with the appointed ARRL TAs (Technical Advisors) when called upon.

11) Is available to give technical talks at club meetings, hamfests and conventions in the Section.

In addition to the above duties, TCs and/or TSs in a given ARRL Section often serve as advisors for radio-frequency interference (RFI) issues. Since RFI can drive a wedge into neighborhood

Fig 2—The ultimate EMI scenario. Sometimes the best option is to cease operation temporarily. *(photo by John Frank, WB9TQG)*

relations, it is frequently a TC/TS with a cool head who resolves such problems.

TCs and TSs may also be asked to work with other ARRL officials, or represent the ARRL at technical symposiums in industry. They may serve on CATV advisory committees, or advise municipal governments on technical matters relating to the Amateur Radio Service.

Chapter 2

Electromagnetic Interference (EMI) Fundamentals

By Bryan Bergeron, NU1N
ARRL Technical Advisor
6 Camelot Ct, #2
Brighton, MA 02135

EMI TECHNICAL FUNDAMENTALS

Knowledge about the fundamentals of EMI is one of the most valuable tools for solving EMI problems. EMI is any electrical interaction that affects normal equipment operation. EMI is present when undesirable signals adversely affect the performance of a device. The term covers interference from radio-frequency sources (RFI), audio-frequency sources (that is, induced hum from a transformer) and electrostatic sources (ESD—electrostatic discharge). Most amateurs are concerned with EMI caused by RF sources, so to hams, the terms "EMI" and "RFI" are usually interchangeable.

Resolving EMI is a three-step process, and all three steps are equally important:

- Identify the problem.
- Diagnose the problem.
- Determine what steps are necessary in order to eliminate the problem.

An accurate diagnosis, regardless of the problem, requires skill with the relevant technology and troubleshooting procedures. The principles of EMI diagnosis are explained in the Troubleshooting chapter.

When dealing with equipment that is not your own (a neighbor's TV set, for example), remember that there is a certain amount of liability attached to any work you might perform. Some states or municipalities have laws that regulate the electronic repair trade. These usually require a repair license and certification in order to service consumer electronic equipment, even if there is no fee for the repair. *Do not open equipment that is not your own!* The owner may hold you to blame for anything that *ever* goes wrong with the equipment after you work on it. The ARRL Technical Information Service staff tell some real horror tales about hams who have gotten into trouble over this sort of thing. Don't be part of the next story!

In general, when dealing with a neighbor, act as a *locator* of solutions, not an *implementer* of solutions. There is no situation where an amateur is required to modify a complainant's equipment. It is wise to remain an advisor.

Manifestations Of EMI

Is it really EMI? Before solving a case of suspected EMI, verify that the symptoms actually result from external causes. A variety of equipment malfunctions can look like interference. It is important to be able to differentiate between EMI and equipment malfunction.

Characterizing EMI is difficult because the symptoms depend on the natures of the emitter and the susceptible device. This complicates the learning process.

The basic causes of EMI can be grouped into several large categories:

- spurious emissions from a transmitter.
- fundamental overload effects.
- intermodulation or externally generated spurious signals.

Transmitters

All transmitters generate some (hopefully few) RF signals that are outside their intended bands. These out-of-band signals are called "spurious emissions." Spurious emissions can be discrete signals or wide-band noise. The most prevalent spurious emissions from transmitters are *harmonics*. A harmonic is a signal that is an exact multiple of the operating frequency. Harmonics are caused by nonlinear operation of oscillator and amplifier stages—no amplifier design can avoid generating *some* harmonics.

Other discrete signals are inadvertently produced as part of the superheterodyne mixing process used in most modern amateur transmitters. Such mixing products can appear above and below the operating frequency. While harmonics are produced in the mixing process, and harmonics play a part in the production of other signals, the term "mixing products" usually is not applied to harmonics. Fig 1 shows the output of a typical transmitter, with harmonics and mixing products present.

Transmitted signals also contain noise. This noise may be broadband composite noise, keying transients or intermodulation products occurring in the linear amplifier stages. Transmitter noise can also cause interference problems.

FCC regulations set two different limits for spurious emissions: an absolute

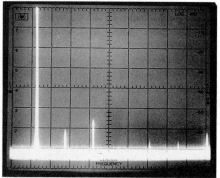

Fig 1—A spectral photo from a typical amateur transceiver. The fundamental (14.2 MHz) appears at left. Harmonics are visible at 28.4 and 42.6 MHz. Amateur stations are subject to stringent FCC spectral-purity regulations (this transceiver complies).

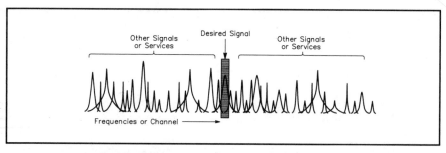

Fig 2—Every electronic appliance must select the signal it needs from all those present in the radio spectrum. The desired signal (in box) may not be as strong as signals that must be rejected.

worst-case limit and an interference limit. The worst-case limits set maximum spurious levels for transmitters. The allowed emission is usually expressed relative to the output (carrier or PEP) power. For example, a spurious emission might be characterized as being –58 dBc, that is, 58 dB "below" (weaker than) the carrier. The worst-case limits range from –30 dBc for 5-W (or less) HF transmitters, to –60 dBc for 50-225 MHz transmitters.

FCC regulations also require attenuation of any spurious emissions that interfere with other radio services. For example, if a spurious emission from a VHF repeater interferes with an FAA communications channel, the emission must be reduced as necessary to eliminate the interference.

Fundamental Overload

Many cases of interference are *not* caused by spurious emissions. The world is filled with RF signals. Properly designed equipment should be able to select the desired signal, while rejecting all others. Unfortunately, because of design deficiencies such as inadequate shields or filters, much equipment is unable to reject strong out-of-band signals. A strong fundamental signal can enter equipment in several different ways. Most commonly, it is conducted along any wires connected to the affected device. TV antennas and feed lines, telephone wiring and ac power-supply leads are the most common points of entry.

The internal wiring, circuit-board traces and components of some poorly shielded equipment can pick up signals

directly. This is known as "direct radiation pickup." In many cases of direct pickup, it is not practical to add shielding, so there is little an amateur can do to fix the problem. In these cases, contact the equipment manufacturer through the Electronic Industries Association (EIA).[1] The EIA has the address and telephone number of the best RFI-problem contact for nearly any consumer-equipment manufacturer.

Amateurs are *not* responsible for the effects of amateur-band fundamental signals, but, of course, practical considerations must sometimes affect our decisions. Fig 2 shows how properly designed consumer equipment should function.

Intermodulation and External Rectification

Most amateurs have heard of "intermod" when operating through repeaters. Because repeaters are often located very close to other strong transmitters, intermod is a common problem. For example, a repeater transmitting on 144.85 MHz might be located near a land-mobile repeater transmitting on 155.55. The 155.55 signal could mix with the amateur fundamental in the output tank circuit and produce a signal at 10.7 MHz. That signal could interfere with the IF of nearby FM-broadcast receivers.

Intermodulation is an unwanted mixing of two or more signals in a nonlinear device (mixer). It is essentially the same process described as "mixing products" under Transmitters, but intermod involves at least one signal from outside the transmitter. In fact, the mixing may not take

place in a transmitter at all.

While "intermod" usually refers to unwanted mixing in electronic equipment, mixing that occurs in nonelectronic components is called "external rectification." The problem is so prevalent and perplexing that there is a chapter dedicated to it later in this book. Although the terms are different, the action and results are essentially identical.

Most any semiconductor junction may act as a "nonlinear device" (mixer). Semiconductor junctions are very common in the form of poor electrical connections. Ac outlets with "push in" connections are good examples, so are corroded connections on TV antennas.

Given the mixer, intermod requires two (or more) strong signals. Since a 0.1-λ wire makes a reasonable receiving antenna, one such conductor for the longer wavelength completes the intermod requirements.

Complete intermod "systems" are common in ac-line wiring, telephone systems and TV antenna systems. Several systems have been particularly troublesome in recent years:

- antenna-switching diodes and feed lines in VCRs, scanners, CBs and other radios.
- unterminated "pairs" in telephone systems.
- LEDs fed by long wires in SWR meters and other accessories.
- ac wiring that uses "push in," rather than screw-terminal connections.

Television Receivers

Although its incidence seems to be decreasing, Television Interference (TVI) remains one of the more common EMI problems. It is perhaps the most troublesome to the amateur. TVs are by necessity broadband devices and are therefore

[1]Electronic Industries Association (EIA), 2001 Pennsylvania Ave NW, Washington, DC 20006, 202-457-4978.

Table 1
Possible Interference Sources

Communications Service Transmitters
CB
Amateur Radio
Police
Cable TV (CATV)
Business or other two-way
Aircraft (near airports only)

Electrical Noise Sources
Doorbell transformers
Toaster ovens
Electric blankets
Fans
Heating pads
Light dimmers
Appliance switch contacts
Aquarium or waterbed heaters
Sun lamps
Furnace controls
Smoke detectors
Smoke precipitators
Computers (and video games)
Ultrasonic pest-control devices
Lights: fluorescent, mercury vapor and
     touch-controlled
Neon signs
Power Company Equipment
     Defective line insulators
     Loose or unbonded hardware
     Discharges from defective lightning
          arrestors
     Defective transformers
Electric fences
Alarm systems
Loose fuses
Sewing machines
Electrical toys (such as trains)
Calculators
Cash registers
Lightning arrestors

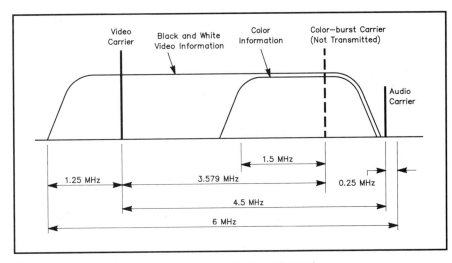

Fig 3—A typical frequency layout of a television channel.

relatively susceptible to interference from external RF signals. The susceptibility problem is compounded because interference makes a much greater impression on the eyes than it does on the ears.

A good TV signal and typical antenna usually result in a TV signal of about –45 dBm at the receiver terminals. Interfering signals that are 40 to 60 dB weaker can cause perceptible interference.

TVI can originate from many sources other than radio transmitter operation. External noise sources such as power lines, electric motors and computer systems (including some video games!) can all interfere with televisions. A complete list of potential interference sources would be quite large. Table 1 is a partial list that may help put the problem into perspective. Obviously, Amateur Radio is only one of many possible RF or noise energy sources. Although Amateur Radio related EMI is our *biggest* concern, remember that amateurs do not comprise the bulk of the EMI problem.

When caused by a radio transmitter, TVI may appear as crosshatching or bars in the picture, a change in contrast or color, complete blanking of the screen, audio interference or any combination of these effects. The manifestation in a specific case depends on the frequency and bandwidth of the interfering signal.

A TV channel is 6 MHz wide, with the picture carrier 1.25 MHz up from the low-frequency edge and the sound carrier down 0.25 MHz from the high-frequency edge (see Fig 3). As the interference frequency approaches that of the video carrier, the interference bars become thicker and more objectionable. As the interference moves away from the video-carrier frequency, the bars become finer until they become a "herringbone" pattern on the screen. This pattern sometimes becomes so fine that it appears as a slight reduction of picture contrast. When these horizontal or diagonal lines are modulated by a voice, they are usually called "sound bars." Fig 4 shows a typical case of TVI.

Television sets are also subject to overload from strong nearby signals. This overload can take several possible forms.

Harmonics or other undesired signals can be generated in the TV tuner circuits. The effect of these tuner-generated signals is exactly the same as the effect from external interfering signals, complicating the task of proper diagnosis.

TV-set circuitry is also subject to *blocking*. Blocking is a reduction of sensitivity (or contrast). The effect can range from a just-perceptible change in color or contrast, to a totally blanked-out picture, color or sound.

The TV audio circuitry can envelope detect an RF signal and amplify the resultant audio signal. This effect is known as *audio rectification*. The resultant audio signal may or may not be affected by the TV volume control. If audio interference is not affected by the volume control, audio rectification is clearly indicated. CW audio interference appears as clicks and thumps, or a reduction in TV audio as the transmitter is keyed. SSB audio interference sounds distorted.

Fig 4—This "herringbone" pattern is typical of a TV receiving interference from a two-way radio.

If the interference is present on all channels, or on channels not harmonically related to the transmit frequency (see Fig 3 in the Televisions chapter), the problem is probably caused by a fundamental signal. If the interference is present only on channels that are harmonically related to the transmit frequency, the problem could be caused by either fundamental overload or transmitter harmonics.

Fringe-area reception can be a two-edged sword. TVI is more of a problem in TV fringe areas, since even low-level RF signals can disrupt a weak TV signal. The good news is that the FCC does not offer protection to fringe-area reception. TV coverage areas are divided into "contours" according to signal strength. Fringe areas are those outside of the "grade-B contour" area. This means that fringe-area viewers are not in the intended audience of the station, and the FCC does not protect them from interference. Of course, hams usually want to be good neighbors and are willing to offer some help in cases of fringe-area TVI.

If TVI is present on all channels, or in channels that are not harmonically related to the transmitter fundamental, the TV may be inadequately filtered or shielded. TVI from harmonics is much more likely when transmitting on 10, 15 or 20 meters than when operating on the lower bands because of the reduced amplitudes of higher-order harmonics.

TVI can usually be cured by the proper application of filters. A low-pass filter installed at the transmitter helps ensure that the station does not emit any spurious signals. High-pass filters and common-mode chokes at the TV system help eliminate interference caused by fundamental overload.

Video Cassette Recorders (VCRs)

The addition of a VCR to an existing TV may increase the susceptibility of the system to EMI. The external tuner and interconnecting cables offer additional entry points for unwanted RF signals. VCR EMI is usually indistinguishable from TVI. The interference ranges from total blanking of the TV picture to bars and lines across the screen.[2] In most cases, TVI cures work for VCRs as well. However, a VCR can contribute to TVI even when switched off. (Diodes within the VCR circuitry can rectify RF energy and generate harmonics.)

The VCR problem is further complicated because the video baseband extends from 30 Hz to about 4.5 MHz, including the FM sound subcarrier. VCRs use a frequency-modulation scheme, with signals up to about 7 MHz. The system contains a magnetic head followed by high-gain amplifiers, with a bandwidth that often exceeds 10 MHz. This, along with plastic cases and marginal filtering, can be a recipe for severe TVI.

In 1982, Public Law 97-259 (there's more on this in the EMI Regulations chapter) gave the FCC authority to regulate EMI susceptibility. The FCC has chosen to rely on voluntary compliance rather than regulation, and the result is the ANSI C-63 Committee. Voluntary manufacturer compliance with the recommendations of that committee has effected a real improvement in TV and VCR susceptibility. VCRs manufactured in the last few years are much better at rejecting unwanted signals than those of several years ago.

When troubleshooting, temporarily remove the VCR from the system. If the TVI disappears at that point, you know that the VCR is the susceptible device. For further information about VCR interference problems, read the VCR section of the Televisions chapter in this book.

Community Antenna Television (CATV)

Interference to CATV is a special case of TVI. The complexity of CATV, however, makes determining the cause of interference more challenging. CATV systems operate from 50 to 450 MHz (higher in some systems). They often use converters to translate each of these channels to a single TV channel (typically channel 2, 3 or 4).[3] A VCR is often used as the cable "converter," or a cable-ready TV may tune the CATV channels directly. Accessories are common in CATV installations, ranging from cable-company installed splitters to customer-installed A/B switches and video games. Such accessory devices are often sources of interference troubles. Some CATV systems make use of "talk-back" channels in the HF range, allowing signals from a subscriber's home to be sent back to the cable company.

CATV is both a blessing and a curse to the amateur. A CATV system is (theoretically) a closed, shielded system, immune to interference from outside sources. FCC regulations limit leakage from CATV systems. Fortunately, leakage is a two-way street, and well-maintained systems exclude interference. Most CATV companies are sensitive to leakage problems and quickly correct any flaws. For more information about TVI in CATV systems, read the CATV section of the Televisions chapter.

In all interference cases, amateurs are responsible only for interference caused by spurious emissions from their stations. Amateurs are not legally responsible for interference caused by fundamental overload. Nonetheless, you may want to help. Try basic TVI cures first: high-pass filters, common-mode chokes and ac-line filters. These devices often cure CATVI, and they are more convenient than securing an appointment and waiting for CATV service.

Some TVs and VCRs respond to signals picked up on the coaxial-cable shield or are subject to direct radiation pickup. The CATV company is not responsible for these two kinds of problems. In cases of direct pickup, contact the TV or VCR manufacturer through the EIA. To eliminate problems from signals on the coaxial-cable shield, install a common-mode choke. Common-mode chokes are discussed in the TVI chapter.

Since CATV systems are not readily accessible, it may be difficult to locate the point of failure. EMI may enter the system via the flexible "drop" cable at a subscriber's home, a poor connection, an unterminated coupler that feeds a TV, external devices or ac-power lines, to name just a few.

Many CATV interference problems can be traced to poorly installed illegal hookups. These are often performed by untrained personnel. There is little regard for proper installation techniques, such as good shielding and secure connections. Some of the worst illegal installations even use twin-lead instead of coaxial cable.

Radio Receivers and Audio Devices

AM, FM and FM-stereo EMI manifests itself as buzzing or voice sounds superimposed on the desired audio. The

---

[2]D. DeMaw, "Joe ham versus VCR RFI," July 1988 *QST*, pp 34-44.

[3]R. Dickinson, "Entertainment and Interference: The Two Faces of CATV," Feb 1982 *QST*, pp 11-15.

unwanted signal may enter through the antenna, feed line, power or control leads, earphone or speaker leads, receiver enclosure or joints and openings in the enclosure. Six-meter (50 MHz) operation is most likely to affect FM receivers (88-108 MHz) because the second harmonic from any transmitter is usually the strongest. Interfering RF is commonly conducted via the speaker leads, which act as antennas. With this kind of EMI, the receiver audio-gain control has no effect on the interference. (The RF enters the set after the audio control, in the output stages.)

All kinds of audio appliances—from alarm systems to computers—are subject to overload by strong radio signals. The radio signal is usually amplitude detected then amplified just as if it were the desired signal. The FCC *Interference Handbook* clearly states that any audio device that receives interference from a radio transmitter "improperly functions as a radio receiver." The FCC does not offer protection to audio devices such as amplifiers or telephones.

EMI CAUSES

There is rarely a single, well-defined cause for a given case of EMI. In many cases, there is a complex system of inter-actions and interdependencies. As described below, these include: (a) various mechanisms by which undesirable RF energy is generated, propagated and then received by susceptible devices; (b) inadequate shields and filters that make some electronic devices especially vulnerable to EMI; (c) improper filters, shields or operation of transmitters and other RF energy sources.

The Emitter-Path-Susceptor Concept

All cases of EMI must involve a source of electromagnetic energy (emitter), a device that responds to this energy (susceptor) and a transmission path (either conducted or radiated) that allows energy to flow from emitter to susceptor. Emitters include radio transmitters, receiver local oscillators, computers and computer peripherals, lightning, electrostatic discharge and other natural sources. Susceptors include radio and TV receivers, VCRs, telephones, amplifiers, computers and even devices such as pacemakers or alarm systems.

There are three fundamental means by which EMI can travel from emitter to

susceptor: conduction, radiation and magnetic induction. Possible conductors include antennas and feed lines, interconnect cables, power lines and ground leads. Radiated EMI propagates from some conductor in the emitter system through space to some conductor in the susceptor system. The transmitting and receiving conductors include those already mentioned, with the addition of circuitry *inside* poorly shielded equipment. When magnetic induction occurs, the magnetic field of an inductor (such as a power transformer) produces an unwanted signal in a nearby conductor (such as an ac branch circuit or telephone cord). Relevant characteristics of emitters and susceptors are their emission and reception spectra and their susceptibilities. That is, what undesired signals do they emit or receive, and how do those unwanted signals leave or enter the equipment?

The coupling path between an emitter and receptor can be extremely complicated. For example, there may be conduction from the emitter to a radiator, radiation to another conductor and conduction again to the susceptor. There might be several transitions between radiation and conduction in the EMI path.

Conducted v Radiated Emissions

Conducted emissions originate from a variety of sources, including relay and switch contacts, fan motors, oscillators, analog devices that are operated over non-linear parts of their design curves, digital devices with short rise and fall times, and high-speed switching devices. Radio transmitters can originate conducted emissions.

Conducted interference can be minimized by using filters to channel energy away from sensitive devices, and by using bypass capacitors to decouple devices from the power bus. Although the spectrum of conducted interference can exceed 1 GHz, higher-frequency conduction currents are heavily attenuated by resistive losses, wire inductance and shunt capacitance. In addition, the higher-frequency signals have a tendency to be radiated and coupled to nearby wiring.

Radiated emissions can result from leakage through coaxial cable shields, corroded surfaces, insulation breakdown, discontinuities in component housing, or inadequate grounding that compromises shield effectiveness. Whereas the same conditions that cause conducted interfer-

ence may also result in radiated interference, the source of radiated interference is much more difficult to diagnose. The direction of maximum radiated energy may not correspond to the exact beam heading of the source antenna because reflections mask the direct path.

The coupling of interfering signals from emitter to susceptor may involve common wiring, mutual capacitance or inductance, or direct radiation. Common wiring includes shared ground and power-supply leads as well as signal cables. For example, consider a 40-meter transmitter grounded to the telephone-entrance ground. If the ground point is 12 ft from the effective earth ground, it presents a high impedance to the third harmonic at 21 MHz. The third-harmonic energy then enters the telephone system, is rectified by a transistorized phone and appears as audio throughout the telephone system.

The degree of capacitive coupling depends on the amount of capacitance, the impedance levels of the emitter and susceptor circuits and the amplitude and frequency of the signal involved. High-impedance, high-frequency circuits favor capacitive coupling. Most amateurs who have ever tried to use an antenna system that placed a high-voltage point in the shack have experienced capacitive coupling. "RF in the shack" results when RF energy capacitively couples from the transmitter and transmission line into metal objects at the operating position.

The degree of inductive coupling between adjacent wires is a function of the frequency and magnitude of current flowing in one of the wires, the spacing of the wires and their common length. The most common amateur example of inductive coupling is the Yagi antenna. The current in the driven element induces currents in the parasitic elements that produce a directive radiation pattern.

Direct radiation pickup generally involves an intermediate element that supports reradiation of the unwanted signal. For example, a fundamental emission could be picked up by a resonant length of ac wiring, which would concentrate and reradiate it for pickup by consumer equipment.

Differential- v Common-Mode Currents

It is important to understand the differences between differential-mode and common-mode conducted signals. Each of these coupling modes requires different EMI cures. Differential-mode cures (the typical high-pass filter, for example) do not work on common-mode signals. On the other hand, a typical common-mode choke does not affect interference resulting from a differential-mode signal. Fig 5 shows differential-mode and common-mode signals.

Differential-mode currents occur between two conductors with no ground reference. There is a 180° phase difference between the currents in the two conductors. In a two-wire transmission line, for example, the signal arrives on one line and returns on the other.

In comparison, common-mode currents are in phase on each conductor of a multiwire cable. Common-mode currents return to their source through some conductor common to both the source and affected circuit (usually the system or earth ground). In a common-mode situation, all wires of a multi-wire system act as if they were a single wire. Since common-mode currents flow in the same direction through all conductors in a cable, little field cancellation takes place. The result can be a good radiator. For example, a coaxial transmission line can act as a long-wire antenna "worked" against earth ground.

The magnitude of the common-mode current induced in each wire of a cable is a function of the cable design. In a balanced two-wire system, such as 300-Ω twin-lead, the conductors carry equal induced currents, just as though the two conductors were wired in parallel. In an unbalanced system, such as coaxial cable, the induced current magnitude is different in each conductor. The induced current is much greater in the coax shield than in the center conductor.

If not recognized, EMI from common-mode currents can be especially troublesome to eliminate. The usual fundamental-overload and harmonic-radiation EMI cures (a low-pass filter for the transmitter, a high-pass filter for the receiver, a ground connection for the transmitter coax shield and power-line filters for the transmitter and TV) do not reduce common-mode interference.[4] The conductors carrying unwanted common-

[4]B. Orr, "Ham Radio Techniques," Feb 1989 *Ham Radio*, pp 18-23.

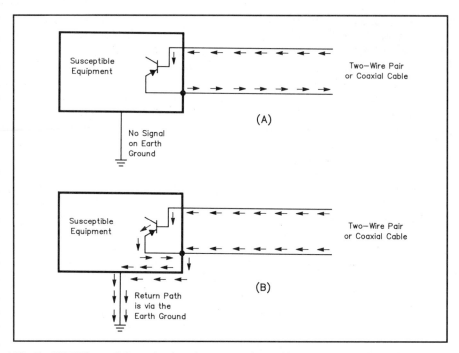

Fig 5—(A) Differential-mode signals are conducted between two wires of a pair. This signal is independent of earth ground. (B) A common-mode signal is in phase on all wires that form the conductor (this includes a coaxial cable). All wires act as if they are one wire. The ground forms the return path, as with a long-wire antenna.

mode currents usually carry desired differential-mode currents as well. Therefore common-mode cures must block in-phase currents without affecting differential-mode signals. Fortunately, this is easily accomplished with common-mode chokes.

Fundamental Overload v Spurious Emissions

Most receivers, including consumer-electronic receivers, have some form of band-pass filter in the front end. Strong signals that lie outside of the passband may, however, pass through to the first stage (fundamental overload). Once there, they may: (1) be passed directly to the output, (2) overload the receiver and make it inoperative (desensitization), or (3) combine with other signals in a nonlinear element to produce spurious signals within the receiver passband. Fundamental overload may occur when the affected device is inadequately shielded, has unfiltered leads, or is in close proximity to an RF source. *Fundamental overload is the result of design deficiencies in the affected system, not improper transmitter operation or design.*

In a high-density RF environment such as a repeater tower, one transmitter can overload a second repeater's receiver. A third transmitter can mix with the first signal, causing intermodulation. Naturally, each user tends to blame the others for the problem.

Spurious emissions include harmonics and parasitic oscillations. Whereas harmonics are always exact multiples of a fundamental signal, parasitic oscillations usually bear no direct relationship to the fundamental frequency. The FCC requires that all commercially made amateur HF transmitters suppress spurious energy at least 40 dB below the peak output level, without exceeding 50 mW. For example, all spurious energy must be 10 mW or less for a 100-W HF transmitter. At VHF, the spurious energy must be at least 60 dB below the peak output power. Although all transmitters generate harmonics, harmonics can be reduced to the required levels by proper transmitter design and filters.

Nonlinear Junctions

Harmonic and intermodulation interference can be caused by RF current flowing through any nonlinear junction. Colloquially referred to as the "rusty-bolt effect," the nonlinear characteristics of corroded joints (oxidized guy wires, coaxial connectors, antennas, feed lines and such) can cause harmonic, cross-modulation and intermodulation products that result in EMI. Even when the transmitter output is clean, EMI may result if there is a poor electrical joint somewhere in the antenna or feed line. A poor solder or mechanical joint can act as a rectifier, generating harmonic currents that may be radiated by the antenna. Poorly conducting joints in nearby conductors (rain gutters, for example) can generate and radiate harmonics when excited by RF energy.

RF current flowing through nonlinear electronic components, including transistors, diodes and ICs, can also result in EMI. Semiconductor junctions in such equipment as audio amplifiers, mast-mounted preamplifiers and SWR bridges may all rectify RF energy to produce spurious signals. Receivers can be EMI sources. When the first stage overloads, the resulting intermodulation signals may be radiated from the receiving antenna. Sensitive preamplifiers, with limited dynamic range and poor selectivity, are especially vulnerable to overload problems.

Intermodulation products can be generated by transmitters when other strong signals are coupled to the final amplifier. There they mix with the fundamental and its harmonics. The resulting signals may be coupled to the antenna and radiated. Intermodulation problems are especially common in the VHF and UHF bands because amateur repeaters tend to cluster around commercial and government VHF and UHF radio services.[5] Wide-band, solid-state amplifiers with low-Q circuits are more likely to generate spurious signals than are narrow-band, high-Q configurations.

CURING EMI

Identification of the EMI path is a key step in resolving EMI. The frequencies common to both the emitter and susceptor are critical factors. The times of emitter operation and the relative location of emitter and susceptor are also important. Only with this data in hand is it possible to select the most effective means of combating the EMI. In most instances, a combination of techniques must be used. The effectiveness of any one method is greatly enhanced when additional methods are also employed. For example, a low-pass filter does not operate effectively unless the transmitter is well shielded.

Shielding

Shielding is perhaps the most universal concept in controlling radiated EMI. It is used to set boundaries for radiated energy. Thin conductive films, copper braid and sheet aluminum are the most common shield materials. Shield effectiveness is expressed as the number of decibels by which a shield reduces the field strength of radiated energy. Effectiveness is a function of shield composition and thickness, the frequency of the radiation and the quantity and shape of any shield discontinuities (seams or holes). For example, RG-59U with 51% braid coverage provides a relative isolation of 18 dB; a cable with 98% braid coverage provides 52 dB of relative isolation.

Maximum shield effectiveness demands solid sheet metal that completely encloses the emitter or susceptor. Solid materials, such as sheet aluminum or copper, attenuate RF signals through reflection and absorption. The shield surface reflects incident energy because of the impedance discontinuity at the air-shield boundary. Similarly, there is internal reflection of RF energy that reaches the opposite face of the shield. The RF energy is further attenuated by absorption as it passes through the shield material.

Because of the expense of solid shields and the popularity of plastic enclosures and composite materials, many modern electronic devices rely on thin film shielding (shielding with a thickness of less than $1/4 \lambda$ at the propagation velocity) to control EMI. Thin film shielding typically consists of a metallic (silver, copper or zinc) film deposited on the nonconductive support via vacuum metallization, flame spraying, plating or metal-filled paints. Pressure-sensitive foils and laminates can also add shielding capabilities to nonconductive materials. Unlike solid materials, thin films provide negligible absorption loss; attenuation is primarily from reflection.

Discontinuities in shield materials decrease shield effectiveness. For this reason, shield seams should be bonded so that the RF impedance of the seam is the same as the material being joined. Otherwise, RF voltages may develop across the seam (which possesses a transfer impedance, Z_T) and allow RF energy to penetrate the shield. Pressure-sensitive foils, also called EMI tape (a form of thin film shielding), can be used across shield breaks to maintain shield effectiveness.

Panel openings (for displays, controls and ventilation) often compromise shield effectiveness. Panel openings for meters and other displays can be protected with conductive glass or a wire mesh across viewing surfaces. A honeycomb construction can be used for features like ventilation ducts, where it is not necessary for the shield to be transparent. Thin film or solid shields behind displays, combined with feed-through capacitors, also minimize the flow of RF energy through panel openings.

Cable shields are typically a compromise between effectiveness and practical considerations. Flexibility, low-weight and low-cost requirements lead owners to choose cables that are not completely shielded. (The main exception is rigid or flexible conduit.) Most cables are shielded with copper braid. The shield effectiveness of braid increases with weave density and decreases with increasing frequency. When braid-shielded coaxial cables are bundled together, leakage from one cable can cause interference to adjacent cables.[6] Leakage of electromagnetic fields through the braid causes a current to flow on the outside of the braid. This produces currents on the outside of other cables in the bundle.

Filters

A major means of separating signals relies on their spectral differences. Filters make this possible because they can offer little opposition to certain frequencies while blocking (by reflection and/or absorption) others. A properly designed and installed filter can reduce the levels of conducted interference, so long as the spectral content of the interference is different from that of the desired signal. Filters are often used to resolve problems resulting from design compromises. For example, harmonic filters would not be needed if circuit linearity were absolute.

[5]D. Potter, "Intermodulation Reviewed," May 1983 *QST*, pp 17-18.

[6]J. Tealby, P. Cudd and F. Benson, "Coupling Between Braided Coaxial Cables," *Fifth International Conference on Electromagnetic Compatibility*, 1986, pp 121-127.

Filters vary in attenuation characteristics (high pass, low pass, bandpass and notch), power-handling capabilities and in their passband and stopband frequencies. For example, power-line filters are low-pass filters that provide little attenuation to 60-Hz energy but substantial attenuation to RF energy. Power-line filters are useful for suppressing conducted emissions that may enter equipment from the power line and vice versa.

The application of standard filter configurations is the same, regardless of operating frequency. Low-pass filters are used at the output of any device that generates unwanted harmonics. Depending on the desired frequencies, however, receiver front ends may require low-pass, high-pass, bandpass or notch filters at the antenna input.

From the perspective of EMI control, the most important attribute of a filter is its frequency characteristic—the relationship between insertion loss and frequency. In choosing a filter, both the desired and unwanted frequencies must be considered. If the frequencies are relatively close, then a filter with a large attenuation-v-frequency slope may be needed. Such filters are expensive to construct, however, because of the precision component values and mechanical accuracy required.

Reflective filters function as reflectors in their stopband, sending energy back to the source rather than allowing it to reach the load. These filters usually consist of a capacitor-inductor combination configured to present a high series impedance with a low shunt impedance in the stopband and a low series impedance with a high shunt impedance in the passband. A variation of the reflective filter that works primarily by diverting energy away from the load is the common power-line surge suppressor. Power-line surge suppressors do not suppress, absorb or otherwise dissipate line noise. Instead, they divert energy from one path to another. Surge suppressors use metal-oxide varistors (MOVs) between the hot and neutral lines to dump high-energy noise from the hot line to the neutral line, creating a potential difference between neutral and ground (and common-mode noise).

Since reflective filters simply divert undesired signals, the potential for interference remains. When undesired signals must actually be eliminated, a filter that provides attenuation via absorption is required. These so-called lossy filters are typically constructed in the form of a transmission line that has a dielectric of ferrite or some other RF-absorbing material. Lossy filters are especially effective when combined with reflective filters; the combination is capable of providing steep cutoff slopes and high stopband attenuation.

The reliability, stability, size, weight, efficiency and effectiveness of a filter is ultimately defined by the nature of the components used in its construction. For example, filter performance is limited because perfect components do not exist. Capacitors provide not only capacitance but also resistance and inductance. The resistance presented by a capacitor can be attributed to dielectric losses, foil resistance and the lead-to-foil contact. The inductance is present in the capacitor plates and leads. Because of self inductance, capacitors exhibit self-resonance at a frequency where the inductive and capacitive reactances are equal. At frequencies above resonance, a capacitor behaves more like an inductor than a capacitor.

The magnitudes of inductance and resistance presented by a capacitor, and therefore the capacitor's suitability for filtering, is a function of capacitor type and construction. Metalized-paper capacitors are poorly suited for use in RF filters because of high contact resistance and a tendency to create RF noise. Large tantalum capacitors are resonant at 2 to 5 MHz, depending on construction and capacitance. In comparison, aluminum-foil capacitors can be used up to 20 MHz, depending on capacitance and lead length. Even low-inductance mica and ceramic-disc capacitors, which are effective up to 200 MHz, are limited by lead inductance. Feedthrough capacitors offer the highest operating frequencies. With their reduced lead inductance, feedthrough capacitors self-resonate above 1 GHz.

Just as there are no perfect capacitors, there are no perfect inductors. In addition to inductance, inductors exhibit resistance, capacitance and self-resonance. At resonance, the reactance of the interwinding capacitance is equal to that of the inductance. To increase the self-resonant frequency (and therefore the useful range of the inductor), inductors can be wound on separate cores and connected in series. This decreases the total interwinding capacitance.

Even the best components and filter designs are fruitless without good shielding and grounding practices. Improper shielding and grounding of a transmitter cabinet allows EMI to radiate around the filter. In addition, improper shielding and grounding make it possible for unfiltered RF currents to flow unimpeded on the outside of the coaxial cable braid and around low-pass filters and traps.

Finally, filters must be properly terminated. The input impedance of a filter, at any frequency, depends on its load impedance. Since the ability of a filter to pass or reject energy results from its input impedance, a suitable load impedance is critical to filter performance.

Ferrite Beads and Toroids

Ferrite is a ceramic containing granulated iron compounds. It comes in toroids, beads and bars that can be used as low-pass filters at RF. When current flows in a wire passing through a ferrite bead, magnetic flux circulates inside the bead. In ferrites formulated for EMI control, most of the magnetic flux is dissipated in the material as heat. The bead forms an absorptive filter with energy absorption proportional to frequency and bead length. The optimum bead size and composition is determined by the application.

Ferrite cores are uniquely suited for attenuating the flow of common-mode currents. Wrapping a cable through a lossy toroid (that is, one with a high-permeability mix) forms an RF choke for common-mode currents, without attenuating differential-mode signals. With unbalanced cables (coax), common-mode current in the shield sets up a field that dissipates the energy as described for beads. With balanced cables (speaker cables, twin lead and such), the differential-mode currents produce magnetic fields of opposite polarity, which cancel so that there is no net field and negligible signal loss. Common-mode currents are dissipated as described for beads.

Speaker leads may be wrapped through a ferrite toroid core to reduce RF energy entering an audio amplifier via the speaker leads. Similarly, common-mode RF current flowing through an ac-line cord can be attenuated by winding a single layer of the cord on a ferrite rod. When both the antenna and power leads carry common-mode current, the attenuation afforded by a toroid can be multiplied by winding both leads (in opposite directions) on the same core.[7]

Ferrite cores are effective filters at higher frequencies than wire-wound inductors because of their low associated capacitance. Ferrite cores are also easy to use. By simply slipping one or more cores onto an antenna lead, speaker cable, pick-up lead, power cable or multiwire cable, RF current on the wire is attenuated. A single core slipped over a wire is equivalent to a single-turn RF choke. Increasing the series inductance is simply a matter of adding more turns, more or longer beads, in direct proportion to the additional inductance required.

Bypass Capacitors

Like ferrite beads, bypass capacitors are commonly used to decouple circuit elements at RF. A bypass capacitor with reactance about 20% of the circuit impedance forms an effective low-pass filter. By shunting higher frequencies to ground, bypass capacitors offer an inexpensive means of reducing RF energy in a circuit. Bypass capacitors (typically 0.001- to 0.01-µF ceramic discs) can be used at panel meters to reduce emissions, from each side of the ac line to the chassis to reduce ac-line pickup, and on speaker connections (*on vacuum-tube amplifiers only,* more on this later) to shunt the RF energy to ground before it enters audio circuitry.

Grounds

Do not think of ground as a huge sink that somehow swallows noise. A properly implemented ground, however, can increase shield and filter effectiveness. Grounding, the establishment of an electrically conductive path between two electrical systems or between an electrical system and a reference point (ground plane), is a circuit concept. The ideal reference point (a zero-potential, zero-impedance body) can only be approximated. In practice, ground (perhaps better called a universal reference) often consists of a metallic automobile or building structure, plumbing, earth ground, steel-reinforced concrete floors, station ground, signal-control cables or telephone lines.

Most amateur installations make use of two ground systems: a safety ground and an EMI ground (a few also use a system of conductors and ground rods for lightning protection). Safety ground is the ac-line lead, which is connected to earth at the service entrance. At power frequencies, the safety ground provides a low-impedance path for fault currents. At higher frequencies, however, the impedance of the ac ground lead may be much greater than that provided by stray capacitive paths between cables, circuit boards and equipment enclosures. Power ground is especially ineffective when the ground wire approximates an odd multiple of 1/4 λ for unwanted signals. Then, the ground lead transforms a low-impedance ground to a high impedance at the grounded equipment.

The EMI or station ground is used for: (1) cable-shield and equipment grounding, (2) EMI filter referencing, (3) noise and interference control (by providing a low-impedance sink for noise currents), and (4) circuit referencing (by allowing signals between equipment to be properly interpreted). The common practice of using house plumbing for EMI ground is not a good idea. The result could be a "shunt feed" system with the entire building behaving as a radiator.

Three fundamental grounding techniques are *floating, single-point* and *multipoint* ground.[8] A floating ground is used to isolate circuits electrically from a common ground plane or from common wiring that might introduce circulating currents (ground loops). Floating ground is not used much in EMI prevention because it is ineffective at frequencies greater than approximately 1 MHz. At higher frequencies, capacitive coupling paths bypass isolation transformers and other isolation mechanisms. This allows RF currents to flow from one point to another on the ground plane. In addition, the electrical isolation may allow static charges to accumulate, resulting in random discharges of EMI-producing current.

Single-point grounds rely on a single physical point in each circuit that is defined as the ground reference point. All ground connections are made directly to that point. In a single-point ground system with multiple cabinets, cabinet and electronic ground are kept separate, and a single ground is used inside of each cabinet. Cabinet grounds are in turn connected at a single reference point. Single-point grounds are effective at higher frequencies than floating grounds. As ground lead length approaches 1/4 λ at the unwanted frequency, however, the impedance of the ground connection increases to an unacceptably high level.

In multipoint grounding, each ground connection is made directly to the ground plane at the closest available point. The advantages of this approach include easier circuit construction and higher operating frequencies. In addition, multipoint ground reduces electrostatic coupling between shielded cables. Unfortunately, multiple-point grounding permits ground loops, a potential source of interference.

The relative advantages and limitations of the three basic grounding techniques should be considered whenever a ground connection is required. For example, if ground lead length approaches 1/4 λ for any potentially interfering signals carried on the lead, multipoint grounding should be used. Other good grounding practices include:

- Keep ground leads as short as possible.
- Ground all equipment to safety ground for shock protection.
- Avoid use of twisted-wire grounds on cables, especially on those carrying signals above 1 MHz.
- Insulate cable shields to prevent undesired grounding. Random contact between shield braid and chassis can result in noise.
- Cable shields should be grounded, at a minimum, at both ends. Optimally, cable shields should be grounded every 1/20th or 1/10th of the wavelength of signals carried by the shield.
- Shields should not be used for signal return, unless the shield is part of a coaxial cable carrying that signal.
- Use floating ground if interference caused by ground loops are a problem.
- Use separate circuit grounding systems for signal returns, signal shield returns, the power system and chassis grounds. Tie all of these grounds together at a single reference point.
- Isolate the grounds of low-level signals.
- Use multipoint grounding of the shield of

[7]J. Wick, "Doughnuts for the Tennessee Valley Indians," March 1982 *QST*, pp 16-19.

[8]J. Osburn and D. White, "Grounding: A Recommendation for the Future," *1987 IEEE International Symposium on Electromagnetic Compatibility*, pp 155-60.

coaxial cables used for high-frequency circuits.

Operating Practices

Although it is no substitute for proper shielding, grounding and filtering, modified operating practices can be an effective means of avoiding EMI. Abstain from transmitting, or operate at relatively low power levels, during prime-time TV viewing hours. Also avoid operating on frequencies with a high potential for EMI. For example, the second harmonic of 10 meters falls within TV channel 2 (54-60 MHz). Since the lowest harmonic frequency is 2 MHz above the low edge of the TV channel, fine cross-hatching of the TV picture can be avoided by restricting operation to 29 MHz and above. The only restriction is that the second harmonic must be kept 200-500 kHz from the TV sound carrier, which is up 5.75 MHz from the low end of the TV channel. Thus, avoid 29.875 MHz (± 250 kHz) as well.

Other operating practices that can reduce the potential of EMI include properly adjusting the drive and tuning of final-amplifier circuits. Mistuning the output of a tube-type amplifier or overdriving the final amplifier can result in very high harmonic output. Similarly, an improperly neutralized transmitter final amplifier may produce parasitic oscillations. A Transmatch can increase the selectivity of a transmitter output circuit, thereby reducing harmonics and parasitics at the antenna.

Antenna Management

Most amateurs select an antenna based on power gain figures, radiation pattern, beamwidth, space requirements, weight, wind survivability and cost. Many of us fail to recognize that antenna polarization, physical configuration, height and orientation affect not only communications, but also EMI. For example, single-band antennas, tuned and resonant at a single band, are much less efficient radiators of harmonics than are multiband antennas.

Often what is best for communications is also optimum from an EMI-prevention standpoint. For example, a narrow beamwidth is not only effective against QRM, but it also restricts the area potentially affected by EMI. A pair of stacked antennas compress the vertical radiation pattern, relative to a single larger

one. A four-bay antenna array, with four antennas stacked two high and two wide, compresses the radiation in both the vertical and horizontal planes.[9]

Increased antenna height not only improves communications, but also reduces the field intensity of transmitted signals at nearby homes. Place the transmit antenna as high above ground as possible to increase the separation between the antenna and neighbors' equipment. Keep the transmit antenna away from CATV and power lines. Energy coupled into these lines may be conducted into neighboring houses. On the way, rectification may occur at corroded connections and junctions of dissimilar metals, resulting in EMI. A beam directed at a CATV system may result in EMI, simply because the connectors or housings of amplifiers and taps are inadequately shielded.

The optimum antenna polarization for EMI control depends on the situation. A switch from vertical to horizontal polarization has been effective in some TVI cases.[10] Vertically polarized transmit antennas, however, induce stronger common-mode signals in nearby cables than do their horizontally polarized counterparts (D. Potter). Experiment to determine what is best for a particular case. Also consider the physical configuration of the susceptor. Horizontally oriented CATV wiring, for example, picks up more RF energy from horizontally polarized antennas.

Most of us strive for a low SWR in our antenna systems. This is reasonable for increased communications efficiency, but a high SWR is often erroneously presented as an intrinsic cause of EMI. While several factors in the antenna system have specific EMI significance, high feed-line SWR does not create EMI problems. For example, it is important that any transmit filters have an appropriate load. That end, however, is easily achieved by placing the filter between the transmitter and Transmatch. An antenna system that is unmatched at the fundamental frequency, however, does not significantly increase feed-line radiation and

[9]S. Katz, "The Interference Survival Guide," Jan 1988 *CQ*, pp 64-66.
[10]C. Schauers, "Ham Clinic," July 1961 *CQ*, pp 75-77.

has little or no bearing on EMI.

Design Practices

Amateurs have an obligation to reduce harmonics and spurious emissions in accordance with good engineering practice. This responsibility not only encompasses the operation of commercial equipment, it should also govern the design and construction of home-built equipment. Because of the typical amateur's limited resources, it is much better to provide for EMI control in the design stages than to correct EMI problems once they appear. For example, although a spectrum analyzer is the best way to ascertain that transmitter output is clean, few amateurs have access to one. In contrast, the following design guidelines for minimizing EMI can be followed by virtually all amateurs:

- Use N or SMA connectors, protected by several layers of good quality vinyl tape to assure clean, oxide-free connections. UHF and BNC connectors tend to leak RF energy.
- Avoid long cable runs in close proximity to one another (to minimize crosstalk).
- Minimize wire length whenever possible to reduce the potential for RF radiation and pickup.
- Isolate return lines of noisy components from the return lines of sensitive components.
- Use thick wire insulation to reduce capacitive coupling between wires.
- Choose minimum signal levels that are consistent with the needed signal-to-noise ratio.
- If signal and power leads cross, make the crossing perpendicular to minimize coupling.
- Use decoupling capacitors or a voltage regulator to decouple circuits from the power supply.
- Use wired returns rather than structure returns to ensure a clean, single-point ground that is free of ground loops. Structure returns (for example, chassis ground) may be more convenient, but voltage drops may occur in the structure used for the return, which may in turn induce voltages into sensitive circuits. (This does not apply to PC-board ground planes.)
- When working with digital circuits, use slow, low-power ICs where possible. The EMI potential of a device is directly proportional to its operating frequency

and output current, and inversely proportional to its rise and fall times.[11] When the output transition times are much shorter than required by the load input, the transition can be slowed by adding a small series resistor (25 Ω) at the output pin, and a small capacitor (50 pF) between the output pin and ground.

- Minimize radiated EMI from PC-board traces by positioning each RF load as close to its signal source as possible. Radiation is a direct function of the RF current path length (Cooperstein).
- If a noisy line must cross a PC board, locate the input and output connectors as close to each other as possible, and run separate ground traces for each line.
- Connect unused IC inputs and outputs to ground or V_{cc}, according to the manufacturer's recommendations.
- The ideal PC-board aspect ratio is 1:1 (Cooperstein). The less a printed circuit board approximates a square, the more difficult it will be to design a good EMI-proof layout.
- Lay out op-amp circuits so that the input and output traces are as far from each other as possible (to prevent amplifier oscillations).
- Don't cover unused circuit board areas with ground plane. The area enclosed by return currents constitute a radiating loop, and additional ground plane area may increase EMI by increasing the effective loop areas.[12]
- Route all conductors subject to common-mode pickup tightly in cabinet corners or otherwise close to a ground plane (to minimize radiation and coupling).
- Shielded twisted-pair wiring can help protect susceptible circuits.

Side Effects

Some might argue that a good working knowledge of the manifestations and

[11]B. Cooperstein, "Design Techniques to Minimize Electromagnetic Emissions from PCBs," 2(1): 1991 *EMC Test Design*, pp 38-40.
[12]A. Swainson, "Radiated Emission and Susceptibility Prediction on Ground Plane Printed Circuit Boards," *Sixth International Conference on Electromagnetic Compatibility*, 1988, pp 295-301.

root causes of EMI is unnecessary, that a random trial-and-error approach can be equally effective. As an illustration, examine the following list of actions known to be effective against EMI:

- Install a filter in the ac line to the affected device.
- Install a 0.01-μF capacitor across the speaker terminals. *(Don't do this!)*
- Install a common-mode choke in the speaker leads.
- Connect the affected device to a good RF ground.
- Use grounded, shielded cable for speaker leads.
- Install a 0.01-μF capacitor between the affected device chassis and ground.
- Ground the device coax shield.
- Install an RF choke in the antenna coax.
- Install a shielded, grounded high-pass filter on the device antenna terminals.

All of these suggestions would likely control most cases of fundamental overload. However, each of these actions may also have unplanned and undesirable effects on the equipment involved. The more troublesome side effects, including parasitic oscillations and high-frequency attenuation in audio amplifiers, hum in phonograph preamplifiers and the "hot-chassis" syndrome, may cripple or destroy the affected device.

Bypass capacitors installed on speaker leads of transistor amplifiers may cause inaudible feedback. The result is oscillations that may destroy the audio amplifier. The Stereos chapter fully discusses this problem and appropriate cures.

Reduced high-frequency response can result when a phono preamplifier for a magnetic cartridge is bypassed. This happens because the bypass increases capacitance in the signal path beyond the maximum load capacitance of the cartridge. Similarly, RF chokes in phono preamplifiers (which are high-gain, high-impedance circuits) may cause hum when nearby magnetic fields induce voltage in the choke. Shield the choke or replace it with a ferrite bead (over the input lead of the first amplifying transistor in the preamplifier circuit) to eliminate the hum.

An RF choke in series with the speaker leads not only breaks the RF path, but also attenuates higher audio frequencies. Even if the fidelity loss is not noticeable, the choke may cause the amplifier load to

vary dynamically. Since some receivers and amplifiers can't tolerate loads of varying impedance, a ferrite bead on each speaker lead, or a toroid common-mode choke may be a better solution.

Bypass capacitors at the primary of a power transformer prevent RF from entering via the ac line, but they may create a shock hazard. The capacitors form a voltage divider that places the chassis above ground potential. Underwriter's Laboratories specifies that the maximum leakage current through any such capacitor should be less than 0.5 mA. Hence, use 0.01-μF capacitors or less. Also, any capacitors installed in ac-line circuits should be rated for that use (1.4 kV is the typical ac rating). An external power filter is a better alternative in most cases. External filters usually work, and anyone can install them.

Given these and other possible side effects of EMI control, it is best to use only minimum control methods after a detailed examination of all emitters, transmission paths and susceptors involved. Examine each in terms of possible EMI control techniques and potential side effects. While certain exploratory methods can and should be used in determining these techniques, the most appropriate method of EMI control varies from one situation to the next. That is where knowledge of EMI fundamentals becomes indispensable.

GLOSSARY

BCI—broadcast interference; interference to broadcast receivers.

Bonding—the establishment of a low-impedance path between two metal surfaces; the physical implementation of grounding.

Broadband emission—an emission that has a spectral energy distribution sufficiently broad, uniform and continuous so that the response of the measuring receiver does not vary significantly when tuned over a specified number of receiver bandwidths.

Bypass capacitor—a capacitor used to provide a comparatively low-impedance ac path around a circuit element.

Common-mode signals—signals that are in phase on both (or several) conductors of an signal lead (often the antenna lead).

Conducted emission—a signal that

propagates through an electrical conductor or any conductive structure. The level of conducted emissions is usually expressed in terms of voltage or current (for example, V or dBV).

Counterpoise—the reference-plane portion of an unbalanced antenna.

Crosstalk—an electromagnetic disturbance introduced by unwanted coupling between conductors.

Cross modulation—modulation of a desired signal by an undesired signal.

Decibel—a logarithmic unit of relative power measurement that is used to express the ratio of two power levels. It is equal to 10 times the common logarithm of this ratio.

Desensitization—a reduction in receiver sensitivity caused by RF overload from either noise or a nearby transmitter.

Electromagnetic compatibility (EMC)—the capability of electronic equipment or systems to be operated with a defined margin of safety in the intended operational environment, at designed levels of efficiency, without degradation from interference.

Electromagnetic interference (EMI)—any electrical imposition that may interfere with the normal operation of equipment. EMI is said to be present when undesirable voltages or currents adversely influence the performance of a device. This term encompasses interference from radio-frequency sources (RFI), audio-frequency sources (that is, induced hum from a transformer) and electrostatic sources (ESD—electrostatic discharge).

Electrostatic discharge (ESD)—a flow of current that results from static electrical charges. This term covers events from doorknob shocks to lightning strikes.

Emission—electromagnetic energy propagated from a source by radiation or conduction.

Emitter—a source of electromagnetic energy.

Filter—a network of resistors, inductors or capacitors that offers comparatively little opposition to certain frequencies while blocking or attenuating other frequencies.

Functional interference—occurs when the normal functions of one system part directly interfere with those functions of another part. Functional interference is generally easy to resolve because the frequencies and power levels are established by design.

Fundamental overload—unwanted desensitization or generation of spurious responses in a receiver that is caused by large amounts of RF energy from a nearby transmitter fundamental output signal.

Grounding—the establishment of an electrically conductive path that connects electrical and electronic elements of a system to one another, or to some reference point that may be designated as "ground."

Ground plane—a metal sheet or plate used as a common circuit return or reference point for electrical or signal measurements.

High-pass filter—a filter designed to pass all frequencies above a certain cutoff frequency, while rejecting those below the cutoff frequency.

Intersystem interference—when the source-coupling-receptor routes of interference include two or more separate and discrete systems.

Intrasystem interference—when the source-coupling-receptor routes of interference are located within a system.

Intermodulation distortion (IMD)—the undesired mixing of two or more frequencies in a nonlinear device, which produces additional sum and difference frequencies.

Low-pass filter—a filter designed to pass all frequencies below a certain cutoff frequency, while rejecting those above the cutoff frequency.

Narrowband emission—an emission that has its principal spectral energy within the passband of the measuring receiver.

Noise—anything that interferes with the exchange of intelligence in electronic communications.

Nonlinear—having an output that does not rise or fall in linear proportion to the input.

Notch filter—a filter that rejects or suppresses a narrow band of frequencies within a wider band of desired frequencies.

Passband—the band of frequencies that a filter conducts with essentially no attenuation; the frequency range in which a filter is intended to pass signals.

Peak output—RF output power (averaged over a single carrier cycle) at the maximum amplitude that can occur with any combination of signals that may be transmitted; the maximum instantaneous power output from the transmitter.

Radiated emission—RF energy that is coupled between circuits, equipment or systems via electromagnetic fields. Radiated energy leaves the source and spreads out in space according to the laws of wave propagation. The strengths of radiated emissions are usually expressed in terms of power density or field strength (for example, V/m and dBV/m).

Receptor—the generic class of devices, equipment and systems which, when exposed to conducted or radiated electromagnetic energy, either lessen performance or malfunction.

RFI—radio frequency interference. Interference to AM and FM radios as well as to various appliances, such as computers, audio systems and telephones, from a source of RF energy. This is a specific case of EMI.

Selectivity—the ability of a receiver to reject unwanted signals. Selectivity, a measure of equipment quality, is a critical factor in most interference cases.

Service entrance—the point where some utility, usually ac power or telephone, enters a building.

Spurious emission—any electromagnetic emission, from the intended output terminal of an electronic device, that is outside of the designed emission bandwidth.

Spurious response—any response of an electronic device to energy outside its designed reception bandwidth through its intended input terminal.

Stopband—that part of the frequency spectrum that is attenuated by a filter.

Susceptibility—the characteristic of electronic equipment that permits undesirable responses when subjected to electromagnetic energy.

Susceptor—a device that responds to unwanted electromagnetic energy.

TVI—interference to television receivers.

Chapter 3

Troubleshooting

By Ed Doubek, N9RF
Illinois Section Technical Coordinator
25 W 062 Wood Ct
Naperville, IL 60563

N ow that you have learned the important fundamentals about EMI, it is time to discuss troubleshooting techniques that will help you locate the source of a specific problem. Consider the various ways an RF signal can get from your transmitter to a neighboring piece of equipment. By the time you factor in all of the variables and the possible cures, the possible combinations can number in the millions. A systematic approach to troubleshooting should be used.

Is interference being caused by harmonics from the transmitter? Are harmonics being generated outside the amateur station? Is it a case of fundamental overload? Is the RF energy conducted along the power lines or picked up directly by the affected equipment? These are the kinds of questions that you must answer as you troubleshoot a case of EMI.

Many things other than amateur transmission can cause EMI. Fluorescent lamps, neon signs, electric motors and transmitters licensed and unlicensed to other services can all contribute to the problem. (See Fig 1.) A TV set can even interfere with itself.

The process of EMI elimination is composed of two phases. Phase one identifies the root cause of the EMI problem. Once the source and victim are identified, and the means of interference is understood, phase two begins. Phase two determines solutions to the problem. This chapter deals mostly with phase one problem causes and suggests some basic solutions. Several easily performed tests are described that make the cause analysis

easier. More detailed explanations of specific cures appear in other chapters of this book. The main steps to troubleshooting any EMI problem are outlined in Table 1.

Many of the techniques discussed in this chapter are most appropriate to an HF transmitter. This is not meant to exclude VHF (and UHF) transmitters. Most VHF operation uses 25 W or less power, which reduces the likelihood of problems. If a VHF transmitter reached the market since 1978, the FCC VHF spectral-purity regulations (spurious emissions less than − 60 dBc at power levels of 25 W or greater), provide greater protection than is afforded HF transmitters. In addition, only a few TV channels are harmonically related to the VHF fundamental signal. All these factors tend to make television interference (TVI) a smaller problem at VHF than at HF.

Of course, a 6-meter operator who lives in an area with channel-2 TV service might see things differently. Most TVs don't have enough front-end selectivity to reject a 6-meter fundamental signal.

Often the best troubleshooting technique is to determine what remedy eliminates or reduces the interference. Thus the distinction between troubleshooting and an actual cure is often blurred. This chapter concentrates on determining the

Fig 1—A few possible RF (EMI) sources. Each transmitter or noise source can interfere with nearby equipment. The Amateur Radio Service is *not* the only possible source of EMI.

Communications Service Transmitters
CB
Amateur Radio
Police
Cable TV (CATV)
Business or other two-way
Aircraft (near airports only)

Electrical Noise Sources
Doorbell transformers
Toaster ovens
Electric blankets
Fans
Heating pads
Light dimmers
Appliance switch contacts
Aquarium or waterbed heaters
Sun lamps
Furnace controls
Smoke detectors
Smoke precipitators
Computers (and video games)
Ultrasonic pest-control devices
Lights: fluorescent, mercury vapor
    and touch-controlled
Neon signs
Power Company Equipment
    Defective line insulators
    Loose or unbonded hardware
    Discharges from defective
        lightning arrestors
    Defective transformers
Electric fences
Alarm systems
Loose fuses
Sewing machines
Electrical toys (such as trains)
Calculators
Cash registers
Lightning arrestors

Table 1
An EMI Troubleshooting Procedure

The Amateur Station

1. Learn about EMI.
2. Obtain outside assistance.
3. Check the amateur station.
4. Install a transmit filter at the amateur station.
5. Try an ac-line filter and common-mode choke at the amateur station.
6. Install a good RF ground at the amateur station.
7. Check the integrity of station connections. Corrosion and arcs can cause interference.
8. Cure interference to home electronics equipment in the amateur household.
9. Maintain an antenna-connected TV monitor at the amateur station.
10. Simplify the problem—disconnect (completely) all accessories and all but one transmitter and determine if the problem still exists. Reconnect one unit at a time.
11. Change one, and *only* one, thing at a time.
12. Keep test conditions (frequency, power level) constant.
13. Consider ways to reduce the signal strength at the affected equipment: Reducing transmit power and altering antenna placement are two possibilities.
14. Eliminate any problems from the amateur station first before working on the consumer equipment.

The Consumer Equipment

1. Maintain a spirit of cooperation between the amateur and neighbor. Nothing gets solved with anger and bad feelings.
2. Simplify the affected system as much as possible.
3. Install appropriate filters on the consumer equipment. This can include differential-mode high-pass filters, common-mode chokes for signal and power-supply leads and ac-line filters.
4. Try one, and *only* one, thing at a time.
5. Disconnect the equipment from long leads to check for direct radiation pickup.
6. Inspect the consumer-electronics ground connection. *(Obey all necessary safety precautions.)* This may include grounding the equipment or feed lines or breaking ground loops with common-mode chokes.
7. Ensure that the consumer equipment is installed in accordance with good EMC engineering practice.
8. If possible, increase the strength of the desired signal. A better TV antenna or *good* distribution amplifier may help in some cases.
9. CATV does *wonders* for TVI!

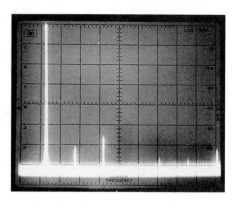

Fig 2—A spectral photo from a typical amateur transceiver. The fundamental (14.2 MHz) appears at left. Harmonics are visible at 28.4 and 42.6 MHz. Amateur stations are subject to stringent FCC spectral-purity regulations (this transceiver complies).

approach your neighbor and responsibility issues are discussed more fully in the First Steps chapter.

When there is interference to audio devices (phonographs, tape decks, most intercoms, burglar alarms and so on) you are most certainly dealing with fundamental overload. The FCC *Interference Handbook* pamphlet indicates that when these devices are subject to interference, they are *improperly functioning as radio receivers.* (See Fig 4.) Thus you need not troubleshoot the station ground, look for harmonics or other causes. In the case of fundamental overload, the only things you can do at your station are those that reduce your field strength at the affected equipment. This includes reducing power and relocating antennas as far as possible from the susceptible equipment.

The cures to fundamental-overload problems are found at the other end—with the faulty equipment. The necessary filtering, shielding or grounding must be applied *there*.

ENSURE THAT YOUR OWN HOUSE IS CLEAN!

To show that your station is free of spurious emissions that cause harmful interference, fix your own consumer-electronics equipment. If an antenna-connected television in your own house is not subject to interference on any channel while you are operating, you can be sure that your station is not a source of TVI problems (see Fig 5). The FCC will see it that way, too!

interference causes, but must often (by necessity) discuss techniques that are covered in more detail in other chapters.

SPURIOUS EMISSIONS V FUNDAMENTAL SIGNALS

It is important to understand amateur responsibilities. There are two ways a station can affect electronic equipment. All transmitters generate small signals outside of the intended bands (see Fig 2). If these *spurious* emissions exceed the FCC limitations for out-of-band signals or cause interference to another service, it is the amateur's responsibility to reduce

these signals as necessary to eliminate the interference.

Many cases of interference result from the inability of consumer equipment to reject strong, out-of-band signals. This is known as *fundamental overload* (see Fig 3). This fundamental RF energy is the signal from your transmitter that is *supposed* to be there—the one you are trying to make as strong as possible. It is, by far, the strongest signal emitted from your station. You are not *legally* responsible for interference resulting from your fundamental signal, but practical considerations can be important. The best way to

It may take a bit of work—installing high-pass filters on the TV, debugging telephone interference, getting the voices out of your stereo—but it will be time well spent. It is very persuasive to show neighbors that if they can't watch their television while you are on the air, they can watch at your house. It usually brings home the point that the problem is in the affected TV.

THE PRELIMINARY STEPS

If you want to locate the cause of an EMI problem quickly, first *simplify the problem.* Every EMI problem is a puzzle, and puzzles with fewer pieces are easier to solve (see Fig 6).

For example, when troubleshooting a case of telephone interference, first disconnect all but one line and one telephone, preferably near the telephone service entrance. Each line, instrument or accessory in a system may cause EMI, and several causes may interact. It is much easier to fix a simple system. Once a minimum system is fixed, reconnect components one at a time. That way causes and effects are more easily recognized and problems can be handled individually.

At the transmitter, remove all non-essential accessories. Disconnect all but one transmitter feed line and antenna (remember to include VHF transceivers, scanners, and so on). See if the problem still exists. If it does, troubleshoot and eliminate the problem, and reconnect equipment one unit at a time, eliminating EMI as you go.

In the case of television interference (TVI), temporarily disconnect any extra TVs, VCRs, games or splitters. (In cable systems, use only one cable-ready TV or the converter supplied by the cable company.) Once the interference is eliminated from a single TV, reconnect equipment one unit at a time, debugging EMI as it recurs. With luck, you will cure all of the problems. If not, at least you will find the specific equipment that causes the problem.

Try one (and only one) thing at a time. If you change too many variables at once, you won't know which change cured the problem. It is useful to know specific causes if you want to make future changes to your station. (You will know the most prevalent mechanism of EMI at your location.)

When testing to determine the cause of interference, or to see if an attempted

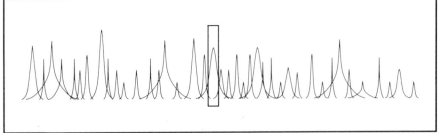

Fig 3—Fundamental overload results when equipment fails to reject signals it was never intended to receive. Here we see a desired signal (in box) that must be selected from many in the spectrum.

PART II

INTERFERENCE TO OTHER EQUIPMENT

CHAPTER 6

TELEPHONES, ELECTRONIC ORGANS, AM/FM RADIOS, STEREO AND HI-FI EQUIPMENT

Telephones, stereos, computers, electronic organs and home intercom devices can receive interference from nearby radio transmitters. When this happens, the device improperly functions as a radio receiver. Proper shielding or filtering can eliminate such interference. The device receiving interference should be modified in your home while it is being affected by interference. This will enable the service technician to determine where the interfering signal is entering your device.

The device's response will vary according to the interference source. If, for example, your equipment is picking up the signal of a nearby two-way radio transmitter, you likely will hear the radio operator's voice. Electrical interference can cause sizzling, popping or humming sounds.

Fig 4—Part of page 18 from the FCC *Interference Handbook* (1990 edition) explains the facts and places responsibility for audio rectification.

Fig 5—Here we see the difference a few filters can make. As an RFI troubleshooter and responsible ham, make sure your TV picture looks like the one on the left when you transmit.

cure has had any effect, keep the test conditions constant. It is difficult to notice improvements if you are constantly changing power levels, frequency or other test conditions. If you need to test at multiple operating frequencies and power levels, eliminate the problem at one set of conditions before you move on to the next set.

The Amateur Station

Although it is not exactly a "troubleshooting" step, the installation of a transmit filter and adequate ground at the amateur station is an important first step. It may eliminate the interference—and the need for further investigation. The effects of RF grounds on EMI is a complex subject. Even the experts don't always agree on how RF grounds actually accomplish EMI control.

A Transmit Filter

The transmit filter for an HF transmitter is usually a low-pass filter, rejecting frequencies above its cut-off frequency (40 MHz, or so). An antenna tuner can also function as a transmit filter to a limited degree, but under some circumstances it may offer only a few dB of attenuation to VHF signals.

Commercial transmit filters for VHF (and up) are not common, mostly because they are not often needed. Amateurs sometimes construct transmit filters for VHF transmitters. These are usually bandpass filters, although a low-pass filter can be used effectively on a VHF transmitter. As you read on, keep in mind that a VHF transmit filter is required only when there are symptoms of harmonic interference or IMD in the transmitter output stage.

Every amateur who expects to encounter interference problems (and this probably includes nearly every amateur) should have a filter on the transmitter or transceiver antenna lead (see Fig 7). It is not a cure-all for interference, but if an interference case is ever brought before the FCC they will want to know that a transmit filter has been used. When confronted with interference, an amateur can point to the filter with pride.

A transmit filter should be installed with short leads, between the transceiver and the antenna or Transmatch. (Filters require a matched load to function properly, so use a Transmatch if needed.) If an amplifier is used, the filter should be installed after the amplifier, between the

"Y'know OM, you really ought to check that line sometime"

A GOOD TVI EXAMPLE

If an amateur can demonstrate that a TV set reasonably close to the transmitter has no TVI at the same time that the affected device shows interference, the amateur station emission is clean. (Any spurious emissions affect *all* TVs within range.) Nothing further can be done at the transmitter to help.

The example TV need not be the ham's own set, but it is a good idea for demonstration purposes. Many TVs are quite susceptible to fundamental overload, so it is a good idea to carry a portable TV for tests.

Filters should be installed on all of the amateur's TV sets, whether needed or not. This demonstrates that filters are effective and do no harm.

Corollary:
If the amateur station is interfering with a known clean TV set, the station must be cleaned up before anything is done with the TVs. The susceptible TVs may also need work of some sort, but the transmitter *must* be clean before any sensible diagnosis is possible,

These principles may seem obvious, but often hams whose stations are already "clean" put wasted effort into adding more transmit filters, shields, grounds and so on. Of course the results are often "no improvement." Sometimes even skilled troubleshooters neglect the obvious.—*Dave Heller, K3TX, ARRL Assistant Director Atlantic Division, Yardley, Pennsylvania*

amplifier and antenna tuner. (In some cases a second filter between the exciter and amplifier helps.)

If there is any suspicion that the transmitter might be coupling harmonic or fundamental energy into the ac-power lines, install an ac-line filter at the transmitter. This may cure the EMI problem, and it eliminates the transmitter as a direct source of power-line RF energy. This is discussed again later in this chapter and in the electrical-interference chapter.

Ground

The first reason to incorporate a good

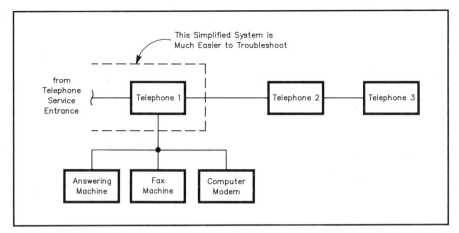

Fig 6—Simplify the systems you must troubleshoot. If interference comes from a system accessory, it might take months to find the culprit with all accessories connected.

ground in the amateur station is for the safety of the operator and the building housing the station. Virtually all local building codes and the *National Electrical Code* require this. If an amateur station is located in the basement of a building, a ground installed through the basement floor may conduct unwanted spurious energy harmlessly away from the transmitter to ground.

In amateur stations above ground level, the ground lead becomes longer and may, itself, radiate energy. The rule to remember is that an RF ground is not an RF ground when the lead connecting it to a transmitter is over about 1/16 λ long. If it is longer than 1/16 λ, at the fundamental or *at the frequency of any spurious emission,* the ground-lead impedance increases and the ground lead may radi-

ate significantly. This is acceptable, in some cases, but in others (especially when the ground lead passes near the affected equipment), a long ground lead can actually worsen the interference.

The discussions of grounds in the Fundamentals and Transmitters chapters explain how grounds contribute to the effectiveness of transmit filters. Those chapters also discuss how grounds may increase radiation from the transmitter cabinet and accessory wiring. Station grounding is a complex subject that is not very well understood. If the transmitter is not connected to a low-impedance RF ground, spurious energy may be radiated directly from the chassis or return to earth ground through ac power system. The FCC expects you to have a good station ground.

A good RF ground is difficult to attain. Even an 8-ft ground rod can have a few dozen ohms of contact resistance in poor soil. Most ground systems use a shorter rod. (Have you ever driven an 8-ft ground rod into even the softest soil?) Remember that at VHF, the frequency range at which spurious emissions cause the most problems, even a few feet of wire is a significant part of a wavelength. This means that a few feet of wire is also a fairly efficient radiator at VHF. When a ground wire passes close to a television-set antenna (not uncommon if the station is on the second floor of an apartment building), that TV may pick up quite a bit of harmonic energy from the ground lead.

At VHF, the length of a ground lead is almost always longer than 1/16 λ. At higher frequencies there is a "skin effect" whereby the RF energy flows only on the surface of a conductor, increasing the resistance. These two effects make it difficult to achieve a good RF ground at VHF.

The EMI Fundamentals chapter discusses the principles of a good ground. Also see "Eliminate TVI with Common-Mode Current Controls" (May 1984 *QST*, pp 22-25) for a discussion of lossy dielectrics as effective RF grounds.

Quick Fixes For The "Other End"

Another diagnostic trick: Install appropriate filters on the equipment that is receiving interference. A high-pass filter (common-mode and differential-mode), telephone filter or three-wire ac-line filter with a separate common-mode fer-

---

AMATEUR STATION TVI CHECKLIST

Have proof that your station is clean. Be sure to separate wishful thinking from reality. Thinking or saying that station output is clean may sound good, but it doesn't help the neighbors. Children have become the innocent victims of neighborhood TVI disputes. The following check list should make it easier to objectively examine a station and determine the source of a problem.

- _ Is the station grounded with a short low-inductance ground lead?
- _ Is the SWR low at the transmit filter?
- _ Does TVI increase when the transmitting antenna points toward the TV? (Fundamental Overload)
- _ Always use the same frequency for TVI tests.
- _ Is the TVI level constant as the antenna rotates? (Harmonics)
- _ Does TVI cause the TV raster to change with no TV antenna connected? (Direct Radiation Pickup)
- _ Are the drive level and any narrow-band stages tuned for minimum harmonic output?
- _ Does TVI decrease when any VHF radio equipment is disconnected from its antenna?
- _ Observe a monitor TV while transmitting.
- _ Is all station equipment bonded to a common ground?
- _ Are all station cables and connectors in good shape?

After completing the amateur station TVI checklist proceed to the TV receiver checklist. Then make a complete analysis of the transmitter, affected equipment and interference path to solve the EMI problem.

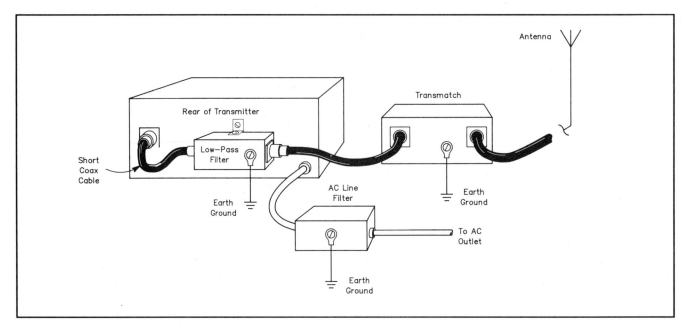

Fig 7—Properly installed low-pass and ac-line filters.

rite-core choke can usually be applied quickly and often results in an instant cure.

Every TVI troubleshooter should keep a variety of filters on hand for test purposes. Several filters appear in Figs 8 through 12. ARRL HQ receives a lot of EMI letters that indicate there are no filters installed on the consumer equipment. Always try the easy things first—if they are real easy, try them twice!

This does not *directly* apply to VHF transmitters. A VHF signal is within the passband of a TV high-pass filter. A VHF signal picked up by a TV antenna passes through the high-pass filter to the TV set. A band-reject filter can help reject VHF fundamental energy. Put it right at the TV terminals. If a mast-mounted TV preamp is used, it may be necessary to install an additional filter at the preamp input. Resonances in the TV feed line may affect filters. Sometimes filters work better at a low-impedance point in the line. It is often difficult to locate a low-impedance point, so experiment with filter position when necessary.

There are two quick fixes for interference from a VHF transmitter. A band-reject filter, tuned to the appropriate VHF amateur band, may eliminate VHF fundamental overload from differential-mode signals that are picked up by the antenna system. TVs and VCRs can also respond to the VHF common-mode signal present on the coax shield. A VHF common-mode

choke can be made using a type-43 ferrite toroid. Wrap about four to ten turns of the coaxial cable through the core and install as close to the TV or VCR (or both) as practical. An example of this filter is shown in Fig 11.

If a VHF signal is apparently entering a CATV system (the Televisions chapter has charts that relate TV channels to amateur bands and their harmonics) try a VHF common-mode choke. If the TV is still receiving the interference you may

---

ADDITIONAL EMI HINTS

Here are a few stock cures that have solved a lot of RFI cases:

1. When using a transmit filter, use a split ferrite bead at each end of the filter. Common-mode signals are then effectively attenuated.
2. Even if there is no transmit filter in the line, a couple of split beads should be installed. If coax is used to feed a balanced antenna, it is a good idea to place a couple of ferrite toroids at the antenna end too.
3. AC power cords should also have toroid chokes, with the wire wrapped through the core several times (in the same direction).
4. Keep all cables (microphone, key and so on) short. If a particular cable must be extended, install a toroid choke in it (close to the rig).
5. Disconnect all unnecessary items from the feed line. Unnecessary items include antenna relays and switches, SWR bridges and monitors. If an SWR bridge must be in the line it should be on the transmitter side of the transmit filter.
6. Ground leads should be short, direct and *thick* (large surface area means low inductance).
7. Commercially available ground rods are usually copper-plated steel, and the simple act of driving them into the earth removes the copper. It is far better to use standard (approximately 1-in.) pipe, available from plumbing supply stores. A ground wire is then easily soldered to the pipe for a corrosion-free connection.
8. Filters must only be used in lines that are terminated in their characteristic impedance. If not, a Transmatch must be installed between the filter and load.—*Howard Liebman, W2QUV, NYC-Long Island Section Technical Coordinator, New Hyde Park, New York*

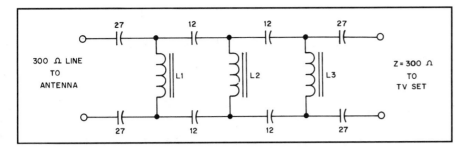

Fig 8—A differential-mode high-pass filter for 300-Ω twin-lead. When installed in the feed line at a TV or accessory input, it rejects differential-mode HF signals picked up by the TV feed line or antenna.

L1, L3—13 turns no. 24 enameled wire on T-44-10 toroid core (0.67 μH).

L2—12 turns no. 24 enameled wire on T-44-10 toroid core (0.57 μH).

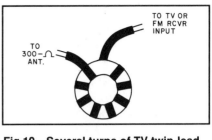

Fig 10—Several turns of TV twin-lead cable on a toroid ferrite core can eliminate HF and VHF common-mode signals from a twin-lead TV system.

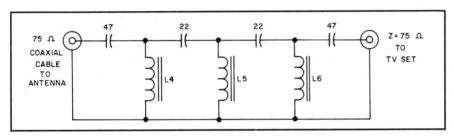

Fig 9—A differential-mode high-pass filter for use with 75-Ω coax. It rejects HF signals picked up by a TV antenna or that leak into a CATV system. It will not reject common-mode signals, HF or VHF, on the outside of the coax. All capacitors are high-stability, low-loss, NP0 ceramic discs. Values are in pF.

L4, L6—12 turns no. 24 enameled wire on T-44-0 toroid core (0.157 μH).

L5—11 turns no. 24 enameled wire on T-44-0 toroid core (0.135 μH).

Fig 11—Several turns of coax on a ferrite core eliminate HF and VHF signals from the outside of a coaxial feed-line shield. This is the only filter that protects against such VHF signals.

have a case of CATV leakage. Contact the CATV company for assistance. This subject is also covered in the CATV section of the TVI chapter.

Harmonic Radiation

Harmonics are a common cause of interference, usually to televisions. A harmonic is an integer multiple of the operating frequency. For example, the second harmonic (f × 2) of a 14.2-MHz signal is 28.4 MHz; the fourth harmonic is 56.8 MHz and so on. The 56.8-MHz harmonic from the last example falls within TV channel 2. Harmonics are caused by insufficient selectivity in transmitter output stages. This can result in a harmonic strong enough to cause interference.

If a harmonic falls within the passband of a TV channel and is not at least 40 dB below the received TV signal, it may cause a herringbone pattern to be superimposed on the intended TV picture. This interference to television reception is commonly referred to as TVI. An example of what a herringbone

pattern looks like appears in the Fundamentals chapter. Charts in the Televisions chapter show amateur harmonics that can affect TV channels.

Transmitters that use a mixing process to obtain the desired output frequency (as do most modern amateur transmitters) can also produce unwanted outputs as a result of the mixing process (see Fig 13). These emissions can cause

interference to other services. Any emission from a transmitter other than the fundamental signal, including mixing by-products and harmonics, is referred to as a *spurious emission* or signal. All transmitters produce some spurious signals. The FCC regulations specify the permitted levels of *spurious* emissions for

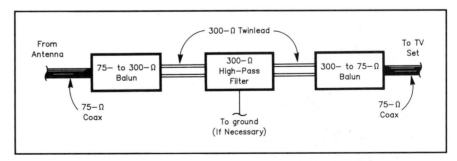

Fig 12—This filter arrangement—two transformers with a 300-Ω high-pass filter—rejects HF signals in both common- and differential-mode interfering signals. It is the best option in coax-fed TV systems. Some 300-Ω high-pass filters (refer to Fig 8) have inductor center taps that are grounded. Such filters are ineffective in this application unless the inductor ground connections are removed.

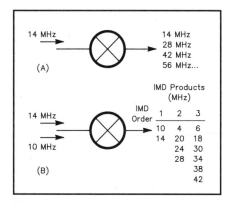

Fig 13—Mixers produce many spurious signals. When there is a single input signal, harmonics result (A). Multiple input signals produce both harmonics and many mixing products known as IMD (B).

"RF Sniffer"

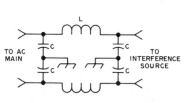

(A)

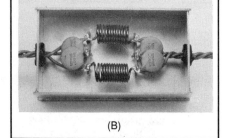

(B)

Fig 14—AC-line filters. A shows a commercial filter available from many sources. B shows a home-built version widely known as a "brute force" line filter.

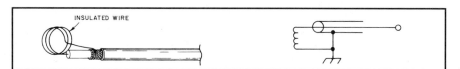

Fig 15—A pickup loop for detecting cabinet radiation and IMD from external rectification. Use the loop with an appropriate radio or TV receiver.

different power levels and operating frequencies. (See the chapter about Transmitters.) An amateur is responsible for interference caused by spurious emissions from the amateur transmitter.

The first test for "spurious" should determine whether the interference is the result of spurious radiation from the transmitter chassis. Connect a dummy load to the transmitter output (in place of the antenna) using coaxial cable. If any interference is present when transmitting into a shielded dummy load, it indicates that the shielding and lead filtering of the transmitter is not adequate. Improved shields and filters or replacement of the

transmitter will be required to eliminate the interference. The techniques for properly shielding and bypassing transmitter key and microphone lines are covered in the chapter about Transmitters. That chapter also discusses techniques used to filter ac-power leads, as does the Power Lines and Electrical Devices chapter. Fig 14 shows ac-line filters (refer to Fig 7 for an installation).

Spurious emissions can leak out of a transmitter cabinet. Even if there are no shielding defects that result in leaks, the equipment cabinet itself can radiate undesired signals. Meter holes, seams, and large gaps are possible sources of cabinet leakage.

While you are transmitting into a dummy load, cabinet leakage can be located by means of a sensing loop and receiver. Make a loop (2 t, no. 20 insulated wire) 3 inches in diameter and mount it at one end of a short piece of coaxial cable (see Fig 15). Connect the cable to a VHF receiver or a TV tuned to the channel of the suspected spurious emission.

As the loop passes near any opening (such as a meter, control knob, or wire that exits the transmitter enclosure), the signal level measured on the receiver S meter, or seen on the TV screen may increase. This indicates hot spots (shield deficiencies).

"EMI tape" can be used to cover seams that allow radiation. Copper screening can be used to create screen cages for meters, and bypass filters can reduce RF energy present on the wires exiting the rig. These techniques are discussed in the Transmitters chapter.

Many accessories such as SWR indicators, external electronic TR switches and monitor scopes all rectify a small amount of RF energy in order to operate. This rectification can generate harmonics. Fig 16 shows test points in a transmitting system. To test, insert a dummy load at each point (sequentially). When the interference returns, the last device added to the system is suspect. In some cases the transmit filter may be needed between the offending device and the antenna.

A good rule to remember: Use as few accessories in the transmitting feed line as possible. Each accessory can pump both spurious and fundamental energy into the system (although this is not very likely).

Harmonics can be locally generated by things like rotator controllers, corroded tower joints and guy wires, and unpowered solid-state equipment (such as a VHF transmitter connected to an antenna). This subject is covered in the chapter on External Rectification.

Another test for spurious radiation in the transmitter output: Insert a transmit filter between the transmitter output and the feed line or antenna coupler. If the interference level is reduced via a low-pass filter at the HF transmitter, spurious radiation is indicated. If no apparent difference is observed, spurious signals may be present, but generated after the filter. Even more likely, the interference is caused by something other than transmitter spurious energy.

Harmonics can be generated outside an amateur station. The tuners in many TVs and VCRs can generate harmonics and other mixing effects when overloaded by strong signals. These harmonics

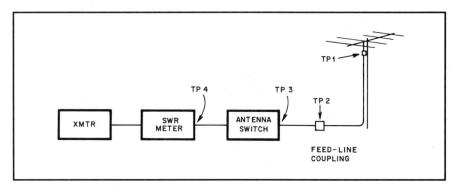

Fig 16—Test points used when troubleshooting a transmitting system. Sequentially place a dummy load at each test point and check to see if the interference is eliminated.

look the same on the TV screen as harmonics generated by the transmitter. Try a differential-mode high-pass filter, with a separate ferrite-core common-mode choke at the TV set. If the harmonics are self-generated, the filter will give a marked improvement.

Even if no difference is noted with this test, it is imperative to retain a transmit filter because it may significantly reduce spurious interference at a distance from your station. (Spurious energy may be masked at your home by some other form of interference such as fundamental overload.)

When installing a coaxial transmit filter, connect the filter to the linear or transceiver with a short feed line. A right-angle connector with a single double-male connector at one end does a good job. If not installed this way, some spurious energy may be radiated from the coaxial cable between the transmitter and filter.

Since modern transmitters must meet FCC regulations (see the Transmitters chapter), the probability of spurious radiation EMI in a reasonable-broadcast service area is low if the transmit equipment is properly operated. The likelihood of interference caused by spurious radiation does increase with power level, so it may be necessary to add filtering, usually a commercially available low-pass filter, to high-power amplifiers.

Incidentally, some exciters exhibit a relatively constant spurious output regardless of the drive-control setting. They can cause a problem if drive is reduced to accommodate an amplifier.

There are many myths about the effect of SWR on interference. By itself,

a high SWR on a feed line will *not* cause interference. However, if you install a coaxial transmit filter it is imperative that a low SWR (1.5 to 1 or so) exists on your feed line *at the filter.*

A typical transmit filter may reduce the spurious output reaching the antenna by 60 to 70 dB when used in a matched transmission line. With a high SWR, the filter effectiveness may be reduced. In addition to increased spurious output, high SWR can cause filter failure from increased feed-line voltage. You may want to read "SWR and TVI," Jan 1954 *QST*, pp 44, 45, 128, 130. Information about line conditions with high SWR appears in *The ARRL Antenna Book.*

Other Station Problems

Harmonics can be generated after RF energy passes through the transmit filter. This can happen at any system hardware or connections that are loose or corroded. For example, if a commercial or home-built spark-gap lightning protector is inserted in the line, bugs, corroded copper or brass pieces can generate harmonics via rectification.

Metallic structures in the near field of your antenna are also subject to these effects. For example, when copper water pipes touch any other metal object (including other pipes or the electrical system ground), corrosion can cause rectification at the points of contact, thus generating harmonic energy after the illuminating RF energy leaves the amateur antenna. This problem can be detected with an absorption frequency meter (see sidebar) tuned to the second harmonic. The RF indicator will peak when passed over the harmonic source. Do not tune the meter to the trans-

ABSORPTION CIRCUIT

A typical absorption frequency-meter circuit is shown in the figure. In addition to the adjustable tuned circuit, L1-C1, it includes a pickup coil, L2, wound over L1, a semiconductor diode, D1, and microammeter or low-range (usually not more than 0-1 mA) milliammeter. A phone jack is included so the device can be used for listening to the signal.

The sensitivity of the frequency meter depends on the sensitivity of the dc meter movement and the size of L2 in relation to L1. There is an optimum size for this coil that must be found by experimentation. An alternative is to make the rectifier connection to an adjustable tap on L1, in which case there is an optimum tap point. In general, the coupling should be a little below (that is, less tight than) the point that gives maximum response, since this makes the indication sharper.

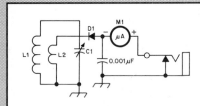

Absorption frequency-meter circuit. The closed-circuit phone jack is optional. Ground the positive meter terminal if the jack is omitted.

Calibration

The absorption frequency meter must be calibrated by taking a series of readings from circuits carrying RF power at known frequencies. The frequency of the RF energy may be determined by means such as a marker generator and receiver. The setting of the dial that gives the highest meter indication is the calibration point for that frequency. This point should be determined by tuning through it with loose coupling to the circuit being measured.—*from The ARRL Handbook (1991 edition)*

---

mitter frequency because any metal in the wall normally results in higher levels of fundamental RF energy. If significant harmonic radiation is indicated by this test, either bond the pieces of metal together or separate them, as convenient. See the External Rectification chapter for a full discussion of this problem.

In an RF system, corrosion means non-linearity, which means harmonics or other undesired mixing effects (intermodulation—IMD). Remember that corroded connections can generate harmonics and other undesired mixing effects. Include a check of these things in your troubleshooting procedures, especially in difficult cases. (Corroded connections in your station installation should be fixed anyway, as part of good engineering practice.)

Wire antennas are often connected to trees instead of towers. If the antenna arcs to tree branches or the trunk, the arcs can disrupt home-electronic equipment. The effect of these arcs normally shows up as a series of black dots on all TV channels or as broadband noise on audio devices. You are responsible for interference caused by arcs that occur in your station equipment, so ensure that the wires of your transmitting antenna cannot arc to nearby objects.

Antenna Support Structures

Here are some guidelines to use when installing a tower:

Make sure all mechanical joints are both mechanically and electrically secure! If the tower is new, the bright galvanize makes a good connection. If the tower is old, shine up the galvanizing with very fine sandpaper at the joints (don't forget the insides). Just brighten it up, don't sand through the coating! Guy lines should be tight, so that they don't move or rub where they are attached. Rohn specifies that guy wires should be tightened to about 10% of their ultimate strength. Tree branches, wire antennas

and other material should not touch the tower as they may arc and cause noise.

Joints between dissimilar metals should be eliminated. For example, never allow bare copper to contact the zinc (galvanize) coating on a tower or guy wire. Not only will the joint quickly form a diode (a rectification interference source), but the zinc will be removed, allowing the steel to rust. Use either bronze or stainless-steel clamps to attach metal to a galvanized tower (or weld the connection).

For lightning and noise protection (and also to comply with the electrical code) ground the tower to a good ground system (Fig 17). The ground system should be specifically designed for soil conditions at the site. Locate the tower ground rods as close to the tower as possible, but at least 2 ft from any concrete base. Connect ground rods to the tower with standard copper aluminum wire with cross-sectional area equivalent to no. 2 (for copper) or no. 0 (for aluminum) solid wire. The radii of any bends in these ground leads should be at least 8 inches. Connect the ground system to your station with no. 6 copper wire; make all runs as short as possible, with smooth bends and never in *metallic* conduit. For additional lightning protection, bond the shields of all cables to the tower where they leave it for the shack. (Be sure to weatherproof the connections.) Avoid braid as a station ground.

There is some controversy about insulating guy wires. If there is concern about resonant guys disturbing the pattern of an HF antenna, use non-conductive guys. If resonances are not a problem (say, for a VHF/UHF-only array), leave the guys continuous and ground them at the anchors for good lightning protection. When using metal guys broken by insulators, ensure that the joints are in good condition without touching or rubbing metal. Keep tree limbs, wire antennas and such well away from guy lines.

The above are general guidelines. In any case, be sure that all aspects of station and antenna installation comply with local electrical codes.

Maintenance

Loose hardware such as clamps, bolts and screws can also contribute to interference problems because they can contribute to arcs. Dissimilar-metal junctions can also cause rectification (harmonics).

Aluminum oxidizes and can cause arcing and high SWR in a system. The near-field intensities of an amateur antenna that is radiating high power can induce voltages in other wires such as guys and metal clotheslines. These can, in turn, arc to tree branches, thus causing interference. This is an uncommon, but easily overlooked, EMI problem.

It is folly to think that an antenna requires no maintenance. Tighten hardware and trim trees, as needed (at least annually). Damage from thunderstorms must also be repaired. The ground for an antenna support structure also needs a yearly inspection.

Summary

Harmonic or spurious-emission interference problems are among the easiest to cure. The proper installation of transmit filters and a proper ground eliminates EMI problems in most cases. The ARRL Technical Information Service doesn't receive many letters about interference that is apparently caused by spurious emissions from transmitters. An amateur is responsible for the effects of spurious emissions from the station equipment and understands how to cure them. The Transmitters chapter discusses cures that should be applied to transmitters.

TELEVISION INTERFERENCE (TVI)

Television interference is mostly a problem of HF operators. This results from a number of factors. HF operation is more apt to use high power. (Even a 100-W "barefoot" rig uses more power than the typical VHF 10- or 25-W FM base station.) The FCC requirements for spectral purity at VHF are more stringent than those at HF. (No spurious emission from a 25-W VHF station can exceed 25 µW.) This results in significantly reduced TVI potential.

TV RECEIVER CHECKLIST

The VHF bands are harmonically related to fewer TV channels than are the HF bands. Note that only the 50-MHz band has harmonic relationship to VHF broadcast TV channels: (1) The second harmonic falls within the FM broadcast band. (2) The fourth harmonic affects channels 11 through 13 (depending on the frequency of 6-meter operation). The fundamental signals and most VHF harmonics fall within various CATV channels.

The most common VHF TVI problems result from CATV leakage, direct pickup by television-set circuitry or TV-set response to a common-mode fundamental signal on the TV coax shield. A 146-MHz fundamental signal affects

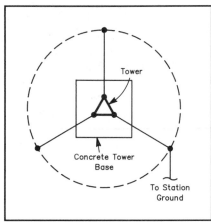

Fig 17—Schematic of a properly grounded tower. All conductors, except the station-ground lead, should be stranded, with cross-sectional area equal to no. 2 (copper) or no. 0 (aluminum) solid wire. The strands must be at least: no. 10 for copper, no. 8 for aluminum or no. 17 for copper-clad steel or bronze. The radius of any conductor bends should be at least 8 inches. All connectors should be compatible with the tower and conductor materials to prevent corrosion. The dashed line represents a buried (1 ft deep), bare, tinned, no. 2 copper wire ring that connects the tower ground rods. Ground rods should be at least 1/2 inch by 8 feet long. More, or fewer, ground rods may be needed for lightning protection depending on soil conditions. (See "How Safe Is Your Ham Shack," Oct 1978 *QST*, pp 39-40 for more information.) Locate ground rods on the ring, as close as possible to their respective tower legs. The station ground leads may be a single no. 10 copper or no. 8 aluminum wire.

cable channel E (channel 18 on most converters or cable-ready TVs); a 222-MHz signal affects channel K (channel 24 on converters and cable-ready TVs). VHF transmitters can also suffer intermodulation problems. A good discussion of intermodulation problems appears in the chapter on External Rectification.

In all cases of TVI, use a TV set to monitor TVI from the shack operating position. If any antenna-connected TV set in the immediate vicinity of the station is interference-free on all channels, then the station is not generating TVI from harmonics or other spurious emissions, or from external effects related to the station (see the chapter on External Rectification). If an interference case is ever brought before the FCC, they are able to quickly determine that the interference is not caused by the

BALUNS

The balun is one of the most misunderstood components in an amateur station. It can help minimize feed-line radiation. Ferrite transformer baluns can saturate under high-power or high-SWR conditions. When ferrite saturates it causes distortion that can generate harmonic energy. An air-core balun is not subject to saturation.

The common question about baluns is, "Do I need one?" The best answer is "Maybe." If the affected equipment is much closer to the feed line than the antenna, radiation received from the feed line might be stronger than the energy received from the antenna. This might be an important consideration in some installations, especially for amateurs who live in apartments.
—*Ed Doubek, N9RF, Illinois Section Technical Coordinator, Naperville, Illinois*

station if a station monitor is clean. If the station has been determined to be clean, then continue trouble shooting at the equipment that is experiencing interference.

The first step in troubleshooting TVI at the TV receiver (after simplifying the system as described earlier) is to install a high-pass filter in the TV feed line. Filters that are effective for TVI are described in the TVI chapter and the filter-study chapter. If an interfering signal enters the TV through the feed line, a filter will eliminate or minimize the problem. A conventional high-pass filter works against differential-mode interference from an HF signal. The two-transformer with 300-Ω high-pass filter in Fig 12 works against HF common- and differential-mode signals. A ferrite-toroid filter as seen in Figs 10 and 11 works against either HF or VHF common-mode signals, but has no effect on HF differential-mode interference. You must decide which filter or combination of filters is best suited for each interference problem (based on the operating frequencies and suspected nature of the interference).

An ac-line filter should be installed on the TV-line cord, as close to the TV as possible. If the interfering signal is conducted via the ac power system, this filter will eliminate or reduce the severity.

If the problem is still evident, the tele-

TV RECEIVER CHECKLIST

Complete the following checklist for each TV with interference.

- Is the received TV signal STRONG?
- Use common-mode, high-pass filters. (See Figs 10 through 12)
- Is there an audio-rectification problem?
- Does the TV antenna have good low-channel gain?
- Could the TV system benefit from a properly installed distribution amplifier?
- Is the TV antenna and coaxial feed line properly grounded?
- Check the area for rectifying devices.
- Is there a line filter at the power plug?
- Is the TV antenna and feed line in good condition?
- Is the TV-antenna feed-line horizontal-run at ground level?

vision circuitry may be picking up signals directly. See the section on direct radiation pickup. (This could also be a case of locally generated harmonics or some form of intermodulation.)

Since the TV set may exhibit varying sensitivity to TVI from different amateur frequencies, keep the transmit frequency constant when testing cures. This frequency-dependent TVI sensitivity can be caused by several factors: The presence of a COMB filter in a TV set causes variable gain across the channel, which can reduce interference at some frequencies. When a harmonic is near either the picture or sound carrier, or near the color subcarrier, the beat frequency is lower and appears as a more coarse (and more visible) interference pattern. As the frequency of the HF transmitter increases relative to the TV video carrier, the video interference pattern is normally less objectionable. Color or audio interference may increase, however. Because of this behavior, video interference caused by operating on phone frequencies is not normally as visible as when operating in the CW portion of a band (but color or audio EMI may increase). Remember, when testing a station for TVI, always listen before transmitting to prevent QRM, and always identify properly.

Direct Radiation Pickup

The term "direct radiation pickup" refers to interference that results when fundamental RF energy is picked up directly by the circuitry of susceptible equipment. As fundamental energy is picked up by the circuitry of a TV (or other piece of equipment), it can result in severe interference.

Measurements in an anechoic chamber indicate that several kilowatts are required to generate this type of interference at the normal distances between an amateur antenna and a TV set. It is highly unlikely that direct radiation pickup by TV circuitry will be encountered unless operating in an apartment building with an indoor or attic amateur antenna. However, this mode of EMI may affect TV remote-control circuitry. Such interference to a remote control is usually eliminated by shielding the leads to the infrared detector on the front of the set.

To test for this type of interference to the picture, replace the TV antenna with an appropriate 300- or 75-Ω termination and note the level of interference to the set raster or remote IR controller. If the raster shows no flicker or drastic change and the IR controller doesn't false, there is no direct-radiation-pickup problem. There must be some other mode of interference such as fundamental overload. With the TV signal removed from the TV set, the TV gain drastically increases (because of the TV circuitry). Hence, this test is a sensitive indicator.

A second test for the presence of direct radiation pickup can be made at amateur stations that have rotatable antennas. Temporarily remove the antenna from the TV set and rotate the amateur antenna while transmitting. If the interference level increases while the amateur antenna is pointing at the house (especially the TV), it is likely that the set has a direct-radiation-pickup problem.

If direct-pickup interference is indicated, contact the TV manufacturer through the Electronic Industries Association (EIA).[1] Additional shielding such

as EMI tape may eliminate most of the chassis pickup. Leave the installation of internal shielding, especially in a neighbor's TV, to qualified service personnel. When shielding is added, devices in a set may overheat if air flow around the devices is changed. Added shielding may also cause short circuits; it is best to turn to the equipment manufacturer for assistance.

In some states, a state-issued electronic-service license is required to work on consumer-electronics equipment. When working inside of a TV set, remember that high voltages remain after the set is turned off. Also remember that in some sets the circuitry and some hardware is dc connected to one side of the ac line (these sets contain no isolation or power transformer). The *117-V ac line can kill!* Please be careful—the Silent Key is not an award. Leave the repair of consumer

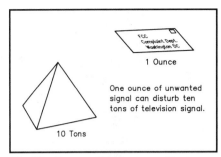

Fig 18—For good picture quality, a television signal must be much stronger than any competing signal (on the same frequency).

equipment to qualified service personnel!

Some hams have learned this the hard way—they have been held responsible for defects that occurred years after they modified a neighbor's set. Allegedly, the failure was caused by the modifications.

ANOTHER HARMONIC TEST

Here is another test to confirm the presence of a harmonic interference problem. Temporarily connect a TV set to the amateur beam antenna normally used with the HF transmitter. Connect a dc voltmeter to the TV AGC line, or disable the TV AGC and use an attenuator between the antenna and TV. Plot a curve of the amateur antenna gain on the channel exhibiting TVI. With an HF beam, you should see a pattern such as that shown in the figure. The pattern shows almost constant gain (loss) in any direction. If you have an old 6-meter converter you can convert it to channel 2 and run a plot of your antenna gain. A TV field-strength meter or a VHF/UHF receiver with an S meter can be used as well. Once you have this plot, connect your amateur antenna to the transmitter and transmit while rotating the amateur antenna and watching a monitor TV. You will see a TVI level corresponding to either the plot of the harmonic response of your amateur antenna or an interference level corresponding to the HF response of your amateur antenna. The presence of an interference level that corresponds to the normal HF directivity (gain) of your ham antenna indicates that TVI does not come from transmitter harmonics.—*Ed Doubek, N9RF, Illinois Section Technical Coordinator, Naperville, Illinois*

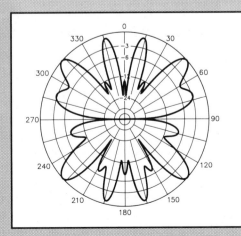

An amateur HF antenna is many wavelengths long at VHF. Here is a typical VHF radiation pattern for an HF Yagi antenna. If interference results from VHF spurious signals created in the transmitting system, the interference magnitude should display similar peaks and nulls as the HF antenna is rotated while transmitting.

---

[1]Electronic Industries Association (EIA), 2001 Pennsylvania Ave NW, Washington, DC 20006, 202-457-4977.

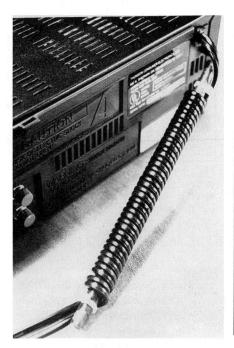

Fig 19—Wrap the ac-line cord around a ferrite rod to break ground loops that include the ac line.

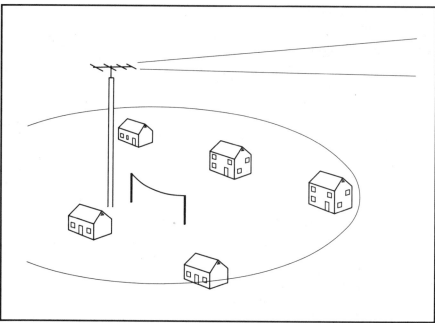

Fig 20—Some towns try to limit antenna height thinking that they will prevent EMI problems. It's easy to see that a higher amateur antenna is better, from both amateur and EMI viewpoints. Here the higher antenna directs energy away from houses and equipment. The lower antenna places its field at ground level.

Fundamental Overload

Well, by now you probably think you shouldn't be having any EMI problems, but this may not be so. Although the internal circuitry of consumer-electronics equipment is fairly immune to strong local signals, the equipment is connected to 117-V ac lines, and antenna or cable systems, that can couple a tremendous signal into the set. Fundamental overload usually generates a herringbone pattern similar to that caused by transmitter harmonics. (The TV cannot distinguish between transmitted harmonics and harmonics generated inside the TV.) Fundamental overload is the most common form of TVI.

Fundamental overload TVI results from the inability of a TV to reject an amateur's transmitted signal. Anechoic-chamber tests show that some TVs have only 30 dB of 20-meter signal rejection. There may be even less rejection of 10-meter and VHF/UHF signals. When compared to a received TV signal, this sensitivity to amateur frequencies is potentially one of the most severe EMI problems (see Fig 18).

Fortunately, a simple test can tell whether a problem is caused by fundamental overload. Refer to the test described for direct radiation pickup, where the amateur antenna was rotated while transmitting and

the TV set was observed. If the interference level increases greatly as the antenna is turned toward the house (the TV set, its antenna and feed line) and decreases in other directions, it is a good indication that the interference is fundamental overload. At VHF, a typical amateur HF antenna is many wavelengths long, resulting in a radiation pattern similar to that shown in the sidebar. As this antenna is rotated through 360°, if the interference problem is caused by VHF transmitter harmonics, the interference will have a series of peaks corresponding to the lobes in the pattern.

The first step in troubleshooting (and often eliminating) a fundamental-overload problem is to install an appropriate filter in the consumer equipment feed line and power leads. Figs 8-12 show examples of TVI filters, for common- and differential-mode HF and VHF signals.

Install the filter as close to the TV antenna terminals as is possible. If you own the TV, you may want to install it inside the cabinet, at the tuner. If the interference is to an audio device, such as a stereo or alarm system, install the filters in the speaker leads or audio input cables, as close to the amplifier as possible. Telephone interference filters should be installed as close to the telephone as possible. Refer to the appropriate chapters in this book for more details about filter

selection. Also, don't forget that dangerous voltages are present inside most consumer-electronics equipment.

You may also want to consider the consumer-electronic equipment ground. A good RF ground may not always be practical, especially in a neighbor's house, but an understanding of the basic principles can sometimes help. A CATV system should be connected to a good earth ground, preferably *outside* the house. This shunts unwanted RF energy on the outside of the coax shield (common-mode) to ground. The electric-power mains and telephone lines coming into the house also have grounds that should be inspected. All utility grounds should be serviced only by qualified personnel.

The consumer-electronics equipment is usually RF "grounded" through the ac-power system. In most cases, this is not a good RF ground. An interference problem can sometimes be cured by improving this ground with an actual earth connection, either to a ground rod or a cold-water pipe that is continuous to earth. In other cases the situation can be improved by RF isolating the equipment ground connection, using 10-20 turns of the ac-line cord on a ferrite toroid or rod, thus breaking the ground loop (see Fig 19). Remember that some equipment has dangerous voltages on parts of the chassis. If you are not *sure*

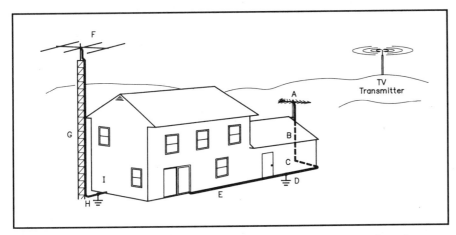

Fig 21—EMI is minimized when antenna and feed-line locations are based on sound EMI thinking. Note the EMI advantages.

A) TV antenna located far from ham antenna
B) TV feed line is run vertically to ground level
C) TV distribution amplifier mounted at ground level
D) TV feed-line shield connected to ground rod
E) TV feed-line horizontal run at ground level
F) Ham antenna does not block TV-reception path
G) Ham feed line runs vertically to ground level
H) Ham feed-line shield connected to ground rod
I) Ham shack located close to station ground

that it is safe to connect an earth ground to any piece of equipment, contact the manufacturer.

BASIC INTERFERENCE REDUCTION TECHNIQUES

Fundamental overload can be reduced by either decreasing the strength of the amateur signal at the affected equipment or increasing the strength of the desired signal. The following techniques often help diagnose or fix problems:

1. Install a TV antenna with greater gain, hopefully with improved directivity.
2. Increase the TV antenna height, thus increasing the desired signal.
3. Increase the height of the amateur antenna, thus moving it farther away from the susceptible equipment (see Fig 20).
4. Move the TV antenna above the roof from under the roof. (Many roofing materials contain aluminum and carbon and attenuate TV signals by up to 20 or 30 dB when wet.)
5. Install a TV distribution amplifier to increase the strength of the desired signal. Some distribution amplifiers have better EMI immunity than a typical TV or FM receiver. They are often available from electronic distributors who sell components to television-service shops. Be wary however: (1) Poor-quality distribution amplifiers can worsen the problem. (2) Distribution amplifiers boost unwanted, as well as wanted, VHF signals.

These signal-increasing fixes all seek to increase the desired-signal to undesired-signal ratio received by the TV set or radio receiver. Another method that may improve the situation is to use an exciter alone or QRP power levels. The few dB lost at the fundamental is not noticed in most contacts. A station power-output increase from 100 to 1000 W typically changes the received signal typically by only 1 to 2 S units, depending on the calibration of the receiver. That same change, however, may increase the interference potential ten-fold. Some amateurs use an amplifier when making the first call and then bypass it to continue. This mode of operating fulfills the FCC regulations and reduces QRM levels on the bands. QRP is not the best EMI-reduction solution, but hams should understand the relationship between transmitter power and EMI problems like fundamental overload.

The TV Installation

When installing a TV antenna system or connecting a TV to a CATV system, it is best to run the TV feed line as directly to ground level as possible and then run the line horizontally in the basement or near the ground surface. This should increase the vertical separation between the amateur antenna and coaxial cable. When a feed line is shorted at one end it acts like a long-wire antenna. The TV antenna and feed line are not a good termination for HF and therefore act as a relatively good antenna at amateur frequencies. The TV feed line can pick up a tremendous number of amateur signals.

Even though a common-mode filter cannot reduce the pickup of your amateur signal by an antenna and feed line (only reducing the total length of the feed line can), it can reduce the coupling of the amateur signal to the TV. Remember that the TV antenna is not effective at HF, but the feed line to the TV is an effective HF antenna. The TV antenna can pick up amateur VHF/UHF signals (usually in a differential mode), and the TV feed line can pick them up in the common mode.

The amateur signal at the TV can be reduced greatly by proper location of the amateur and TV installations. When planning your antenna placement, consider the horizontal and vertical distances between the amateur and TV antennas and feed lines (see Fig 21). Place the amateur antenna as far as possible from the TV system. If you can locate the ham antenna so that there are no TV antennas or feed lines between it and favorite operating directions, the field strength at the TV will be reduced.

With proper antenna placement and orientation, including antenna polarization, harmonic and fundamental overload conditions can be reduced. You may even be able to give nearby neighbors a break with this trick. Since harmonic energy will be picked up to a lesser degree off the back side of a TV antenna about 20 to 40 dB less signal will be received if the ends of each antenna can be pointed away from each other.

Vertical separation between antennas or between the amateur antenna and horizontal runs of the TV feed line is more effective than horizontal separation. Raising the height of the amateur antenna not only reduces take-off angle, but also the field strength at the TV (thereby reducing fundamental overload).

The TV Preamplifier Problem

In theory, a TV preamplifier at the TV antenna should boost the signal in relation to the interference picked up by the feed line. In practice, this does not work because the selectivity of these preamps is poor, and they eagerly amplify amateur signals as well (especially VHF signals). In general, avoid the use of such preamps.

If a TV distribution amplifier has a good noise figure and is built into a metal box, it can go a long way toward solving the problem. The amplifier should be mounted at ground level (where the TV feed line changes from a vertical run to horizontal). This location minimizes stray pickup by the distribution amplifier. A good common- and differential-mode high-pass filter should be placed before the distribution amplifier input (antenna) terminal and at the TV (output) terminal of the distribution amplifier. Ground the

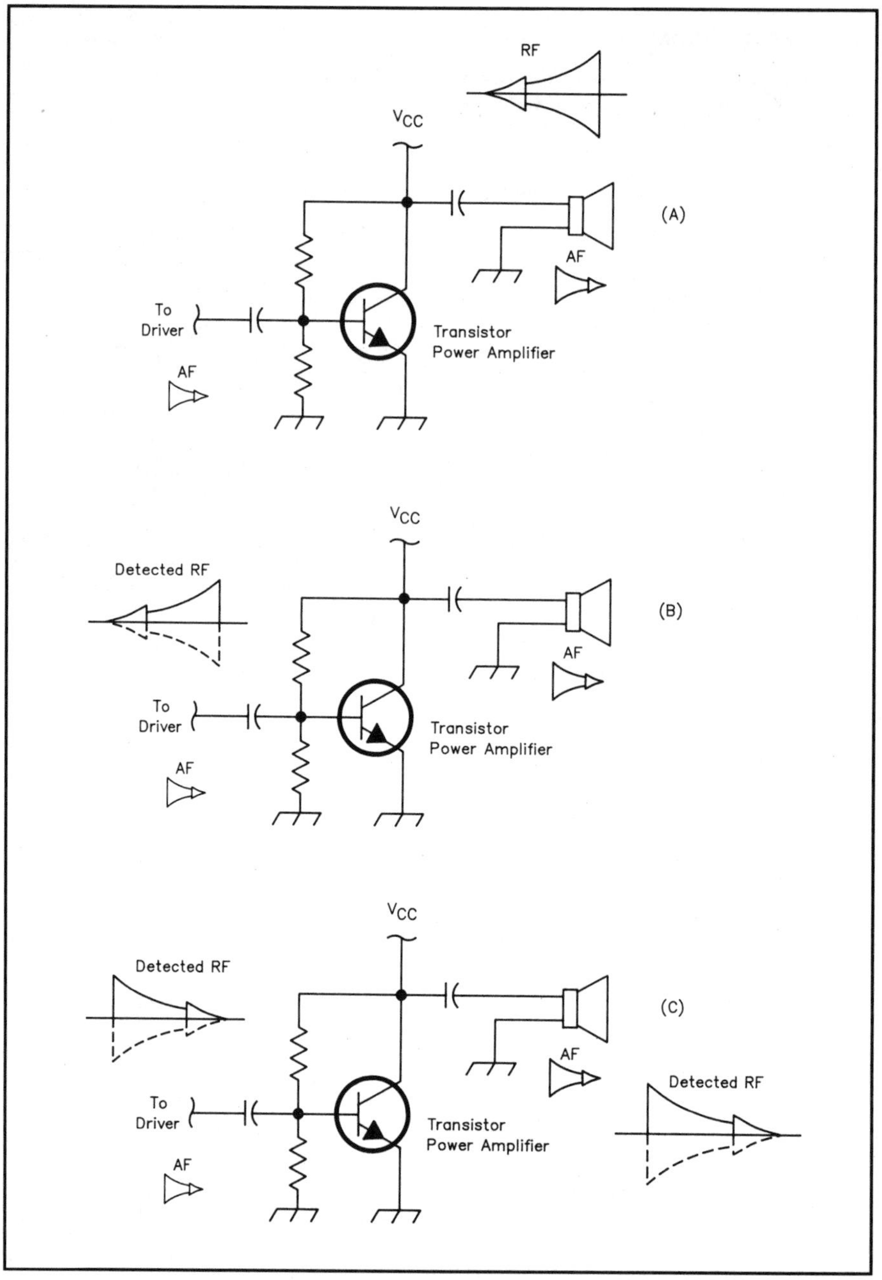

Fig 22—In audio-rectification interference, some device that is not designed to act as an RF detector does so. RF on the output lines (A) is detected by the output device (B). The detected signal is then amplified and applied to the speaker (C).

amplifier chassis to a nearby ground rod. An additional filter may be required at each TV as well. Unused amplifier outputs should have appropriate resistive terminations. In general, a system as just described increases the transmitter level at which TVI begins by up to 10 dB or so. This is an expensive but excellent fix. TV distribution amplifiers are usually available from electronic distributors who sell components to television service shops.

A fringe-area (over 150 miles) TV antenna increases received TV signal strength, especially at channels 2 to 4. Even though a smaller TV antenna may provide an acceptable picture, there is no signal-to-noise margin to cope with near-field amateur transmissions. Buy a long-range TV antenna once to avoid replacing a smaller antenna later.

Passive signal splitters result in signal loss and many units are poorly shielded (plated plastic). Splitters may be adequate for a two-TV installation, but larger splitters introduce too much loss.

Video Cassette Recorders (VCRs)

It is often difficult to protect a VCR from TVI. Older models are more susceptible because of poor shielding. Newer models meet the 1 V/meter (signal strength) immunity recommendation of the ANSI C.63 Committee (see the chapter on EMI Regulations). Although the cases are usually plastic, newer VCRs are internally shielded to meet the standard.

There are several modes of VCR interference. If a VCR degrades the TV picture even when the VCR is off, filters are needed (see Fig 12). If the TV picture is not affected with the VCR turned on, it will probably record well because the record-head amplifiers are not usually sensitive to RF fields.

The most difficult VCR problem results from poorly shielded heads. They tend to pick up interference from nearby amateur stations. To improve shield quality, try grounding the VCR cabinet or place the VCR inside a grounded enclosure (a sheet-metal or screen cage). The VCR manufacturer (who can be contacted via the EIA) may be of some assistance.

Wooden entertainment-center cabinets can be shielded, but it may be more practical to purchase a metal cabinet. Such additional shielding can improve the EMI performance of the TV-VCR system in some cases.

The VCR section of the TVI chapter describes VCR problems in detail. For the most part, treat a VCR as if it were a TV set and apply appropriate common- or differential-mode filters to the inputs, outputs and ac line.

Audio Rectification

Audio rectification occurs because a transistor can act like a diode when power is removed or it is driven into nonlinear operation (saturation) by a strong RF signal (see Fig 22). Thus, the "diode" envelope detects RF, and the AM-detected result appears in the affected device.

Audio rectification is a form of fundamental overload. It is not the fault of the amateur station that the affected equipment was inadequately designed to reject unwanted signals. Most audio rectification results from RF pickup on wires connecting stereo components. It can also occur in telephones, answering machines, and even things like alarm systems that

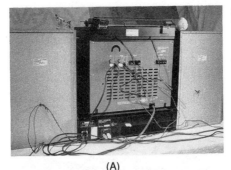

Fig 23—A typical stereo installation at A, and its EMI-resistant counterpart at B.
Photos by Karen Sullivan

produce no audio.

The interference magnitude is affected by many factors: the length of wires connecting components, component shield quality and component lead filtering. RF pickup of long wires may be reduced by using shielded cable. It may help to bypass the base lead of transistor amplifiers with 0.001-µF capacitors, but it is better to apply filtering techniques, usually common-mode, as described in the chapter about stereos. *It is usually not necessary to apply internal modifications.*

It is not difficult to diagnose audio rectification in a device that contains no RF components. The FCC RFI pamphlet clearly states that audio devices that pick up RF signals are improperly functioning as radio receivers. When this occurs, it is clearly a case of audio rectification. Steps that reduce the strength of the amateur's fundamental signal may help the situation, but the fault and solution both lie at the audio-equipment end.

In a radio or television system it is more difficult to differentiate between fundamental overload, harmonic interference and audio rectification. Whenever a TV or radio is involved, the amateur should assume that the amateur station *could* be at fault. The neighbor should

realize that the amateur station *may not* be at fault. There are two easy ways to determine whether audio rectification causes a TV or radio interference problem. If the interference is audio-only and not affected by the volume control, the problem is almost certainly rectification in the receiver audio stages.

Amateurs should be familiar with the unique sound of AM-detected SSB signals. If your receiver has an AM mode, listen to the SSB signals in that mode. If the audio interference to a TV or radio sounds similar, the problem is probably audio rectification.

Next determine, if possible, how RF energy is entering the affected equipment. The basic step of simplifying the problem is a good start. For example, if dealing with a multi-component audio system, disconnect all of the input cables from the audio amplifier or tuner and see if the interference is present with only the amplifier and speakers. If so that leaves only direct pickup, the ac line, or the speaker leads as suspects.

The solutions to audio rectification involve filters and shields. At this point you could try a common-mode ac-line filter and a common-mode speaker-lead filter. The ac-line filter has been mentioned earlier in this chapter and is discussed in detail in the electrical-devices chapter. A common-mode filter should be installed on the feed line of a TV set as well.

Stereo installations often have a rat's nest of wires, cables and antennas forming large loops behind the equipment. Clean up this mess by using minimum-length cables. Bundle signal-input cables, speaker wires and antenna leads away from each other as well (see Fig 23).

A quick fix: Shorten the speaker leads. Many times speakers are located next to the stereo, with several feet of speaker wire coiled into a large loop behind the equipment. Either shorten the wires, or wrap the excess into a small loop that fits around your hand and secure it with a wire tie. This technique forms the extra wire into an inductor (common-mode choke).

Another simple audio-rectification test is performed by turning off the affected device and listening for detected RF (clicks from CW or distorted audio from sideband) from the speakers while operating the amateur transmitter. If clicks are heard, audio rectification is taking place in the final audio transistors. Another

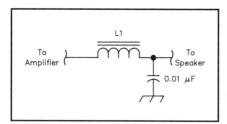

Fig 24—A low-pass filter for use at speaker leads. The inductor prevents the capacitor from sending transistor output stages into fatal oscillations. L1 is 15 turns no. 24 enameled wire on an FT-37-43 core (100 μH).

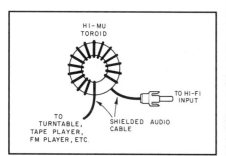

Fig 25—A ferrite toroid common-mode choke for use on speaker leads and interconnect cables.

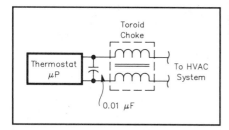

Fig 26—A choke-capacitor combination for use with thermostats and alarm systems.

good clue is if the interference level is *not* affected by any volume control.

Contrary to previous ARRL literature, it is *not* a good idea to bypass speaker leads with a capacitor! This works just fine for tube amplifiers, but solid-state amplifiers are apt to break into full-power oscillation with capacitive loads. Damage to the amplifier is the probable result. Fig 24 shows a low-pass filter suitable for speaker leads.

A good audio common-mode filter suitable for speaker or input leads can be made from an FT-240-61 ferrite core (see Fig 25). Wrap approximately 10 to 20 turns (a single layer) of shielded speaker

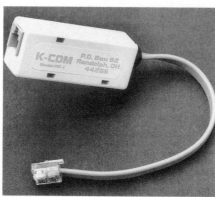

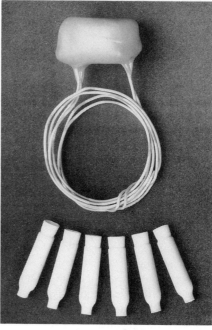

Fig 27—Some filters for use with telephones.

or inter-connect cable. In most cases this filter greatly attenuates the level of audio rectification.

If external audio amplifiers are used to amplify stereo or TV audio signals, a much more difficult situation is present. Only judicious filtering and bypassing of the audio leads can reduce audio rectification to an acceptable degree. See later chapters of this book for more ideas.

A simple test for audio rectification for a TV set that uses external amplifiers, is simply to switch the TV set off and leave the audio amplifiers on. When the amateur transmitter is operating, clicks or pops indicate the presence of audio rectification. The same test applies to a home stereo system.

Intercom Systems

Some of the most difficult audio-

rectification problems involve interference to wired intercom units. Susceptible audio equipment is often connected to hundreds of feet of wire that acts as an immense antenna.

Intercom units are seldom well-shielded, often use no shielded wiring, and may be designed by engineers who do not fully understand the EMI/EMC field. The nature of the connectors used makes lead filtering almost impossible (because of mechanical considerations). External common-mode filters formed from ferrite cores may give some degree of suppression, but unfortunately they don't work well in this application.

Perhaps it is best to contact the intercom manufacturer. This problem is potentially severe because intercoms are never switched off, and they aggravate pets as well as people. Affected systems can awaken people. They can also contain electronic door bells that tend to go off with amateur transmissions.

Setback Thermostats

Microproccessor-controlled "setback" thermostats are a problem as well. They respond to RF energy picked up on their wires. In most cases, the interference shows up as a random reset of the thermostat program.

Thermostats, however, respond well to common-mode ferrite chokes. A single-layer winding on a 3-inch core is usually sufficient. In difficult cases, wire a 0.01-μF capacitor across the leads to suppress differential-mode signals (see Fig 26). Cut a hole in the wall behind the thermostat to hide the choke. Make the hole big enough for the wound core, and pass the core through the hole to hide it. Then remount the thermostat.

Telephones

Audio-rectification interference to telephones occurs in the amplifier and level-controlling circuitry. It results from RF pickup on the telephone wiring or inadequate shields and filters within the telephone. Filters, such as those shown in Fig 27, can protect phones from RF on the wires.

In their *Interference Handbook*, the FCC states that a telephone that picks up radio signals is improperly functioning as a radio receiver. Cordless telephones are Part 15 devices, which are not (by law) entitled to protection from interference. Cordless telephones even interfere with each other.

The Telephones chapter has all of the necessary information to diagnose and correct telephone EMI.

A Word About Ferrites

Ferrite cores are made by several suppliers. There are two groups of ferrites. The first group acts inductive over its specified frequency range. Ferrites in the second group act like a resistor in parallel with an inductor over the specified frequency range. For these reasons, it is difficult to buy ferrite material at hamfests; the material and supplier are frequently unknown. When misapplied, ferrite seldom works; this breeds mistrust of a valuable EMI suppressor.

The inductive ferrites are the most apt to solve interference problems because they work to oppose RF currents within their windings. Information on ferrite is most easily obtained by writing for information and catalogs from advertisers in *QST* and other publications.

Many hams use ferrite obtained from old TV yokes and flyback transformers. These ferrites are usually rated for 15 to 20 kHz and may not work well in HF applications.

Ferrite beads typically work only in low-impedance circuits. They provide little (if any) attenuation when used in FET-gate or vacuum-tube-grid circuits.

The following types of ferrite materials are the most useful:

Material 43: μ = 850. Wide-band transformers to 50 MHz. Optimum frequency attenuation from 40 MHz to 400 MHz. This material makes the best common-mode filter against HF signals. It is also used for ferrite beads, with a peak impedance of about 30 Ω per bead at 200 MHz.

Material 61: μ = 125. Wide-band transformers to 200 MHz. Optimum attenuation above 200 MHz. This material makes the best common-mode filter against VHF signals.

Material 64: Ferrite bead material. Its peak attenuation is approximately 40 Ω per bead at 400 MHz. This is the best bead for VHF and UHF amateur signals.

Material 73: Ferrite bead material. Its peak attenuation is approximately 30 Ω per bead at 25 MHz. This is the best bead for the upper end of the HF range.

Material 75: Ferrite bead material. Its peak attenuation is approximately 30 Ω per bead at 6 MHz. This is the best bead for AM broadcast interference and MF/lower-frequency HF amateur signals.

In general, a ferrite filter formed with a toroid is more effective than one formed with beads. Ferrite filters, especially those formed with beads, are usually only effective in low-impedance circuits. Of course, the impedance of any particular circuit (at the interference frequency) is often difficult to predict.

Ferrite material is available from Amidon, Palomar, RADIOKIT and many electronic-parts distributors. See the ARRL Parts Suppliers list in the components chapter of the *ARRL Handbook* for an up-to-date list.

Interference Caused By Someone Else

The world is filled with radio transmitters and noise sources. When these operate in proximity to susceptible equipment, interference often results. Fig 1 lists just a few possible RF sources. Each transmitter or noise source can interfere with nearby equipment. The Amateur Radio Service is *not* the only possible source of EMI.

The station log is one valuable tool to prove that a station is *not* involved in a particular interference case. The FCC has greatly relaxed logging requirements over the past few years, but it is still important to maintain a logbook. The log can quickly relate any reported interference to station operation. If an amateur station is involved, the log shows the frequency, power level, antenna orientation and equipment in use at the time. The logbook often contains these details, or serves as a memory aid for other aspects of station operation that may be important.

When another source of interference is suspected, it may be necessary to locate that source. Read the EMI Direction-Finding chapter of this book for several interference-locating techniques.

Commercial paging transmitters located in many metropolitan areas cause many urban interference problems. They transmit 250 W, or so, for much of the day. Since many pagers operate between the 10-meter band and TV channel 2, they may cause both IMD and harmonic problems. Other pagers, at about 158 MHz, easily enter a CATV system because of their high power level. In most cases the CATV company responds rapidly when you locate the source and report the problem.

Conducted interference signals fol-low the power or telephone line and find their way into a TV set or some other piece of equipment. This interference takes two forms or modes. When all lines carry the interference in phase, the interference is referred to as common-mode. When the signal phase is opposite between the line wires, it is in the differential mode. Filters are available to handle one or both types of interference, although the common-mode problem is more prevalent. It is unfortunate that most ac-line filters do not filter the common-mode problem because they don't filter the ground lead or shield. Unless all three leads are filtered, the filter *will not work* against ac-line common-mode interference. *Remember that the ground lead is an important safety feature. Do not remove the ground lead from an electric circuit or a piece of equipment.*

Line-to-line interference is referred to as differential-mode interference. Motors and other similar noise sources usually couple a differential-mode signal into the power line. When RF energy strikes an unshielded, elevated power line, some signal is induced into the power system. That signal frequently appears as common-mode interference in the power system. This energy then propagates down the power line to do its damage at some nearby TV or other electrically powered device.

A simple (but not 100% accurate) test for common-mode line interference is as follows: Connect a battery/ac powered TV to the TV antenna and compare the interference levels when battery powered with those when operated from the ac power line. If the interference changes significantly, there is probably some power-line involvement. If so, install an ac-line filter. This test is not 100% reliable because the power lines may radiate some of the interfering signal to the battery-operated set's antenna.

If the electrical balance (symmetry) of the power line is degraded at some point (and it probably is), conducted interference can be radiated again. Interference then propagates through the air only to enter the victim at the antenna as well as the power line. If the resultant interfering signal is at the frequency of the desired signal, the problem can only be corrected at the source. For example, it would be necessary to locate a noisy neon sign and repair or replace it.

Shop tools are a frequent cause of

interference to consumer devices and amateurs as well. Most power tools contain a series-wound (universal) motor. These motors contain brushes that make and break connections as the motor turns. This hot switching causes sparking. The spark, like any other arc, creates a signal that can interfere with amateur receivers or consumer equipment.

In many cases, an ac-line filter installed close to the motor eliminates the problem. In extreme cases install 0.01- to 0.05-μF capacitors across the power line inside the tool. The capacitors should be rated for ac-line use (at least 1.4 kV)! In most cases a motor generates differential-mode interference, but there may be some common-mode problems as well. Supplement a commercial ac-line filter with a common-mode ferrite choke when noise persists. Electric-motor interference appears on a TV as a large number of black dots or streaks.

Power-pole hardware such as insulators and tie wires can cause EMI problems. Power-line arcing to tree branches and transformers can do so as well. Since many of these arcs are related to the line-voltage peaks, this interference can be confirmed by listening to it on an amateur receiver while looking at the receiver audio output with an oscilloscope synchronized to the ac line. If the patterns are stable, the interference is not coming from a motor. It is probably an arcing doorbell transformer, oil-burner igniter, a defect in the power grid or other device. If the pattern is not well synchronized on the scope, the interference is probably motor related. For further information on this subject, read the Power Lines chapter. The subject of power-line interference is discussed in detail in the *Interference Handbook* by William R. Nelson, WA6FQG. The book is available from ARRL HQ.

Computers and Peripherals as a Source of TVI

Since all present home personal computers contain clocks that exceed 10 kHz, they fall under the jurisdiction of the FCC. One need only open a computer manual to see an FCC-required statement warning that the computer may cause TVI or other interference. Harmonics of the clock frequency are the most common form of computer interference (see Fig 28). In most cases the herringbone patterns they generate are similar to those from ama-

teur harmonics.

Because computer systems frequently integrate units (peripherals) designed and sold by several manufacturers, the interference problems they generate can be difficult to pin down. Problems may emanate from switching power supplies, peripherals or the monitor.

To find the trouble, simplify the system by removing peripherals where possible. Reconnect one peripheral at a time to determine the interference source. In general, most problems arise from poorly shielded cables and excessive cable lengths. Possible solutions include: shielded cables, split ferrite cores placed over flat cable, relocation of the computer or peripherals. Ac-line filters may also help. Sometimes RF exits the computer chassis via the CRT 117-V power cable. If

so, plug the CRT into a nearby outlet rather than into the computer. Do not use bypass capacitors at computer buses; excessive capacitance can disable the bus. The computer EMI situation is covered in detail in a separate chapter.

Switching Power Supplies

Switching power supplies are more efficient than linearly regulated supplies, and they certainly save weight. The "down side," however, is that they can generate substantial amounts of noise (see Fig 29). This noise is generated by their fast switching characteristics.

Here again, the interference is in the form of harmonics. The resulting herringbone TVI pattern can be mistaken for amateur transmitter harmonics. A review of the Yaesu FT-757GX (in Dec 1984 *QST*) noted that the transceiver generated TVI from the switching power supply.

Some high-power battery chargers

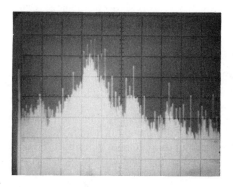

Fig 28—Spectral photo of a computer clock signal. Such signals appear on all computer buses and leads to peripherals such as the monitor and printer.

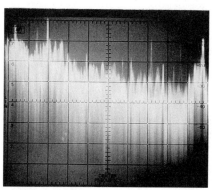

Fig 29—Spectral photo of switching power-supply noise.

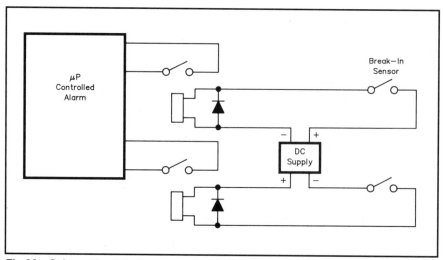

Fig 30—Schematic of dc loops used to EMI harden a microprocessor controlled alarm system.

use switching regulation. They can certainly create TVI and cause interference on the ham bands as well.

Burglar Alarms—A Two-Way Street

Many modern burglar alarms are full of logic gates and microprocessors. Because of this marvelous technology, the burglar alarm can detect RF fields. Alarms have generated serious harmonics when used in the vicinity of amateur stations. Problems include false alarms when mobile radio is used near the alarm.

Police radios as well as amateur radios frequently generate fields strong enough to "false" these electronic wonders.

Microprocessor-controlled alarms may emit interference at the clock frequency and its harmonics. Some electronic alarms fill the ham bands with birdies as well. The most effective way to reduce these problems is to use dc loops and drive relays from the perimeter sensors (see Fig 30). If the relays are mounted close to the burglar alarm, little wire is present to act as an antenna. Burglar alarms hardened in this fashion also are

much more resistant to lightning damage—isn't technology wonderful?

Be Smart

If the troubleshooting methods and cures described in this book are applied, most cases of interference will be reduced or eliminated. There is one other final suggestion, however: when the Super Bowl or some similar big-audience event is broadcast, join the ranks of viewers (just in case one or two of the TV sets in your neighborhood remains unfixed).

Hints & Kinks

from March 1986 *QST*, p 48:
MORE RFI/TVI TIPS

Here is some advice for hams operating in apartments and restricted to indoor antennas. If you are experiencing TVI/RFI, or are aware of such a condition in another apartment, the following techniques may provide some relief:

1) Use the least possible number of ac-powered accessories. Use battery-powered keyers, and install good low-pass and ac-line filters. Sell your high-power amplifier.
2) Eliminate any rat's nest of wire and cable from behind the operating position by keeping cables and leads as short and as neat as possible. Suggest the same technique to a complaining neighbor. It is amazing how many 48-inch cables are used when 18 inches will do. Twenty-foot speaker leads often feed stereo speakers that are only separated by 8 or 10 feet.

I discovered, quite by chance, that TVI/RFI to my downstairs neighbor resulted from my own television when connected to the ac line—even though the set was off during ham operation. Both apartments are served by CATV, and my neighbor has many CATV lines (including one to his stereo system). Nevertheless, I was the source of interference to both TV and stereo. The problem dogged me for nearly a year before I pulled the plug of my

own CATV-connected television.
3) If you are using an indoor antenna, try moving it a bit. A small change in position can make a large difference. Avoid random-length antennas. Run a couple of λ/4 "counterpoise" wires along the baseboard for each band used, and attach them to the ground post of a quality Transmatch.

Do not ignore any possibility when searching for an RFI solution, no matter how absurd it may appear. If operating while standing on your head eliminates interference, you may have to install leg straps on the radio room ceiling!—*Dick Downey, KA2JIZ*

from December 1986 *QST*, p 45:
LIVING WITH TVI

I live in a small apartment building at a summer resort area. During the colder half of the year, I am the only occupant and have no TVI worries. As warm weather approaches, however, the other apartments start filling up. Three tenants have hand-me-down TV sets with poor antennas that are particularly susceptible to TVI. (My own set is free of TVI, even when I use my amplifier. Thus, my station emissions are clean. That doesn't cut any ice with the neighbors, however, who want to see their programs.) For my part, it is good practice to keep my neighbors happy. So, do I go QRT during all TV-viewing hours? Not on your life! I have

set up a "TV detector" to determine when the neighbors are watching TV.

If you live in an apartment building, perhaps you have noticed that your AM broadcast receiver is little better than useless when your (or your neighbor's) TV is on. This is the result of interference from the TV horizontal-sweep oscillator, and it is especially prevalent near the low end of the AM-broadcast dial. Such interference is much worse on longwave frequencies (150-300 kHz). All I do is tune my receiver near 150 kHz (the 10th harmonic of the sweep frequency) and a loud roaring noise can be heard when a neighboring TV is on.

My discovery does not cure TVI, but it does allow me to operate many hours when I would otherwise have to stay off the air.—*Robert J. Panknen, K4SYP/EA5CHT*

from December 1986 *QST*, p 45:
FLASH! VCR CURES TVI!

Here is a tip on the use of a VHS videotape recorder. I live in the weak-reception area of several Los Angeles television stations. When the signals from those stations are very weak, my 7-MHz amateur transmissions produce a light cross-hatch pattern on Channel 5. I have found that the interference is eliminated when the received TV signal is passed through my operating VCR. I do not know the gain of the VCR front end, but it seems significant.—*K. C. Jones, W6OB*

Chapter 4

EMI Direction Finding

Edited by Robert Schetgen, KU7G
ARRL Assistant Technical Editor

Some EMI can affect relatively large areas. For example, touch-controlled lamps have placed EMI on power lines, where it affected broadcast receivers several miles away. External-rectification sources are often many yards from the RF source and EMI receptor. Similarly, intermodulation distortion (IMD) contributors may be some distance from the affected devices. In treating these and similar situations, direction finding (DF) skills are useful.

An EMI investigator is likely to confront two kinds of signals for DFing. Electric motors, arcs, sparks, external rectifiers and digital equipment all produce broadband noise. Sources of fundamental overload, harmonics and IMD produce narrow-band signals. Let's consider broadband sources first.

BROADBAND DF

Broadband noise is usually an AM phenomenon. Current flow in sparks and arcs varies widely. Digital clocks are square-wave generators with constantly changing current levels. Therefore, AM receivers are best for noise DFing.

Broadband sources release energy all across the spectrum. The strength of that energy, however, varies with frequency. Although there may be peaks at certain frequencies (resulting from resonance of the conductor that radiates the energy), the energy is generally stronger at lower frequencies. Hence, signal strength decreases as frequency increases. This characteristic is useful in DF work.

Distant EMI sources are best detected on a battery powered AM-broadcast receiver. Tune the receiver to a clear frequency somewhere near the low end of the band and listen for the interference.

Most such receivers use a ferrite-loop (loopstick) antenna oriented along the longest dimension of the radio case.

Happily, loopstick antennas are directional. Fig 1 shows the field pattern of a typical loopstick antenna. When the noise is faint, use the loopstick lobes to indicate the source along a line perpendicular to the antenna core. As the search progresses, the signal should strengthen. Then use the antenna nulls to indicate the direction of the EMI source.

If the EMI becomes too strong for DF bearings in the loopstick nulls, switch to a higher frequency and continue the search. A pocket scanner that receives the aircraft band is suitable. (Aircraft use AM for communication.) A monopole or dipole antenna has maximum response perpendicular to the antenna axis, and

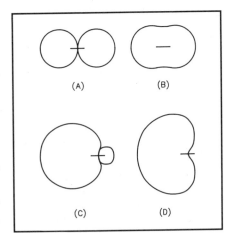

Fig 1—Small-loop field patterns with varying amounts of antenna effect— the undesired response of the loop acting merely as a mass of metal connected to the receiver antenna terminals. The heavy lines show the plane of the loop.

minimum response off the end. Use the maximum response mode when the signal is weak, and switch to the minimum response technique as the signal grows stronger. This technique should be sufficient to locate the building powering electrical noise sources such as thermostats, fluorescent lights and touch-controlled lamps.

Within A Building

Continue listening to the EMI and use the main fuse or circuit breaker to switch the power off and then on, so that the entire building is momentarily without electricity. If the interference disappears, the source is probably in the building.

If the source is in the building, a process of elimination can locate the offending device. Remain at the electrical service box, and cut power to one branch circuit at a time. The EMI should start and stop as power to one of the branches is switched. When the source branch has been determined, unplug each electronic or electrical device on the branch while monitoring the receiver. Once the EMI source is located, refer to the appropriate chapter of this book for suitable treatments.

On The Power Lines

If DFing leads to a power line, the situation gets a little more complicated. EMI radiated from power lines is limited to the general area of the source, while EMI induced into power lines can travel significant distances.[1] It is more appro-

---

[1]W. Nelson, *Interference Handbook* (Wilton, CT: Radio Publications, 1981), p 105.

priate to continue the search by car.

If the car has an AM radio, tune it to a clear frequency at the high end of the band and listen for the noise. Listen to the noise while driving around the general area. Be concerned with general signal-strength trends, not normal variations at down leads, pole grounds and hardware. When the line with the strongest noise is located, drive along that line.

The strength of EMI radiating from a single source connected to a power line usually varies in a pattern of peaks and nulls. The distance between peaks (or between nulls) is minimum at the source (Nelson, p 106). By noting the peak/null pattern, the source can be found. (The signal strength is not significant at this point.)

Do the final location at VHF. Listen for a noise peak with an AM VHF receiver. Once the maximum noise is found, drive past it until the noise becomes very weak. Then, turn around and drive in the opposite direction until it is again weak. Perform this check by driving on both sides of the suspect line if possible.

If the source is at a building, speak with the owner and perform the service-panel tests described earlier in this chapter ("Within a Building"). If there is no building nearby, note the pole number (usually stamped on a metal plate attached to the pole) and call the power company. When reporting the problem, give the pole number, describe the noise and the time of day that it was heard. *Do not, under any circumstances, tamper with the power lines or their attached hardware or equipment!* It is dangerous to strike poles with hammers or automobiles. Doing so could down live wires and cause broken hardware or insulators to fall on passersby.

CATV DF

With a Hand-Held Transceiver

If you believe that you are experiencing CATVI, you can perform your own CATVI-hunt with a hand-held 2-meter transceiver, a detachable antenna, and some basic transmitter-hunting skills. First, tune your receiver to 144.00 MHz. If you hear a loud buzzing (the TV sync pulse) or your receiver quiets, start sweeping the area looking for the strongest signal. (It helps if your receiver has an S meter, but it is possible to gauge signal strength by ear alone.) When you locate the strongest signal, loosen the antenna

connector (to reduce the received signal strength) and continue to hunt for the strongest signal. (It sometimes helps to use your body as a shield for increased directivity.) When the signal strength again reaches the point of receiver overload, remove the antenna entirely, and probe with the open receiver antenna connector for the strongest signal. (You'll probably locate the leak this way.)

With an FM Broadcast Receiver

It is possible to locate distant, strong cable leaks with a mobile FM broadcast receiver.[2] Some CATV systems have a "cuckoo" signal near the high end of the FM broadcast band. The signal is often a double-tone signal, but it may be a single tone. At its strongest, the cuckoo is about 6 dB above the other FM-band signals. The strength of the cuckoo is usually stepped periodically. These steps should be visible on the signal-strength meter of receivers so equipped.

If the CATV system does not use a cuckoo, use a weak FM station that is on the cable, but not received over the air in your area. Since broadcast stations do not use stepped power levels, DFing the station will not work as well as DFing the cuckoo.

Once in the vehicle tune the mobile receiver to the cuckoo. If you are far from the source, only the strongest steps will be audible, and those may not fully quiet the receiver. Drive the area while listening to the strength steps. This method can indicate the general area of leaks. The technique is not completely reliable, however. The CATV shield, telephone cables, guy wires and other nearby conductors can echo the leak. Once you suspect an area, note the pole number and call the CATV company. Let them find the actual leak and fix it.

With a Portable TV

It is also possible to DF CATV leakage with a portable, battery powered TV and a directional antenna. The directional patterns of commercial TV antennas vary, so experiment with a known over-the-air station to determine the pattern. If the pattern is too bad, construct a directional antenna for an appropriate frequency.

[2]J. Moell and T. Curlee, *Transmitter Hunting* (Blue Ridge Summit, PA: Tab Books, 1987).

Suitable directional antennas appear in this chapter, *The ARRL Antenna Book* and *The ARRL Handbook*.

Once the equipment is assembled, tune the TV to the weakest channel that is on the CATV system, but not received over the air. DFing with a TV is easy because multipath signals are visible as "ghosts." TV ghosts are caused by the time delays inherent in multipath propagation. Therefore, the left-most image on the TV screen is always the image produced by the most-direct signal path. Simply follow the path indicated by the maximum strength of the left-most image.

External Rectification Source DF

First, determine the power threshold at which rectification begins. Reduce transmit power gradually until the problem disappears. TVI with less than 5 W indicates the transmit system. Give the transmit antenna, connectors and ground system a detailed inspection. 10 or 20 W is a more common threshold (higher on 80 and 160 meters, where the average bit of metal around the house isn't of significant length) for nearby rectifiers.

It is difficult to exactly locate a rectifier. If the affected device is a TV with a rotatable antenna, make a "first pass" by rotating the TV antenna and looking for a peak in the interference strength. Be careful; it is easy to mistake a TV-signal null for an interference peak. Rotate the transmit antenna (where possible), and look for a TVI peak in the direction of the rectifier. (Do this test with the minimum power required to cause interference when the transmitting antenna is pointed in the most sensitive direction.)

A portable TV or scanner is required to "home in" on the actual interference source. Check suspicious metallic objects by proximity, one at a time. Don't confuse a drop in legitimate signals with a "hot spot" of interference. The usual clanging, banging, twisting, "torquing" and pushing may produce recognizable interference changes that lead to the source. Above all else, check the transmitting and receiving antenna systems thoroughly before beginning a rectifier hunt.

If the transmitter is a commercial broadcast station, the hunt follows a slightly different course. Since the transmitter is always on, and the antenna can't be rotated, the rectifier must be "DFed" from the receiver end. This isn't too difficult—the signal is coherent, narrow band and easily identified. After DFing the general direction, set out on foot with a portable receiver and search by proximity. If the interference seems to be farther afield, use a mobile rig or portable receiver in a car to locate the general vicinity of the source.

When rectification occurs in rain gutters, rusty water pipes and so on, second and third harmonics are usually evident, as well as third- and high-order IMD with broadcast stations. More often than not, the audio is remarkably clear, but scrambled by the simultaneous presence of two sources.

Hum on an interference signal is a possible sign of power-line rectification. A mobile rig and a quick drive around the neighborhood should indicate one or more poles. Report them to the utility company for repair. (Sometimes you can actually hear the broadcast audio while standing near the offending pole.)

Arcs are notorious for producing a wide variety of RF products. Arcs can generate second and third harmonics of the fundamental, as well as second and third order IMD of many broadcast signals near and far.

Rectification may occur in passive conductors (grounds, guys and so on), yet a strong hum component may be evident. This is caused by 60-Hz energy in close proximity; it modulates the current flowing in the rectifying junction.

If IMD products are intermittent but seemingly periodic (seconds), suspect a long expanse of wire. Power lines in residential areas swing like pendulums, with a period of 1 to 2 seconds. As a line sways in the breeze, it tightens and loosens or gently rocks the insulators back and forth. This action may be just sufficient to repetitively break or short the offending rectifier, thus switching the BCI on and off.

Any lengthy conductor can be checked with a portable receiver. Couple the internal loop antenna to the conductor under test. The BCI always gets louder (the metal object is an antenna); but in BCI sources it gets disproportionately louder. Test several objects in the area, using an attenuator (or proximity) to control receiver sensitivity. It doesn't take long to get an intuitive feel for levels that are "normal" and those that aren't.

If the interference is generated in residential wiring or plumbing, check for ground return currents in the water pipe between the street (meter) and dwelling. In temperate climates (where pipes are buried relatively close to the surface) return currents radiate quite well. By simply "sweeping" the sidewalk near the water entrance with a portable receiver, the relative interference level can be detected. Signals are strongest at the house with rectification problems.

A search by foot and by car is usually required to locate IMD sources. If the IMD is weak, it may be difficult to detect on portable receiver. Check other related frequencies for a stronger IMD product. The second and third harmonics should be strongest, followed by third-order IMD with other local stations. The search for TVI and BCI is more art than science. A few simple experiments, however, and some practical experience with a portable receiver quickly makes an expert out of a beginner.

NARROW-BAND DF

Narrow-bandwidth signals appear on a small set of frequencies. Possible sources are two-way radios as used for Citizens Band, police, fire, aircraft, radar, land-mobile, cellular telephone, cordless telephone and Amateur Radio.

Hams often DF narrow-band signals (repeater interference, aircraft emergency-locator transmitters—ELTs—and transmitter hunts) for public service and for fun. Hence, narrow-band DF techniques have been discussed in many magazine articles and a few books.[3]

Is A Search Necessary?

Because the energy of narrow-band signals is concentrated in one area of the spectrum, a narrow-band noise source may be located several miles from the affected device. It is unlikely that distant narrow-band sources will interfere with consumer electronic equipment such as telephones and cassette players. Nonetheless, distant sources may affect sensitive receivers.

[3]Happy Flyers, *Radio Direction Finding*, copies are available from Hart Postlewaite, WB6CQW, 1811 Hillman Ave, Belmont, CA 94002, 415-341-4000 or Paul Hower, WA6GDC, PO Box 2323, La Mesa, CA 92041, 714-465-5288.

Before beginning a search, listen to the target signal on the affected device. Can you discern intelligence? If a voice or Morse code is evident, try to copy the station identification. (Nearly all licensed transmitters are required to identify periodically.) Some digital modes (such as amateur AMTOR, RTTY and ASCII) are permitted to identify in the digital code. If the target signal sounds like a digital code, try to get equipment to read the intelligence. If you can discern a call sign, DF techniques may not be needed. Contact your nearest FCC office to get the owner's name and address. Then contact the owner and discuss the situation. Remember! Cooperation is the key to most EMI problems.

Direction-Finding Systems

A directive antenna and a device for detecting the radio signal are required for any DF system . In Amateur Radio applications the signal detector is usually a receiver; for convenience it should have a meter to indicate signal strength. At very close ranges a simple diode detector and dc microammeter may suffice.

The receiver should be a small portable or mobile instrument. For amateur use, a standard, unmodified, commercially available receiver works well.

Antennas for DF work, on the other hand, are not generally the same kinds used for two-way communications. Directivity is a prime requirement, and here the word directivity takes on a somewhat different meaning than is commonly applied to antennas. Directivity is normally associated with gain: an antenna pattern having a long, thin main lobe. Such a pattern is of value for coarse measurements in DF work, but precise bearings are not possible. There is always a spread of a few (or perhaps many) degrees on the "nose" of the lobe where a shift of antenna bearing produces no detectable change in signal strength. In DF measurements, it is desirable to correlate an exact bearing or compass direction with the position of the target. To do this accurately, an antenna exhibiting a null in its pattern is used. A null can have very sharp directivity, a half degree or less.

Loop Antennas

A simple antenna for DF work is a small loop tuned to resonance with a capacitor. Several factors must be considered in the design of a DF loop. The loop must be small compared with the wavelength. In a single-turn loop, the conductor should be less than 0.08 λ. For 28 MHz, this is less than 34 inches (diameter of approximately 10 inches). Maximum response from the loop antenna is in the plane of the loop, with nulls exhibited at right angles to that plane.

To obtain the most accurate bearings, the loop must be electrostatically balanced with respect to ground. Otherwise, the loop exhibits two modes of operation. One is the mode of a true loop, while the other is that of an essentially nondirectional vertical antenna of small dimensions ("antenna effect"). The voltages introduced by the two modes are not in phase and may add or subtract, depending on the direction from which the wave is coming.

The theoretical true loop pattern is illustrated in Fig 1A. When properly balanced, a loop exhibits two nulls that are 180° apart. Thus, a single null reading with a small loop antenna may not indicate the exact direction toward the transmitter—only the line along which the transmitter lies. Ways to overcome this ambiguity are discussed in later sections.

When the antenna effect is appreciable and the loop is tuned to resonance, the loop may exhibit little directivity, as shown in Fig 1B. However, by detuning the loop to shift the phasing, a pattern similar to Fig 1C may be obtained. Although this pattern is not symmetrical, it does exhibit a null. The null, however, may not be as sharp as that obtained with a loop that is well balanced, and may not be at exact right angles to the plane of the loop.

By suitable detuning, the unidirectional pattern of Fig 1D may be approached. This adjustment is sometimes used in DF work to obtain a unidirectional bearing, although there is no complete null in the pattern. In most cases, however, the loop is adjusted for the best null.

An electrostatic balance can be obtained by shielding the loop, as shown in Fig 2. The shield is represented by the broken lines in the drawing. Shielding eliminates the antenna effect. The response of a well-constructed shielded loop is quite close to the ideal pattern of Fig 1A.

For the low-frequency amateur bands, single-turn loops of convenient physical size for portability are generally unsatisfactory for DF work. Therefore, multiturn loops are generally used instead. Such a loop is shown in Fig 3. This loop may also be shielded, and if the total conductor length remains below 0.08 λ, the directional pattern is that of Fig 1A.

Ferrite Rod Antennas

These antennas consist essentially of many turns of wire around a ferrite rod. They are known as loopstick antennas and also as ferrite-rod antennas. Probably the best-known example of this antenna is that used in small portable AM broadcast receivers. The advantage of ferrite-rod antennas is reduced size. Loopsticks are used almost exclusively for frequencies below 150 MHz.

As implied in the earlier discussion of shielded loops, a true loop antenna responds to the magnetic field of the radio wave (not to the electrical field). The volt-

---

THE PRACTICAL SIDE OF FOX HUNTING

Recently a co-worker, Ed Dabrowski, WB9NLO and I went on a peculiar type of fox hunt. Ed and I are engineers for a Chicago TV station. About two weeks prior to the fox hunt I began noticing a herringbone pattern on TV Channel 2 on my home TV sets. The pattern was relatively weak with no audible modulation and a slight variance in frequency. The herringbone pattern was sometimes on for most of the day and at other times for just a short while. It was enough to make watching Channel 2 frustrating. I asked my immediate neighbors if they were having the same trouble, and they were. I then drove around the neighborhood to see if any new CB or other antennas has been put up recently. Finding none, I checked some of the nearby industries and a nursing home to see if they were using any induction or dielectric heaters or diathermy. Again I drew a blank.

Then, to add to my frustrations, I started getting phone calls blaming me for the disturbance on Channel 2. Five antennas on my tower and three more on my roof make me a prime target for such blame. I checked all of my receivers to see if I could locate a spur that was causing the problem. Next I connected my 2-meter beam to a TV set and made a sweep of the area. The signal seemed to be coming from within my subdivision.

The FCC was contacted and I explained to one of the field engineers just what was happening and what I had done so far. He suggested looking for a TV booster amplifier that had gone bad. He indicated that they sometimes go into self-oscillation and radiate through the TV antenna.

On that advice I called out the reserves. Ed helped me wire up my 5-inch portable so we would be able to use it in the car. We took a bearing on the signal and plotted it on a map of the subdivision. It was dark as we drove around trying to locate the source of the signal. We pinned it down to a few houses, began knocking on doors and explained the situation. We asked the residents if they had recently bought any new electronic devices or if they had a TV antenna booster amplifier. Ed and I also asked the residents to turn off the main power to their houses to see if the herringbone interference would be eliminated. People were very cooperative as we announced ourselves as TV engineers and Amateur Radio operators. One of the homeowners said, "Oh yeah, I know where the trouble is coming from. It's that guy about two blocks west of here with the big tower in his backyard." He was quite embarrassed when I told him that I was the guy he was referring to and that I was trying to locate the source of the signal.

We checked all of the houses in the suspect area, except one where the people were not at home. We were about to pack up when the owners came home. We showed them the portable TV in the car and asked them to turn off their power. When the power was turned off the interference went away. We had found it! The residents indeed had a TV antenna booster amplifier in the attic. It has since been disconnected or repaired, and TV Channel 2 is clear and sharp with no herringbone. Our local community newsletter printed a nice word for us and I got off the hook with the neighbors who suspected me of causing the problem.—*Murray Cutler, W9EHQ*

[Reprinted from November 1979 *QST*.]

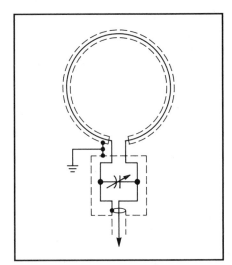

Fig 2—Shielded loop for direction finding. The ends of the shielding turn are not connected, to prevent shielding the loop from magnetic fields. The shield is effective against electric fields.

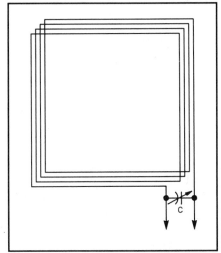

Fig 3—Small loop consisting of several turns of wire. The total conductor length is very much less than a wavelength. Maximum response is in the plane of the loop.

age delivered by the loop is proportional to the amount of magnetic flux passing through the coil and the number of turns in the coil. (The action is much like that of a transformer secondary winding.) For a given loop size, the output voltage can be increased by increasing the flux density (increasing core permeability). A 1/2-inch-diameter, 7-inch rod ($\mu_i = 125$) is a suitable loop core for 1-10 MHz. For increased output, wind the turns on two rods that are taped together, as shown in Fig 4. Loopstick-antenna construction projects are described later in this chapter.

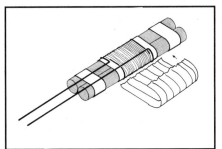

Fig 4—A ferrite-rod or loopstick antenna. Turns of wire may be wound on a single rod or two rods taped together (to increase the output from the loop). The core material must be selected for the intended frequency range of the loop. To avoid bulky windings, use fine wire such as no. 28 or 30, with larger wire for the leads

The maximum response of a loopstick antenna is broadside to the axis of the rod as shown in Fig 5, whereas maximum response of a small loop is in the plane of the loop. Otherwise the performance of loopstick and ordinary loop antennas are similar. A loopstick may also be shielded to eliminate the antenna effect, such as with a U-shaped or C-shaped channel of aluminum or other form of "trough." The length of the shield should equal or slightly exceed the length of the rod.

Sensing Antennas

Because loop antennas exhibit two nulls that are 180° apart, an ambiguity exists as to which null indicates the true direction of the target. For example, assume that a bearing measurement indicates the transmitter lies on an east and west line from your position. With a single reading, there is no way to know whether the transmitter is east or west of you.

If two (or more) receiving stations take bearings on a single transmitter, or if a single receiving station takes bearings on the transmitter from two positions, the ambiguity may be worked out by a method called triangulation. (Further detail on triangulation is given in a later section.) However, an antenna pattern with only one null eliminates ambiguity.

A loop or loopstick antenna can yield a single null if a second antenna element is added. The element is called a sensing antenna, because it gives an added sense of direction to the loop pattern. The second element must be omnidirectional, such as a short vertical. When the signals from the loop and the vertical element are

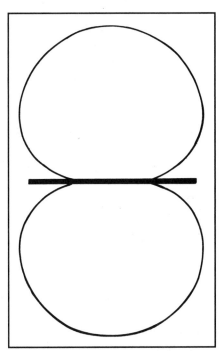

Fig 5—Field pattern for a ferrite rod antenna. The dark bar represents the rod on which the loop turns are wound.

combined (with a 90° phase difference) a cardioid pattern results. Fig 6A shows the patterns of the loop, the sense antenna and the resulting cardioid. (The resultant is not simply the sum of the two patterns.)

Fig 6B shows a circuit for adding a sensing antenna to a loop or loopstick. R1 is an internal adjustment used to set the level of the signal from the sensing antenna. For the best null in the composite pattern, the signals from the loop and the sensing antenna must be of equal amplitude, so R1 is adjusted experimentally during setup. In practice, the null of the cardioid is not as sharp as that of the loop, so the usual measurement procedure is to first use the loop alone to obtain a precise bearing reading, and then to add the sensing antenna and take another reading to resolve the ambiguity. (The null of the cardioid is 90° away from the loop nulls.) For this reason, the sensing element is usually switched in and out of the circuit.

Phased Arrays

Phased arrays are also used in amateur DF work. Two general classifications of phased arrays are end-fire and broadside configurations. Depending on the spacing and phasing of the elements, end-fire patterns may exhibit a null off one end of the axis of the two or more elements. At the same time, the response is

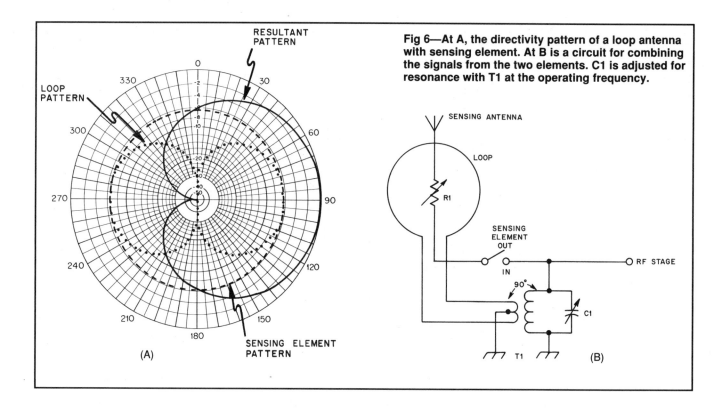

Fig 6—At A, the directivity pattern of a loop antenna with sensing element. At B is a circuit for combining the signals from the two elements. C1 is adjusted for resonance with T1 at the operating frequency.

maximum off the other end of the axis, in the opposite direction from the null. A familiar arrangement is two elements spaced 1/4 wavelength apart and fed 90° out of phase. The resultant pattern is a cardioid, with the null in the direction of the leading element. Other arrangements of spacing and phasing for an end-fire array are also suitable for DF work. One of the best known is the Adcock array, discussed in the next section.

Broadside arrays are inherently bidirectional, which means there are always at least two nulls in the pattern. Ambiguity therefore exists in the true direction of the transmitter, but depending on the application, this may be no handicap. Broadside arrays are seldom used for amateur DF applications.

Adcock Arrays

One of the most popular end-fire phased arrays is the Adcock. This system was invented by F. Adcock and patented in 1919. The array consists of two vertical elements fed 180° apart, and mounted so the system may be rotated. Element spacing is not critical; it may range from 1/10 to 3/4 λ. The two elements must be of identical length, but need not be self resonant. (Elements that are shorter than resonant are commonly used.) Because neither the element spacing nor the length is critical in terms of wavelengths, an

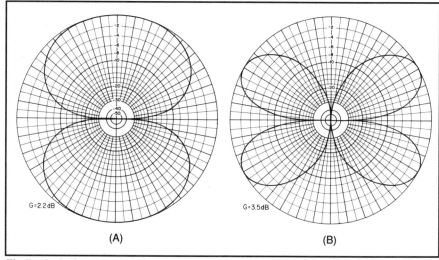

Fig 7—At A, the pattern of the Adcock array with an element spacing of 1/2 λ. In these plots, the elements are aligned with the vertical axis. As the element spacing is increased beyond 3/4 λ, additional nulls develop off the ends of the array. At a spacing of 1 λ the pattern at B exists. Pattern B is unsuitable for DF work.

Adcock array may be operated over more than one amateur band. This is handy for EMI DFing because the user need only change the antenna phasing-line length for each frequency of interest.

The radiation pattern of the Adcock is shown in Fig 7A. The nulls are in directions broadside to the axis of array. They become sharper with greater element spacings. However, as element spacing

exceeds 3/4 λ, the pattern develops additional nulls off the ends of the array axis. At a spacing of 1 λ the pattern is that of Fig 7B, and the array is unsuitable for DF applications.

Short vertical monopoles are often used in what is sometimes called the U-Adcock, so named because the elements with their feeders take on the shape of the letter U. In this arrangement the elements are worked against the earth as a ground or counterpoise. If the array is used only for reception, earth losses are of no great consequence. Short, elevated-vertical dipoles are also used in what is sometimes called the H-Adcock. The Adcock array is often used at HF for sky-wave DF work. In this application, it is not considered a portable system.

The Adcock array, with two nulls in its pattern, has the same ambiguity as the loop and the loopstick. Adding a sensing element to the Adcock array has not met with great success. Difficulties arise from mutual coupling between the array elements and the sensing element, among other things.

Loops v Phased Arrays

Although loops can be made smaller than suitable phased arrays for the same frequency of operation, the phased arrays are preferred by some for a variety of reasons. In general, phased arrays can yield sharper nulls, but this is a function of the care used in constructing and feeding the individual antennas, as well as of the size of the phased array (in terms of wavelengths). The primary constructional considerations are: (1) electrically shield and balance the feed line to prevent unwanted signal pickup, and (2) electrically balance the antenna for a symmetrical pattern.

In general, loops and loopsticks are used for mobile and portable operation, while phased arrays are used for fixed-station operation. However, phased arrays are used successfully above 144 MHz for portable and mobile DF work. Practical examples of both types of antennas are presented later in this chapter.

Direction Finding Techniques

The ability to locate a transmitter with DF techniques is a skill that is acquired only with practice. Familiarity with the antenna being used and especially with its limitations are probably the most important criteria. Of course, knowing the measuring equipment is also a requirement. But in addition to this, one must know how radio signals behave at different frequencies and in different kinds of terrain. Experience is the best teacher, although reading about the experiences of others and talking with others who are active in DF also helps.

Ground-Wave DF

Most amateur DF activity is conducted with ground-wave signals, where the transmitter is seldom more than a dozen or so miles away. The considerations for these measurements are discussed in this section.

Before setting out to locate a signal source, note some general information about the signal itself. Is its frequency constant, or does it drift? Drifting means you may have to retune the receiver often during the hunt. Do transmissions occur at regular intervals, or are they sporadic? Are transmissions very short in duration, modest in length, or is there a continuous carrier? The most difficult sources to locate are those which transmit for very brief periods at irregular intervals. Such transmitters are a real challenge that require extreme patience. They also call for quick action when the transmitter is on the air (to obtain a null reading in those few seconds). Once you've determined the general signal characteristics, begin the search.

Best accuracy in determining a bearing to a signal source is obtained when the propagation path is over homogeneous terrain, and when only the vertically polarized component of the ground wave is present. If a boundary exists, such as between land and water, the different conductivities of the two mediums under the ground wave cause bending (refraction) of the wave front. In addition, reflection of RF energy from vertical objects (such as mountains or buildings) can add to the direct wave and cause DF errors.

The effects of refraction and reflec-

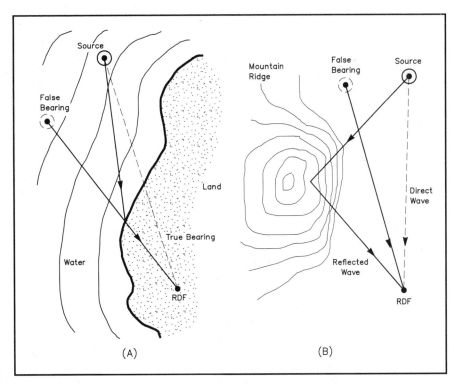

Fig 8—DF errors caused by refraction (A) and reflection (B). At A, a false reading is obtained because the signal actually arrives from a direction that is different from that to the source. At B a direct signal from the source combines with a reflected signal from the mountain ridge. The two are averaged at the antenna, giving a false bearing to a point somewhere between the two apparent sources.

tion are shown in Fig 8. At A, the signal is actually arriving from a different bearing than the true direction to the transmitter. This happens because the wave is refracted at the shoreline. Even the most sophisticated of measuring equipment cannot indicate a true bearing in this situation. (Equipment can only show the direction from which the signal is arriving.) It is important to know how radio signals behave at different frequencies and in different kinds of terrain in this situation. If the measuring equipment is portable, an experienced DF enthusiast will probably move to the shoreline to take a measurement with less chance of error. Even so, a precise reading may still be impossible to obtain if the source lies along the shoreline.

In Fig 8B there are two incoming signals—one on a direct path from the source and another that is reflected from the mountain ridge. In this case the two waves add at the antenna of the DF equipment. A careless or uninitiated observer might follow a false bearing somewhere between the two "sources." An experienced DFer might notice that the null reading in this situation is not as sharp as usual or perhaps not as deep. But these indications would be subtle, and easy to overlook. This is where knowing the antenna and the equipment becomes very helpful.

Water towers, tall radio towers and similar objects can also lead to false bearings. The effects of these objects become significant when they are large in terms of a wavelength. In fact, if the direct path to the transmitter is masked by intervening terrain or other radio "clutter," it is possible that a far stronger signal may be received by reflection than from the true direction to the source. Such towers atop hills are prime sources of strong "pinpoint" reflections. Triangulation may even "confirm" that the transmitter is at that high elevation, unless a reading is taken from in the clear, where the true radio path is unmasked by the clutter.

Local objects also tend to distort the field, such as buildings of concrete and steel construction, power lines and the like. It is important that the DF antenna is in the clear, well away from such surrounding objects. Trees with foliage may also create an adverse effect, especially at VHF and above. If it is not possible to avoid such objects, the best procedure is to take readings at several different posi-

tions, and to average the readings for the result.

Triangulation

Geometry can help locate a target transmitter. Two angles and the side between them define a unique triangle. For long-range DFing, the side and angles can be plotted (on an appropriate map, more on this later) to reveal the approximate location of the target. For close-in work, the location can be guessed by eye or plotted on a roughly drawn sketch of the area.

The procedure for triangulation begins with the first bearing. Choose a spot that is clear of obstructions, power lines and metal objects if possible. Once the bearing is taken, move to another location. The second location should be in the clear like location one, but its bearing from location one should be nearly 90° from the first bearing. If the difference between bearing one and two is less than 10°, travel further along the line from one to two and take another bearing.

The word "approximate" is used here because there is always some degree of

uncertainty in the bearings. This uncertainty arises from equipment limitations (antenna null width, for example). Propagation effects may increase the degree of uncertainty. In order to obtain a more precise fix, a bearing measurement from yet a third position is helpful. When the positions and bearing lines for all three measurements are plotted on a map, the three lines seldom intersect at a single point. Instead, they enclose a small triangle on the map. The unknown source is presumed to be located inside the triangle.

In order to best indicate the probable area of the transmitter location on the map, the bearings from each position should be drawn as a narrow sector instead of as a single line. The sector width should represent the uncertainty. A portion of a map marked in this manner is shown in Fig 9. It is graphically evident how the measurement from DF Site 3 has narrowed the probable area of the transmitter position. The small black section indicates where the transmitter is likely to be found, a significantly smaller area than that of the sectors from DF Sites 1 and 2 alone.

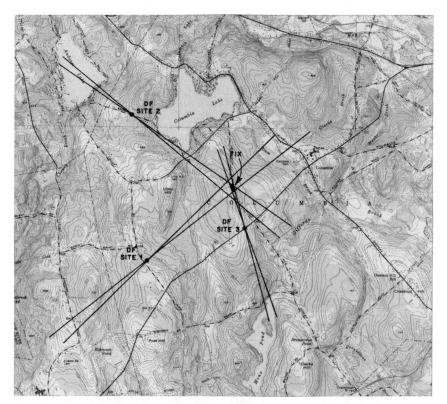

Fig 9—Bearing sectors from three DF positions drawn on a map. This method is known as triangulation. Note here that sensing antennas are not required at any of the DF sites; antennas with two null indications 180° apart are quite acceptable.

Several kinds of maps are suitable for triangulation. Ordinary road maps, however, usually lack sufficient detail, and are not generally satisfactory. Contour maps (like that shown in Fig 9) are preferred for open country. Aeronautical charts are also suitable. City maps and street maps are usually acceptable. Your local city or county engineer can help you find a source for suitable maps.

Closing in on the Fix

In the final phases of locating an unknown source, portable DF equipment is used to pinpoint the exact location. As the equipment is brought near the transmitter, signals become very strong. Receiver front-end overload may lead to confusing and inconsistent readings. An attenuator in the feed line from the antenna to the receiver may help. Attenuators are discussed in *The ARRL Handbook*.

Even with an attenuator in the line, a strong electromagnetic field may couple some energy directly into the receiver circuitry. The effect is an apparent reduction of antenna directivity. In other words, the received signal strength changes only slightly or perhaps not at all as the DF antenna is rotated in azimuth. To cure this, shield the receiving equipment. Simply place the receiver in a bread pan or cake pan covered with a piece of copper or aluminum screening (securely fastened at several points). Difficult situations may call for more sophisticated shields.

As an alternative, the receiver could be replaced by a simple diode detector when approaching the target. The detector should have a sensitive meter indicator with some kind of sensitivity control, and it may be frequency selective, if desired. An optional meter-amplifier circuit could increase its sensitivity. It should be constructed inside a well shielded metal enclosure to avoid unwanted signal pickup.

Final physical location of the target may be easy or difficult. If the emission is a spurious output from a communications transmitter, a transmitting antenna will probably stand out. On the other hand, a hidden transmitter that is jamming a repeater may be very difficult to find. Sometimes such transmitters are buried in the ground, and the antenna may be a thin wire concealed in the bushes. It may not be spotted by an observer standing right next to it. Sharp eyes and some astute observations are helpful.

DF System Calibration And Use

Once a DF system is initially assembled, it should be "calibrated" before use. The balance or symmetry of the antenna pattern is of primary concern. A loop with a lopsided figure-8 pattern, for example, is undesirable; the nulls are not 180° apart nor are they at exact right angles to the plane of the loop. If this fact was not known in actual DF work, measurement accuracy would suffer.

Initial checkout can be performed with a low-power transmitter at a distance of a few hundred feet. It should be within visual range and must be operating into a vertical antenna. (A quarter-wave vertical or a loaded whip is quite suitable.) The site must be reasonably clear of obstructions, especially steel and concrete or brick buildings, large metal objects, near-by power lines and so on. If the system operates above 30 MHz, trees and large bushes should also be avoided. An open field makes an excellent site.

The procedure is to "find" the transmitter with the DF equipment as if its position were not known, and compare the DF null indication with the visual path to the transmitter. For antennas with more than one null, each null should be checked.

If imbalance is found in the antenna system, there are two options available. One is to correct the imbalance. Toward this end, pay particular attention to the feed line. The use of a coaxial feed line with a balanced antenna invites an asymmetrical pattern (unless an effective balun is used). A balun is not necessary if a loop is shielded, but an asymmetrical pattern can result if the shield break is misplaced. A sensing antenna may upset antenna balance slightly. Experiment with its position (relative to the main antenna) to possibly correct the error. Also note that the position of the null shifts by 90° as the sensing element is switched in and out, and the null is not as deep. This is of little concern, however, as the sensing antenna is only used to resolve ambiguities. (The sensing element should be switched out when accuracy is desired.)

The second option is to accept the imbalance of the antenna and use some kind of indicator to show the true directions of the nulls. Small pointers, painted marks on the mast or an optical sighting system might be used. Sometimes the end result of the calibration procedure is a compromise between these two options. A perfect electrical balance may be difficult or impossible to attain.

A Portable Field-Strength Meter

Few amateur stations, fixed or mobile, are without need of a field-strength meter. This instrument serves many useful purposes during antenna experiments and adjustments, and it may be used for closing in on a DF target. When work is to be done from many wavelengths away, a simple wavemeter lacks the necessary sensitivity. Further, its linearity leaves much to be desired.

The field-strength meter described here takes care of these problems. Additionally, it is small, measuring only $4 \times 5 \times 8$ inches. The power supply consists of two 9-V batteries. Sensitivity can be set as desired. The circuit should not be too sensitive, though, or it may respond to unwanted signals. This unit has excellent linearity with regard to field strength, which is very helpful in DF work. (The field strength of a received signal varies directly with the distance from the source, all other things being equal.) The frequency range includes amateur bands from 3.5 through

148 MHz, with band-switched circuits (thus avoiding the use of plug-in inductors). All in all, it is a useful instrument.

The unit is pictured in Figs 10 and 11, and the schematic diagram is shown in Fig 12. A 741 op amp is the heart of the unit. The antenna is connected to J1, and a tuned circuit is used ahead of a diode detector. The rectified signal is dc coupled to the op amp and amplified. Op-amp sensitivity is controlled by switching R3 through R6 into the circuit by means of S2.

The circuit shown can detect an antenna voltage on the order of 100 mV (in its most sensitive setting). Linearity is poor for approximately the first 20% of M1's range, but it is almost perfectly linear from there to full-scale deflection. (Diode nonlinearity causes the poor linearity at lower readings.) For DF measurements, just remember that the indicated strength is not proportional to distance at the low end of the scale.

The 741 op amp requires both positive and negative voltage sources. This is obtained by connecting two 9-volt batteries in series and grounding the center. The instrument can be used remotely by connecting an external meter at J2. This is handy if you want to adjust an antenna and observe the results without having to leave the antenna site.

L1, the 80/40-meter coil, is tuned by C1. The coil is wound on a toroid form. For 20, 15 or 10 meters, L2 is switched in parallel with L1 to cover the three bands. L5 and C2 cover approximately 40 to 60 MHz, and L7 and C2 from 130 MHz to approximately 180 MHz. The two VHF coils are also wound on toroid forms.

Construction Notes

The majority of the components may be mounted on an etched circuit board. Use shielded leads for all connections to S2. Otherwise, parasitic oscillations may occur in the IC because of its very high gain.

For 144-MHz coverage, mount L6 and L7 directly across the appropriate terminals of S1, rather than on a circuit board. (The extra lead length adds too much inductance to the circuit.) It isn't necessary to use toroid forms for the 50- and 144-MHz coils. They were used in the version described here simply because they were available. Air-wound coils of the appropriate inductance can be substituted.

Calibration

The field-strength meter can be used as a relative-reading device for DF work. A linear indicator scale serves admirably. The instrument is much more useful for antenna work when it is calibrated in decibels (enabling the user to check relative gain and front-to-back ratios). If a calibrated signal generator is available, use it to calibrate the meter. Signal-generator voltage ratios can be converted to decibels by using the equation

$$dB = 20 \log \left(\frac{V1}{V2} \right) \qquad \text{(Eq 1)}$$

where V1/V2 is the ratio of the two voltages, and log is the common logarithm.

Let's assume that M1 is calibrated evenly from 0 to 10. Next, assume we set the signal generator to provide a reading of 1 on M1, and that the generator is feeding a 100-mV signal into the instrument. Now we increase the generator output to 200 mV, giving us a voltage ratio of 2 to 1. Also let's assume M1 reads 5 with the 200-mV input. From the equation above, we find that the voltage ratio of 2 equals 6.02 dB between 1 and 5 on the meter scale. M1 can be calibrated more accurately between 1 and 5 by adjusting the generator and figuring the ratio. For example, a ratio of 126 mV to 100 mV is 1.26, corresponding to 2.0 dB. By using this method, all of the settings of S2 can be calibrated. In the instrument shown here, the most sensitive setting of S2 (R3 = 1 MΩ) provides a range of approximately 6 dB for M1. Keep in mind that the meter scale must be calibrated similarly for each setting of S1 (for each band). The degree of coupling of the tuned circuits for the different bands varies, so each band must be calibrated separately.

The instrument may also be calibrated

Fig 10—The linear field-strength meter. The control at the upper left is for C1 and the one to the right for C2. At the lower left is the band switch. The zero-set control for M1 is located directly below the meter.

Fig 11—Inside view of the field-strength meter. At the upper right is C1 and to the left, C2. The dark leads from the circuit board to the front panel are the shielded leads described in the text.

using a transmitter. (Measure its output power with an RF wattmeter.) Since the source is rated in power rather than voltage, this equation applies:

$$dB = 10 \log \left(\frac{P1}{P2} \right) \qquad \text{(Eq 2)}$$

where P1/P2 is the power ratio.

The power output of most transmitters can be varied, so calibration of the test instrument is rather easy. Attach a pickup antenna to the field-strength meter (a short wire a foot or so long will do) and position the device in the transmitter antenna field. Set the transmitter output to 10 W and note the reading on M1. Increase the output to 20 W (a power ratio of 2). Note the reading on M1, and then use Eq 2. A power ratio of 2 is 3.01 dB.

Use this method to calibrate the instrument on all bands and ranges.

With the tuned circuits and coupling links specified in Fig 12, this instrument has an average range of 6 dB for the two most-sensitive positions of S2 and 15 dB and 30 dB for the next two successive ranges. The 30-dB scale is handy for making front-to-back antenna measurements without switching S2.

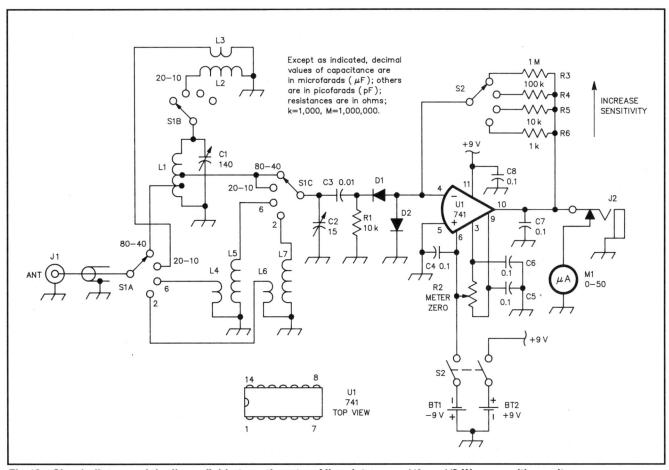

Fig 12—Circuit diagram of the linear field-strength meter. All resistors are 1/4- or 1/2-W composition units.

C1—140 pF variable.
C2—15 pF variable.
D1, D2—1N914 or equiv.
L1—34 turns no. 24 enameled wire wound on a T-68-2 core, tapped 4 turns from ground end.
L2—12 turns no. 24 enameled wire wound on T-68-2 core.
L3—2 turns no. 24 enameled wire wound at ground end of L2.

L4—1 turn no. 26 enameled wire wound at ground end of L5.
L5—12 turns no. 26 enameled wire wound on T-25-12 core.
L6—1 turn no. 26 enameled wire wound at ground end of L7.
L7—1 turn no. 18 enameled wire wound on T-25-12 core.

M1—50 or 100 μA dc.
R2—10-kΩ control, linear taper.
S1—Rotary switch, 3 poles, 5 positions, 3 sections.
S2—Rotary switch, 1 pole, 4 positions.
S3—DPST toggle.
U1—741 op amp. Pin nos. shown are for a 14-pin package.

A Shielded Loop with Sensing Antenna for 28 MHz

Fig 13 shows the construction and mounting of a simple, shielded, 10-meter loop. The loop is made from an 18-inch length of RG-11 (solid or foam dielectric) secured to an aluminum box of any convenient size, with two coaxial cable hoods (Amphenol 83-1HP). The outer shield must be broken at the exact center. C1 is a 25-pF variable capacitor. It is connected in parallel with a 33-pF mica padder capacitor, C3. C1 must be tuned to the desired frequency while the loop is connected to the receiver in the same way as it will be used for DFing. C2 is a small differential capacitor used to provide electrical symmetry. The receiver feed line is 67 inches of RG-59 cable (82 inches if the cable has foam dielectric).

The loop can be mounted on the roof of a car by means of a rubber suction cup. The builder might also fabricate some kind of bracket assembly to mount the loop temporarily in the window opening of an automobile, allowing for loop rotation. Reasonably true bearings may be obtained through a windshield when the car is pointed in the direction of the hidden transmitter. More accurate bearings may be obtained with the loop held out the window and the signal coming toward

that side of the car.

Sometimes the car broadcast antenna may interfere with accurate bearings. Dis-

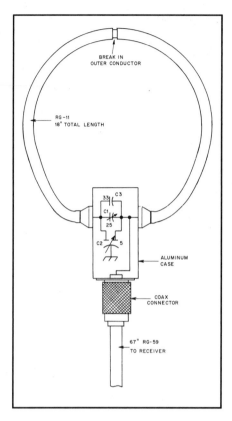

Fig 13—Sketch showing constructional details of the 28-MHz DF loop. The outer braid of the coax loop is broken at the center of the loop. The gap is covered with waterproof tape, and the entire assembly is given a coat of acrylic spray.

connecting the antenna from the broadcast receiver may eliminate this trouble.

Sensing Antenna

A sensing antenna can be added to the loop to eliminate bearing ambiguity. Add a phono jack to the top of the aluminum case as shown in Fig 13. Connect the insulated jack center terminal to the tuning capacitors where they connect to the center conductor of the RG-59 coax feed line. Solder a piece of no. 12 or 14 solid wire to the center pin of a phono plug for insertion in the jack. Plug in and remove the sensing antenna as needed. Start with a sense-antenna length about four times the loop diameter. Prune the sense-antenna length until the pattern is similar to Fig 1D.

A Loopstick Antenna for 1.8 MHz

Fig 14 is a diagram of a ferrite-rod loop. The winding (L1) has the appropriate number of turns to permit resonance with C1 at the operating frequency. Spread L1 over approximately 1/3 of the core center. Litz wire yields the best Q, but enameled wire can be used if desired. A layer of glass tape is recommended as a covering for the core before adding the wire. Use masking tape if nothing else is available.

L2 functions as a coupling link. It is placed over the exact center of L1. C1 is a dual-section variable, although a differential capacitor might yield better balance. The loop Q can be controlled by C2, which is a mica compression trimmer.

Electrostatic shielding of rod loops

can be effected by centering the rod in a U-shaped aluminum, brass or copper channel that extends slightly beyond the ends of the rod loop (1 inch is suitable). This idea is shown in Fig 14B. The open side (top) of the channel can't be closed, as that would constitute a shorted-turn condition and render the antenna useless. This can be proved by shorting across the center of the channel with a screwdriver

Fig 14—Diagram of a ferrite loop (A). C1 is a dual-section air variable. The circuit at B shows a rod loop contained in an electrostatic shield channel (see text). A low-noise preamplifier is shown in Fig 17.

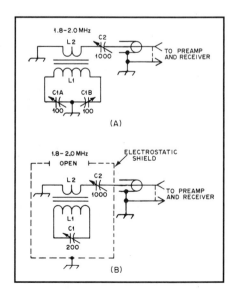

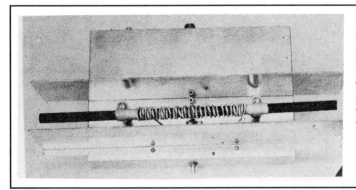

Fig 15—The shielded ferrite-rod loop for 160 meters. Two rods have been glued end to end (see text).

blade when the loop is tuned to an incoming signal.

Fig 15 is a photograph of a shielded rod loop. It was developed experimentally for 160 meters and uses two 7-inch ferrite rods that were glued end-to-end with epoxy cement. The longer core gives improved sensitivity for weak-signal reception.

Obtaining a Cardioid Pattern

Although the bidirectional pattern of loop antennas can be used effectively for tracking transmitters by means of triangulation, a unidirectional pattern reduces the time spent when doing so. It is simple to add a sensing antenna to the loop, and it provides the desired cardioid response.

Fig 16 shows how this is accomplished. The link from the rod loop is connected (via coaxial cable) to the primary of T1, which is a tuned toroidal transformer with a split secondary winding. C3 is adjusted for peak signal response at the frequency of interest (as is C4). Then, R1 is adjusted for minimum back response of the loop. It may be necessary to readjust C3 and R1 several times to compensate for their interaction. Repeat the adjustments until no further null depth can be obtained. Tests show that null depths as great as 40 dB are possible with the circuit of Fig 16 on 75 meters. A near-field

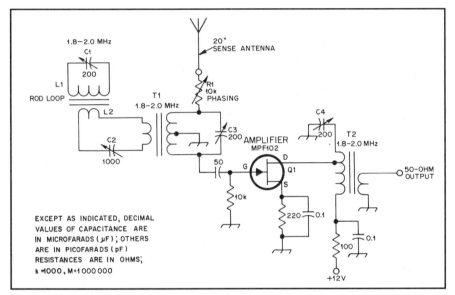

Fig 16—Schematic diagram of a rod-loop antenna with a cardioid response. The sensing antenna, phasing network and a preamplifier are also shown. The secondary of T1 and the primary of T2 are tuned to resonance at the operating frequency of the loop. T-68-2 or T-68-6 toroid cores are suitable for both transformers.

Fig 17—Schematic diagram of a two-stage broadband amplifier. T1 and T2 have 4:1 impedance ratios and are wound on FT-50-61 toroid cores that have a μ_i of 125. They contain 12 bifilar wound turns of no. 24 enameled wire. The capacitors are disc ceramic. This amplifier should be built on double-sided circuit board for best stability.

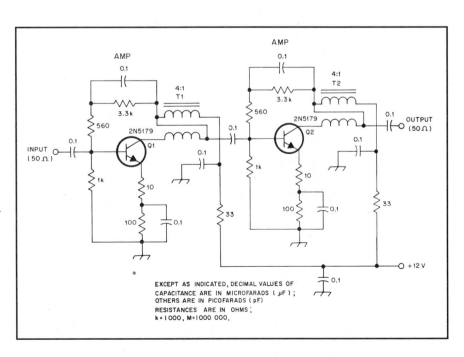

weak-signal source was used for the tests.

The greater the null depth, the lower the signal output from the system. Plan to include a preamplifier with 25 to 40 dB of gain. Q1 of Fig 16 delivers approximately 15 dB of gain. The circuit of Fig 17 can be used after T2 to obtain an additional 24 dB of gain. In the interest of maintaining a good noise figure, even at 1.8 MHz, Q1 should be a low-noise device. A Siliconix U310 JFET would be ideal in this circuit, but a 2N4416, an MPF102 or a 40673 MOSFET would also be satisfactory.

The sensing antenna can be mounted 6 to 15 inches from the loop. The vertical whip need not be more than 12 to 20 inches long. Some experimenting may be necessary in order to obtain the best results. (It also depends on the operating frequency of the antenna.)

A Loopstick for 3.5 MHz

Figs 18 through 20 show a DF loop suitable for the 3.5-MHz band. It uses a construction technique common in low-frequency marine direction finders. The loop is a coil wound on a ferrite rod from a broadcast-antenna loopstick. Because it is possible to make a high-Q coil with the ferrite core, the sensitivity of such a loop is comparable to a conventional loop that is a foot or so in diameter. The output of the vertical-rod sensing antenna, when properly combined with that of the loop, gives the system the cardioid pattern shown in Fig 1D.

To make the loop, remove the original winding on the ferrite core and wind a new coil as shown in Fig 19. Other cores may be substituted; use the largest coil available and adjust the winding so the circuit resonates in the 75-meter band within the range of C1. The tuning range of the loop may be checked with a dip meter.

The sensing system consists of a 15-inch whip and an adjustable inductance that will resonate the whip as a 1/4-λ antenna. It also contains a potentiometer to control antenna output. S1 is used to switch the sensing antenna in and out of the circuit.

The whip, the loopstick, L1, C1, R1 and S1 are all mounted on a 4 × 5 × 3-inch box chassis as shown in Fig 20. The loopstick may be mounted and protected inside a piece of 1/2-inch PVC pipe. A section of 1/2-inch electrical conduit is attached to the bottom of the chassis box and this supports the instrument.

To produce a pattern with only one null, there must be a 90° phase difference between the outputs of the loop and sensing antennas, and the signal strength from each must be the same. The phase shift is obtained by tuning the sensing antenna slightly off frequency by means of the slug in L1. Since the sensitivity of the whip antenna is greater than that of the loop, its output is reduced by adjusting R1.

Adjustment

To adjust the system, enlist the aid of a friend with a mobile transmitter and find a clear spot where the transmitter and DF receiver can be separated by several hundred feet. Use as little power as possible at the transmitter. (Remove your own transmitter antenna before trying to make any loop adjustments and remember to leave it off during transmitter hunts.) With

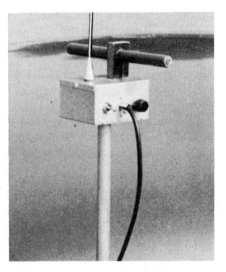

Fig 18—Unidirectional 75-meter DF setup using a ferrite-core loop with sensing antenna. Adjustable components of the circuit are mounted in the aluminum chassis supported by a short length of tubing.

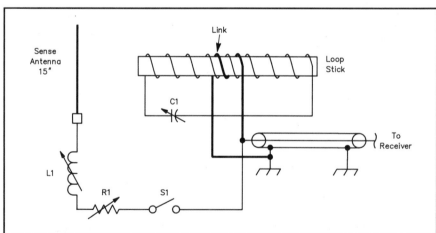

Fig 19—Circuit of the 75-meter direction finder.

C1—140 pF variable (125 pF ceramic trimmer in parallel with 15 pF ceramic fixed).
L1—Approximately 140 µH adjustable (Miller no. 4512 or equiv).
R1—1-kΩ carbon potentiometer.
S1—SPST toggle.

Loopstick—Approx 15 µH (Miller 705-A, with original winding removed and wound with 20 turns of no. 22 enameled wire). Link is two turns at center. Winding ends are secured with electrical tape.

Fig 20—Components of the 75-meter DF are mounted on the top and sides of a channel-lock box. In this view, R1 is on the left wall at the upper left; C1 is at the lower left. L1, S1 and the output connector are on the right wall. The loopstick and whip mount on the outside.

the test transmitter operating on the proper frequency, disconnect the sensing antenna with S1, and peak the loopstick using C1, while watching the S meter on the receiver. Once the loopstick is peaked, no further adjustment of C1 is necessary. Next, connect the sensing antenna, and set R1 to minimum resistance. Then adjust the slug of L1 for maximum S-meter reading. It may be necessary to turn the unit a bit during this adjustment to obtain a larger reading than with the loopstick alone. The last turn of the slug is quite critical, and hand-capacitance may play a part.

Now turn the instrument so that it is broadside to the test transmitter. Turn R1 a complete revolution; if the proper side

was chosen, one particular position of R1 will yield a definite null on the S meter. If not, turn the antenna 180° and try again. This time leave R1 at the setting that produces the minimum reading. Now adjust L1 very slowly for minimum S-meter reading. Repeat this several times (first R1, and then L1) until the best minimum is obtained.

Finally, as a check, have the test transmitter move around the DF and follow it by turning the DF. If the tuning has been done properly the null will always be broadside to the loopstick. Note the proper side of the DF for the null, and the job is finished.

A Phased Array For 144-MHz DF Work

Although there may be any number of different antennas that will produce a cardioid pattern, the simplest design is depicted in Fig 21. Two 1/4-λ vertical elements are spaced one 1/4-λ apart and fed 90° out of phase. Each radiator is shown with two radials (approximately 5% shorter than the radiators).

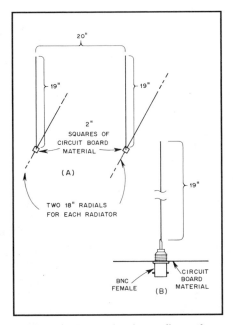

Fig 21—At A is a simple configuration that produces a cardioid pattern. At B is a convenient way of fabricating a sturdy mount for the radiator using BNC connectors.

During the design phase of this project a personal computer was used to predict the impact on the antenna pattern of slight alterations in size, spacing and phasing of the elements. The results suggest that this system is a little touchy and that the most significant change comes at the null. Very slight alterations in the dimensions caused the notch to become much more shallow and, hence, less usable for DF. Early difficulties in building a working model confirmed this.

If you decide to build this antenna, spend a few minutes tuning it for the deepest null. If built with the techniques presented here, this should prove to be a small task that is well worth the extra effort. Tuning is accomplished by adjusting the length of the vertical radiators, the spacing between them and, if necessary, the lengths of the phasing harness that connects them. Tune for the deepest S-meter null, while using a signal source such as a moderately strong repeater. Do this outdoors, away from buildings and large metal objects. Initial indoor tuning on this project was done in a kitchen; reflections from the appliances produced spurious readings. Beware too of distant water towers, radio towers and large office or apartment buildings. They can reflect the signal and give false indications.

Construction is simple and straightforward. Fig 21B shows a female BNC

connector that has been mounted on a small piece of PC-board material. The BNC connector is held "upside down" and the vertical radiator is soldered to the center solder lug. A 12-in piece of brass tubing provides a snug fit over the solder lug. A second piece of tubing, slightly smaller in diameter, is telescoped inside the first. The outer tubing is crimped slightly at the top after the inner tubing is installed. This provides positive contact between the two tubes. For 146 MHz the radiator length is about 19 inches.

Small brass tubing should be available at hobby stores. If none is available in your area, consider brazing rods. These are often available in hardware sections of discount stores. Two 18-inch radials are added to each element by soldering them to the board. Two 36-in pieces of heavy brazing rod were used in this project.

The Phasing Harness

As shown in Fig 22, a BNC T connector is used with two different lengths of coaxial line to form the phasing harness. This method of feeding the antenna is superior to other simple systems for obtaining equal current in the two radiators. Unequal currents tend to reduce the depth of the null in the pattern, all other factors being equal.

With no radials or with two radials perpendicular to the vertical element, a 1/4-λ section of RG-59, 75-Ω coax pro-

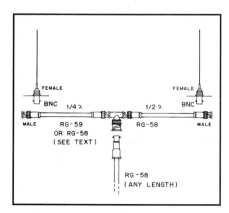

Fig 22—The phasing harness for the 144-MHz DF array. The phasing sections must be measured from the center of the T connector to the point that the vertical radiator emerges from shield portion of the upside-down BNC female connector. Don't forget to consider the length of the connectors when constructing the harness. If care is taken and coax with polyethylene dielectric is used, you should not need to prune the phasing line. With this phasing system, the null is along the boom, on the side of the 1/4-λ section.

duced a deeper null than a 1/4-λ section made of RG-58, 50-Ω line. However, with the two radials bent downward somewhat, the RG-58 section seemed to outperform the RG-59. There is probably enough variation between antennas to make it worth trying both sections. The 1/2-λ section can be made from either RG-58 or RG-59 because it should act as a 1-to-1 transformer.

The coax for the harness should be of the highest quality (well shielded, polyethylene dielectric). Avoid foam dielec-

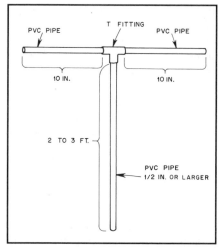

Fig 23—A simple mechanical support for the DF antenna made of PVC pipe and fittings.

tric because its velocity factor may vary. (It can be used if you have test equipment to determine its electrical length.) Avoid coax designed for the CB market or the do-it-yourself CATV market. (Good choices are Belden 8240 for the RG-58 or Belden 8241 for the RG-59.)

Both RG-58 and RG-59 with polyethylene dielectric have a velocity factor of 0.66. Therefore, a 146-MHz 1/4-λ feed line is 20.2 × 0.66 = 13.3 inches. A 1/2-λ section is twice that length or 26.7 inches. Remember that the transmission line is the total length of the cable *and the connectors*. Depending on the construction and connectors you choose, the physical length of the coax may vary somewhat. Determine the length empirically.

Y connectors that mate with phono

plugs are widely available and the phono plugs are easy to work with. Avoid the temptation to substitute these for BNC connectors. Although they are more difficult to find, BNC T connectors provide superior performance. They are well worth the extra effort. If you must make substitutions, use UHF connectors (type PL-259).

Fig 23 shows a simple support for the antenna. PVC tubing is used throughout. (By not cementing the PVC fittings together, you have the option of disassembly for transportation.) Cut the PVC for the dimensions shown with a saw or a tubing cutter. A tubing cutter is better because it produces smooth, straight edges and is less messy. Drill a small hole through the PC board near the female BNC of each element assembly. Measure 20 inches along the boom (horizontal) and mark the two end points. Drill a small hole vertically through the boom at each mark. Use a small nut and bolt to attach the element assembly to the boom.

Tuning

The dimensions given throughout this section are for approximately 146 MHz. Once the antenna is built to rough size, the fun begins. Get a signal source near the frequency of the DF target. Adjust the length of the radiators and the spacing between them for the deepest S-meter null. Make changes in increments of 1/4 inch or less. If you must adjust the phasing line, make sure that the 1/4-λ section is exactly one-half the length of the 1/2-λ section. Keep tuning until you have a satisfactorily deep null on the S meter.

An Adcock Antenna

While loops are adequate in applications where only the ground wave is present, what can be done to improve the performance of a DF system for sky-wave reception? The Adcock antenna has been used successfully for this purpose. There are many possible variations, but the basic configuration is shown in Fig 24.

The operation of the antenna when a vertically polarized wave is present is very similar to a conventional loop. Con-

sequently, the directional pattern is identical to a loop, with a null broadside to the plane of the elements and with maximum gain occurring in end-fire fashion. Element spacing and length are not critical, but somewhat more gain occurs for larger dimensions than for smaller ones. In an experimental model, the spacing was 21 feet (approximately 0.15 λ on 40 meters) and the element length was 12 feet.

Response of the Adcock antenna to a

horizontally polarized wave is considerably different from that of a loop. Good nulls were obtained with the experimental model under sky-wave conditions that produced only poor nulls with small loops (both conventional and ferrite-loop models). Generally speaking, the Adcock antenna has very attractive properties with regard to amateur DF applications. Unfortunately, its portability at HF leaves something to be desired, and it is more suitable

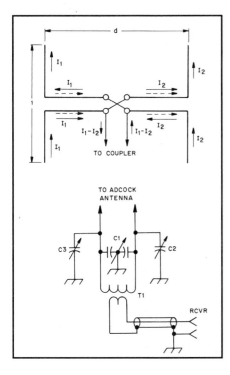

Fig 24—A simple Adcock antenna and suitable coupler (see text).

Less distortion of the pattern should result.

Since a balanced feed system is used, a coupler is required to match the unbalanced input of the receiver. It consists of T1 which is an air-wound coil with a two-turn link wrapped around the middle. The combination is then resonated to the operating frequency with C1. C2 and C3 are null-clearing capacitors. A low-power signal source is placed some distance from the Adcock antenna and broadside to it. C2 and C3 are then adjusted until the deepest null is obtained.

The coupler can be placed on the ground below the wiring-harness junction on the boom and connected by means of a short length of 300-Ω twin-lead. A piece of aluminum tubing was used as a mast, but in a practical application it could be replaced with a length of PVC tubing extending to the ground. This would facilitate rotation and would provide a means of attaching a compass card for obtaining bearings.

to fixed or semiportable applications. While a metal support for the mast and boom could be used, wood is preferable because of its nonconducting properties.

Bibliography

G. Bonaguide, "HF DF—A Technique for Volunteer Monitoring," *QST,* March 1984.

D. Bond, *Radio Direction Finders,* 1st edition (1944), McGraw-Hill Book Co, New York, NY.

D. DeMaw, "Maverick Trackdown," *QST,* July 1980.

D. DeMaw, "Beat the Noise with a Scoop Loop," *QST,* July 1977.

T. Dorbuck, "Radio Direction-Finding Techniques," *QST,* August 1975.

H. Jasik, *Antenna Engineering Handbook,* 1st edition, McGraw-Hill Book Co, New York, NY.

R. Keene, *Wireless Direction Finding,* 3rd edition (1938), Wireless World, London.

J. Kraus, *Antennas* (1950): McGraw-Hill Book Co, New York, NY.

J. Kraus, *Electromagnetics,* McGraw-Hill Book Co, New York, NY.

L. McCoy, "A Linear Field-Strength Meter," *QST,* January 1973.

P. O'Dell, "Simple Antenna and S-Meter Modification for 2-Meter FM Direction Finding," *QST,* March 1981.

Ramo and Whinnery, *Fields and Waves in Modern Radio,* John Wiley & Sons, New York, NY.

F. Terman, *Radio Engineering,* McGraw-Hill Book Co, New York, NY.

Practical VHF Homing Methods

The most common amateur DF occurs on the VHF bands and requires mobile or portable installations. Mobility is important when locating sources of deliberate interference, neighborhood electrical noise, or a hidden transmitter for fun. The following material (contributed by Joseph Moell, KØOV) discusses DF equipment for the popular 2-meter band, but the principles are applicable to other VHF frequencies as well.

Success at mobile DF requires equipment that is effective and reliable. There are three distinct methods of mobile VHF DF commonly used by amateurs: directional antennas, switched dual antennas, and Doppler-shift detectors.

Each has advantages over the others in certain situations. Many DF enthusiasts employ more than one method.

Directional Antennas

Mobile FM or multi-mode transceivers can be pressed into DF service by adding an appropriate directional antenna system. When using directional antennas, the receiver should have a sensitive, easily read S meter. An RF attenuator is needed to keep the S meter usable while "closing in."

Loops

Fractional-wavelength loops are seldom employed above 60 MHz because

they have ambiguous bidirectional patterns and low sensitivity compared to other practical VHF antennas. Their usefulness is limited to close-in on-foot "sniffing" where compactness and sharp nulls are assets—and low gain is of no consequence.

Phased Arrays

The small size and simplicity of two-element driven arrays make them a common choice of newcomers at VHF DF. Antennas such as phased monopoles (see "A Phased Array for 144-MHz DF Work," earlier in this chapter) and ZL Specials have modest gain in one direction and a null in the opposite direction.

The gain is helpful with weak signals, but the broad lobe makes it tricky to take an exact bearing.

As the signal gets stronger, we can use the null for a sharper S-meter indication. Signal reflections, however, may distort the null or obscure it completely. Always take bearings from clear locations for best results with this type of antenna.

Parasitic Arrays

Parasitic arrays are the most common DF antennas in high-competition areas such as southern California. Antennas with significant gain are a necessity because low-power signals are often encountered, and distances of 200 miles or more are covered during some weekend-long events. Typical 2-meter installations feature Yagis or quads with three to five elements, sometimes more. Yagis are commercial or home-made models. Quads are typically home built.

Two types of mechanical construction are popular for mobile VHF quads. Light-wire quads (Fig 25) use thin wire suspended on wood or fiberglass spreaders. They are lightweight and may be turned rapidly by hand. Heavy-wire quads (Fig 26) use thick wire (such as AWG 10) mounted on a plastic-pipe frame. They are more rugged and have a somewhat wider frequency range than light-wire models.

A well-designed mobile Yagi or quad

installation should include a method to select antenna polarization. Poor results occur when a VHF DF antenna is cross-polarized to the incoming signal because multipath and scattered signals (which have indeterminate polarization) are enhanced, relative to the cross-polarized direct signal.

The installation of Fig 27 consists of two identical orthogonally polarized Yagis with separate feed lines on a common boom. Inside the vehicle is a switch box that allows the operator to select horizontal, vertical, 45°, 135°, left-hand circular, or right-hand circular polarization. The Yagi outputs are combined in various phase relationships to produce these polarizations.

The installation of Fig 25 features a slip joint at the boom-to-mast junction, with an actuating cord to rotate the boom and change the polarization. Mechanical stops limit the boom rotation to 90°.

Parasitic-Array Performance for DF

The gain of a parasitic array (typically 8 dB or more) makes it excellent for both weak-signal competitive hunts and for locating low-level interference such as CATV leakage. With an appropriate

receiver, bearings can be taken on any type of signal, including noise. Because only the response peak is used, the null-fill problems and proximity effects associated with loops and phased arrays do not exist with beams.

Multiple incoming signals can be observed as the antenna is rotated, and the operator can make an educated guess as to which is the direct signal. Skilled operators can estimate distance to the transmitter from the rate of signal-strength increase with distance traveled. A DF beam is useful for transmitting, if necessary, but take care to avoid attenuator damage.

The 3-dB beamwidth of a typical mobile VHF parasitic array is about 80°. This is a great improvement over two-element driven arrays, but it is still difficult to get pinpoint bearing accuracy. Careful meter reading can yield errors of less than 10°. In practice, this is no major hindrance to successful mobile DF. The mobile user is less concerned with precise bearings than a fixed station because bearings are used primarily for general direction to "home in" on the signal. Mobile bearings are continuously updated from new, closer locations.

Rapid variations of incoming signal strength are a problem. The target transmitter may be changing output power; the target antenna may be moving or near reflectors (a well-traveled road or airport). Rapid S-meter movements caused

Fig 25—The mobile DF installation of WB6ADC features a strung wire quad for 2 meters and a mechanical linkage that permits either the driver or front passenger to rotate the mast by hand. (K0OV photo)

Fig 26—K0OV uses this mobile setup for DF on several bands, with separate antennas for each band that mate with a common lower mast section, pointer, and 360° indicator. Antenna shown is a stiff-wire quad for 2 meters. (K0OV photo)

Fig 27—Elements from two commercial Yagis on one boom form the heart of WA6DLQ's DF system for 2 meters. A switch box in the vehicle selects from six possible polarization modes. (K0OV photo)

by such conditions make it much more difficult to take accurate bearings with a quad. Furthermore, the process is slow because the antenna must be carefully rotated by hand.

Switched Antenna DF Units

There are three popular DF systems that are relatively insensitive to variations in signal level. Two of them switch the receiver input between a pair of vertical antennas (spaced 1/2 λ or less) at a rapid rate. In use, the indications of the two systems are similar, but the principles are different.

Switched-Pattern Systems

A switched-pattern DF set (Fig 28) alternately creates two cardioid antenna patterns with lobes to the left and the right. The patterns are generated in much the same way as in the phased arrays described above, with PIN RF diodes selecting the patterns. The combined antenna outputs are fed to a receiver with AM detection. Processing after the detector output determines any phase or amplitude differences between the pattern responses to the signal.

Switched-pattern DF sets typically use a zero-center meter as an indicator. The meter swings negative when the signal is coming from the user's left, and positive when it is on the right. When the antenna plane is exactly perpendicular to the signal direction, the meter reads zero.

The sharpness of the zero crossing indication makes possible more precise bearings than those obtainable with a quad or Yagi. Meter deflection indicates which way to turn. For example, a negative (left) reading requires turning the antenna left. This solves the 180° ambiguity caused by the two zero crossings in each complete rotation of the antenna system.

Because it requires AM detection of the switched-pattern signal, this DF system finds its greatest use in the 120-MHz aircraft band, where AM is the standard mode. Commercial manufacturers make portable DF sets with switched-pattern antennas and built-in receivers for field use. These sets can usually be adapted to the amateur 2-meter band. Other designs are adaptable to any VHF receiver that covers the frequency of interest and has an AM detector either built in or added.

Switched-pattern units work well for DF from small aircraft. The two vertical antennas are mounted in fixed positions.

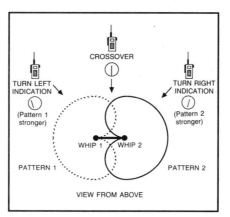

Fig 28—In a switched-pattern DF set, the responses of two cardioid antenna patterns are summed to drive a zero-center indicator.

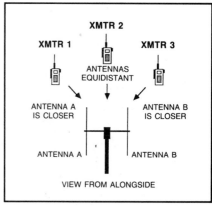

Fig 29—A dual-antenna TDOA DF system has a similar indicator to a switched-pattern unit, but it obtains bearings by determining which of its antennas is closer to the transmitter.

For rapid deployment, they may be simply taped inside the windshield. The left-right indication directs the pilot to the signal source. Since street vehicles generally travel only along roads, mount the antennas on a rotatable mast.

Time-of-Arrival Systems

Another kind of switched-antenna DF set uses the difference of signal wavefront arrival time at two antennas. Narrow-aperture Time-Difference-of-Arrival (TDOA) technology is used for many sophisticated military DF systems.

The rudimentary TDOA implementation of Fig 29 is quite effective for amateur use. The signal from transmitter 1 reaches antenna A before antenna B. Conversely, the signal from transmitter 3 reaches antenna B before antenna A. When the plane of the antennas is perpendicular to the signal source (transmitter 2 in the figure), the signal arrives at both antennas simultaneously.

If the receiver input is switched between the antennas at an audio rate, any difference in arrival time can be detected by an FM discriminator. The resulting short pulses are heard as a tone in the receiver output. The tone disappears when the antennas are equidistant from the signal source, giving an audible null.

Two popular designs for practical TDOA DF sets are the Double Ducky and Handy Tracker (see bibliography). Both employ short vertical whips working against a ground plane, and are intended primarily for short-range on-foot use. For longer range and mobile mounting, vertical dipoles on a metal boom are preferred

for improved performance (Fig 30).

The polarity of the pulses at the discriminator output is a function of which antenna is closer to the source. Therefore, the pulses can be processed and used to drive a zero-center meter in a manner similar to the switched-pattern units described above. "Left" and "right" LED indicators are often used instead of a meter for simplicity.

DF operation with a TDOA dual-antenna DF is done in the same manner as with a switched-antenna DF set. The main difference is the requirement for an FM receiver in the TDOA system and an AM receiver in the switched-pattern case. No RF attenuator is needed for close-in work in the TDOA case, nor is a shielded receiver necessary.

Performance Comparison

Both types of dual-antenna DFs make good on-foot "sniffing" devices and are excellent performers when there are rapid amplitude variations in the incoming signal. They are the units of choice for airborne work. Compared to Yagis and quads, they give good directional performance over a much wider frequency range. They frequently give inaccurate bearings in multipath situations, however, because they cannot resolve signals of nearly equal levels from more than one direction.

The operator can sometimes resolve the problem, however. Because multipath signals are a combined pattern of peaks and nulls, a multipath signal usually fluctuates in strength along the bearing line. You may have observed this phenomenon with FM-broadcast signals in your auto.

Fig 30—A set of TDOA DF antennas is lightweight and mounts readily through a sedan window without excessive overhang. *(KØOV photo)*

If you are receiving a station via multipath and stop at a null, reception improves as you move forward or back 1/2 λ. Similarly, a multipath signal fades as you move the DF antenna along the bearing path (although the fade may be difficult to observe on an FM radio); a direct signal will have constant strength.

The best way to overcome this problem is to take many bearings as you move toward the transmitter. Bearings taken while in motion average out the effects of multipath, making the direct signal more readily discernible.

Switched antenna systems generally do not perform well when the incoming signal is horizontally polarized. In such cases, bearings may be inaccurate or unreadable. TDOA units require a constant-amplitude signal such as FM or CW; they usually cannot yield bearings on noise or pulse signals. Transmitting high power into any switched-antenna system may damage the switching diodes.

Unless an additional method is employed to measure signal strength, it is easy to "overshoot" a hidden transmitter location with a TDOA set. It is not uncommon to see a TDOA fox hunter walk over the top of a concealed transmitter and walk away, following the opposite 180° null, because he had no display of the signal amplitude.

Doppler Shift Detectors

DF sets using the Doppler principle are popular because they are easy to use. Such systems have an indicator that instantaneously displays the target direction relative to vehicle heading, on either a circular ring of LEDs or a digitized readout. A ring of four, eight or more antennas picks up the signal. Quarter-wavelength monopoles on a ground plane are popular for vehicle use, but better performance is obtained from half-wavelength vertical dipoles in free space, where practical.

Radio signals received on a rapidly moving antenna experience a frequency shift known as Doppler effect, a phenomenon well known to anyone who has heard a moving car with its horn sounding. The horn pitch rises as the car approaches, and lowers as the car recedes. Similarly, received radio frequency changes when the transmit or receive antennas move relative to each other. An FM receiver can detect this frequency change.

Fig 31 shows a 1/4-λ monopole antenna moving on a circular track around point P, with constant angular velocity. As the antenna approaches the transmitter, the received frequency is shifted higher. The highest instantaneous frequency occurs when the antenna is at point A, because tangential velocity toward the transmitter is maximum at that point. Conversely, the lowest frequency occurs when the antenna reaches point C, where velocity is maximum away from the transmitter.

Fig 32 shows a plot of the component of the tangential velocity that is in the direction of the transmitter as the antenna moves around the circle. Comparing Figs 31 and 32, notice that at B in Fig 32, the tangential velocity is crossing zero from the positive to the negative and the antenna is closest to the transmitter. The Doppler shift and resulting audio output from the receiver discriminator follow the same plot, so that a negative-slope, zero-crossing detector, synchronized with antenna rotation, senses the direction of the incoming signal.

The amount of frequency shift from the Doppler effect is proportional to the RF frequency and the tangential antenna velocity. The velocity is a function of the radius of rotation and the angular velocity (rotation rate). The radius of rotation must be less than 1/4 λ to avoid errors. To get usable FM deviation (comparable to typical voice modulation) with this radius, the antenna must rotate at approximately 30,000 RPM (500 Hz). This puts the Doppler tone in the audio range for easy processing.

Mechanically rotating a whip antenna at this rate is impractical, but a rapidly rotating antenna can be simulated with a ring of whips, switched to the receiver in succession by RF PIN diodes. A number of commercial Doppler DF models are available on the amateur market. They must be used with receivers having FM detectors. The DoppleScAnt and Roanoke Doppler (see bibliography) are mobile Doppler DF sets designed for inexpensive home construction.

Four whips are typical (see Fig 33), but eight or more may be used. A display unit, which detects the phase of the Doppler tone and compares it with anten-

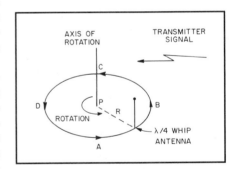

Fig 31—A theoretical Doppler antenna circles around point P, continuously moving toward and away from the source at an audio rate.

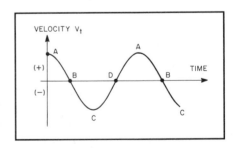

Fig 32—The frequency shift versus time produced by the rotating antenna movement toward and away from the signal source.

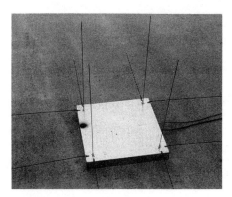

Fig 33—A set of four home-made Doppler DF antennas on an aluminum chassis. The switching circuitry is inside. Whips are made from 3/32-inch bronze welding rod. (K0OV photo)

Fig 34—A window box allows the navigator to turn a mast-mounted antenna with ease while remaining dry and warm. No holes in the vehicle are needed with a properly designed window box. (K0OV photo)

na rotation, completes the setup. Excellent filtering is required in the control unit to separate the added Doppler tone from other modulation and noise on the signal.

Doppler Advantages and Disadvantages

Ring-antenna Doppler sets are the ultimate in simplicity of use for mobile DF. There are no moving parts, and no operator effort is needed to point an antenna. Bearings may be taken on very short signal bursts. Power variations in the source signal cause no difficulties, so long as the signal remains above the DF detection threshold. Doppler systems are excellent for chasing mobile "bunnies," particularly if the hunter is alone.

A Doppler approach does not provide superior performance in all situations, however. If the signal is too weak for detection by the Doppler unit, the hunt advantage goes to teams with beams. Doppler installations are not intended for on-foot "sniffing." The limitations of other switched-antenna DFs also apply: poor results with horizontally polarized signals, no indication of distance, constant-amplitude signals only, and no transmitting through the antenna.

Bearings are displayed to the nearest degree on some commercial Doppler units. This does not guarantee that accuracy, however. A well-designed four-antenna set is typically capable of ±5° accuracy on 2 meters—if the signal is vertically polarized and there is no multi-

path propagation.

Rapid antenna switching can introduce cross-modulation products when the user is near strong off-channel RF sources. This self-generated interference can temporarily render the system unusable. While not a common problem with mobile Dopplers, it makes the Doppler a poor choice for use in remote DF installations at fixed sites with high-power VHF transmitters nearby.

Mobile DF System Installation

Of the mobile VHF DF systems discussed above, the Doppler ring system is clearly the simplest from a mechanical installation standpoint. A four-antenna Doppler DF setup can be readily implemented with magnetic-mount antennas. Alternatively, the four whips can be mounted on a frame that attaches to the vehicle roof with suction cups. The switching circuitry should be installed on the frame, so that only one feed line and one control cable go inside the vehicle (see Fig 33).

A frame also keeps the whips in proper position at all times, with no chance of improperly hooking up the individual antenna lines when the unit is moved between vehicles. It is important that a ground plane is provided, either continuous or with radials as shown, for at least 1/4 λ around the base of each whip.

Rotating Antenna Arrays

Small rotatable antennas can be turned readily by extending the mast through a

window. Installation on each model of vehicle is different, but the mast can usually be held in place with some sort of cup on the arm rest and a plastic tie at the top of the window, as in Fig 30. This technique works best on cars with frames around the windows, which allow the door to open with the antenna in place. Local vehicle codes often limit how far the antenna may protrude beyond the line of the fenders on each side. Large antennas may fit on the passenger side, where greater overhang is usually permissible.

A window box (Fig 34) is an improvement over through-the-window units. It provides a solid, easy-turning mount for the mast, and the plastic cover keeps out bad weather. A custom design is necessary for each model of vehicle. Vehicle codes may limit the use of a window box to the passenger side of the vehicle.

For the ultimate in convenience and versatility, cast your fears aside, drill a hole in the center of the roof, and install a waterproof bushing. A roof-hole mount permits the use of large antennas without overhang violations. The driver, front passenger, and even a rear passenger can turn the mast when required. An installation such as that of Fig 26 should include a pointer and 360° indicator (at the bottom of the mast) for precise bearings.

The installation shown in Fig 26 uses a roof-hole bushing made by mating threaded PVC pipe adapters and reducers. When the fitting is not in use for DF, a PVC pipe cap is a watertight cover for the hole.

Bibliography

J. Moell and T. Curlee, *Transmitter Hunting—Radio Direction Finding Simplified,* TAB Books, Blue Ridge Summit, PA.

D. Geiser, "Double-Ducky Direction Finder," *QST,* July 1981, p 11.

D. Geiser, "Updating the Double-Ducky Direction Finder," *QST,* May 1982, p 15.

J. Moell, "Build the Handy Tracker," *73 Magazine,* September 1989, p 58 and October 1989, p 52.

T. Rogers, "A DoppleScAnt," *QST,* May 1978, p 24; Feedback: July 1978, p 13.

Chapter 5

Transmitters

By Alan Bloom, N1AL
San Francisco Section Technical Coordinator
1578 Los Alamos Rd
Santa Rosa, CA 95409

According to the FCC, most interference from amateur transmitters is not the fault of the amateur equipment. This has been confirmed by the experiences of many local interference committees over the years. Still, transmitters are no more immune to malfunction than any other type of electronic equipment: Perhaps there really is a problem with the rig. If so, read this chapter.

All transmitters emit *some* harmonics, and possibly other spurious frequencies as well. The transmitter may be operating well within specifications, but because it's in a TV fringe area, or because the ham antenna is close to the TV antenna, the specified performance is just not good enough.

The spurious emission levels listed in the FCC amateur regulations [§97.307 (d) and (e)] are not stringent. Below 30 MHz, a properly designed vacuum-tube amplifier can meet these requirements with no more low-pass filter than the pi network used for output impedance matching. For broadband solid-state amplifiers, a simple two-section low-pass filter is sufficient.

Figs 1 and 2 show the allowed spurious levels for HF and VHF transmitters, respectively. Below 30 MHz, when output exceeds 500 W, 0.05 W of spurious radiation is allowed. Now, 1/20 W may seem like real QRP to the average ham, but that much power close to a TV-channel carrier frequency can wipe out fringe-area TV sets for blocks.

Even worse, harmonic attenuation is specified (and tested) into a broadband 50-Ω dummy load. Actual amateur antennas are *not* 50 Ω and resistive at TV frequencies. This can cause filter networks to provide less harmonic attenuation than specified.

The point is, a perfectly legal transmitter is quite capable of causing television interference (TVI). You can calculate the interference level if you like. See the sidebar "Taking the Mystery Out of TVI" for the details.

The calculations show that if you run high power with nearby TV antennas in an area of deep-fringe TV reception, the interference may be difficult or impossible to eliminate. A partial solution is to operate on frequencies where harmonics do not fall within any local TV channels. Fig 3 of the Television chapter should help. As a harmonic approaches the video carrier, picture interference worsens.

ISOLATING THE PROBLEM

Let's assume that you have read the Troubleshooting chapter and have proved that the interference is caused by transmitter harmonics or other spurious frequencies. A simple test can determine whether the unwanted signals are radiated from the antenna or directly from the transmitter chassis and/or cables.

First, disconnect the coax from the back of the transmitter or amplifier, and connect the rig to a shielded dummy load. Use a short piece of good quality (95%

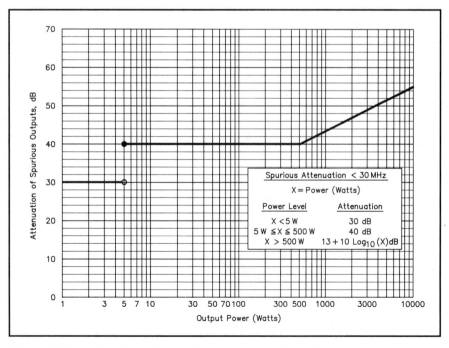

Fig 1—FCC specifications for minimum harmonic attenuation of amateur transmitters operating below 30 MHz.

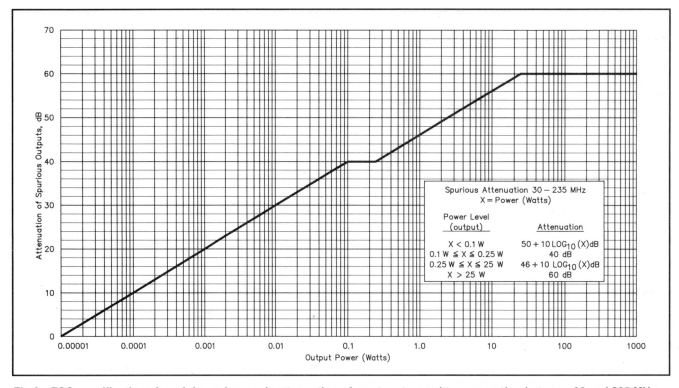

Fig 2—FCC specifications for minimum harmonic attenuation of amateur transmitters operating between 30 and 225 MHz.

braid coverage) coaxial cable with properly installed connectors. (This means the coax shield is soldered to each connector around the entire circumference of the connector shell.)

Transmit with full power into the dummy load. Use all bands, and a selection of frequencies on each band. "Play" with the transmitter tuning controls and attempt to generate TVI. If there is *any* interference on the TV set, there is radiation either from the transmitter chassis (poor shielding) or from the power cord or other leads. If the transmitter is "clean" into the dummy load, but not into the antenna, and if the problem is not in the TV set or caused by external rectification, then the transmitter must be emitting signals outside the amateur bands via the coax feed line.

If there is no interference with the dummy load connected directly to the transmitter output, try moving the load to the antenna end of the coax. If this point is not accessible, move it to the point where the coax leaves the shack, after any low-pass filters, wattmeters, antenna tuners, and so on. If the interference returns, examine the connecting coax cables for good solid connections between the cable shields and connector bodies. This is especially important for the cable between the transmitter and

low-pass filter. It should be as short as possible and have the shield solidly soldered around the full circumference of the connector shell. Even better: use a double-male coaxial adapter and mount the filter directly to the transmitter output connector. Coax-shield problems may only show up with the antenna feed line connected, because the feed line can act as an antenna for the interference.

Poor Shielding

If there is interference with a dummy load connected, look for poor shielding in the transmitter. Are any shields missing? Are there any missing or loose screws? Corrosion or paint at the seams can let a surprising amount of VHF energy escape. Make sure all joining surfaces are clean and bright. A salty or acidic atmosphere can corrode plated metal surfaces. So can cigarette smoke. If the radio has been used much by a heavy smoker, a major cleaning of all contacting surfaces may be necessary to restore good connections.

The ac-line cord is another common source of incidental radiation. Wrap several turns of the cord around a ferrite rod or through a ferrite toroid core. If there is *any* change in interference, it is a sure sign that there is RF on the line cord. Other connecting leads, such as the

microphone or key cords, merit the same attention. A good device for the purpose is a toroid with a permeability greater than 800 such as an FT-82-43.

Crud Out The Coax

If the interference only occurs with the feed line actually connected to the antenna, the transmitter output signal must contain interference that is reaching the antenna. First, try a low-pass filter. Locate the filter directly at the transmitter output, before any SWR meter, Transmatch or other accessories. If there already is one filter, add a second in series. One may not be enough in areas of weak TV reception.

Check used transmitters for modifications or improper alignment by a previous owner. The proper alignment procedures appear in the manufacturer's service manual (often an extra-cost item, separate from the owner's manual).

Vacuum-tube transmitters are especially susceptible to misadjustment because the operator can adjust the tuning controls. Be sure that the final amplifier is not overdriven, especially with older rigs that do not have automatic level control (ALC). Reduce the drive to the lowest level that yields full power output. Also be sure that the final amplifier bias is set correctly (check for the proper *resting*

TAKING THE MYSTERY OUT OF TVI

Why do some hams have no trouble with television interference when other amateur stations, seemingly just as "clean," turn the neighbors' TV screens into multi-color light shows? The mystery can be solved with a single equation:

$$H_R = (T_P + T_A + G_T + G_R + S_A) - S_S + S_R$$

H_R is the additional harmonic attenuation (in dB) required for the transmitter. T_P is peak transmitter power in dBm, T_A is harmonic attenuation (dB) of the "bare" transmitter, G_T is transmit antenna gain (dB) at the TV frequency, G_R is TV antenna gain (dB), S_A is path loss between the transmitter and TV antennas (dB), S_S is TV signal level (dBm), and S_R is the signal-to-interference ratio required for an acceptable picture (dB). Here's how to estimate these numbers:

Transmitter power, T_P, is measured in dBm, or decibels referenced to 1 mW. One watt is +30 dBm, 100 W is +50 dBm, and 1500 W is +62 dBm.

Transmitter harmonic attenuation, T_A, is typically somewhat better than required by FCC regulations, especially for the higher-order harmonics. The specifications in the transmitter manual should give a conservative estimate. Older transmitters, built before FCC harmonic regulations, may have only –30 to –35 dB suppression of the second and third harmonics.

The transmitting antenna may have up to –10 dB or so of gain, G_T, at TV frequencies, depending on the type of antenna and its orientation. A conservative estimate is 0 dB.

A good (fringe-area) TV antenna may give up to +10 dB of gain, G_R, in its main lobe. Figure about –10 dB at right angles to the main lobe and 0 dB off the rear.

Path loss can be estimated from the procedure described below, or calculated using the following equation:

$$S_A = 10 \log_{10} \left(\frac{\lambda^2}{(4 \pi D)^2} \right)$$

D is the distance between antennas in the same units as the wavelength, λ. To estimate S_A, note that if D is 10 meters (33 feet), then S_A varies from –27 dB on TV channel 2 to –31 dB on channel 6. Add –6 dB each time the distance doubles.

The received TV signal level, S_S, depends on the TV-station effective radiated power, distance to the TV antenna, terrain, and other factors. If the proper equipment is available, measure the signal level right at the TV end of the coax. Otherwise, figure about –60 dBm for fringe-area reception (more than 50 miles) and –35 dBm for local stations.

The signal-to-interference ratio, S_R, required for a decent TV picture varies. It's about 40 dB for interference close to the video carrier. For signals more than 2 MHz from the video carrier and not too close (a few hundred kHz) to the sound and color subcarriers, it's about 20 dB. 35 dB is a reasonable value.

Sample Calculation

Let's perform the calculation for an amateur station operating on 15 meters with 1.5 kW of output power. The third harmonic (in TV channel 3) is down 45 dB, just within FCC specifications. The TV antenna is 40 meters (130 ft) away, pointing at the ham antenna. The TV station is more than 50 miles away. The numbers are:

T_P = +62 dB,
T_A = – 45 dB,
G_T = 0 dB,
G_R = +10,
S_A = – 28 –12 = – 40 dB,
S_S = – 60 dBm,
S_R = +35 dB.

$$H_R = (+62) + (– 45) + (0) + (+10) + (– 40) – (– 60) + (+35)$$
$$= 82 \text{ dB}$$

It would probably take two low-pass filters in series to get this kind of attenuation. Other factors, such as power- and control-wire radiation and shielding, not to mention fundamental overload of the TV set, would also be tough "nuts" to crack. Fortunately, few of us are faced with such a worst-case scenario. This information was derived from "Harmonic TVI: A New Look at an Old Problem," by D. Rasmussen, W6MCG, and D. Gerue, K6YX (Sep 1975 *QST*, pp 11-14).—*Alan Bloom, N1AL, San Francisco Section Technical Coordinator, Santa Rosa, California*

---

current with the transmitter on, but with no RF output), as specified in the owner's manual.

Neutralization

Some tube amplifiers include a special circuit that prevents oscillation by *neutralizing* the tube internal (feedback) capacitance. If the neutralization circuit is not adjusted correctly, there may be evidence of oscillation: erratic jumps in plate or grid current while tuning. In a properly operating amplifier, maximum power output, minimum plate current and maximum control- and screen-grid current should all occur at about the same control settings. If not, the transmitter needs neu-

tralization. This procedure should be described in the transmitter service manual; otherwise use the procedure described in the *ARRL Handbook* (look in the index under "neutralization").

Don't overlook the transmitter manufacturer as a source of help for spurious-emission problems. If there is a stubborn problem, the service department may be willing to offer suggestions.

CIRCUIT MALFUNCTIONS

This section covers transmitters that are operating outside their design limits: There is something wrong with the rig. The simple fixes mentioned above didn't work, and now it's time to dive into the

circuitry to find out what is wrong. The two most common problems are excessive harmonics and spurious oscillations.

Final amplifier overdrive is a common cause of harmonics. Most modern transmitters have an ALC circuit to limit the peak power level. If a slight reduction in drive level causes a dramatic reduction in interference, the ALC circuit may be faulty. There is usually an internal adjustment to set the ALC level, and readjustment of this control may be all that is needed. A previous owner may have "optimized" this setting to increase power output.

Both solid-state and tube transmitters can have defective tank-circuit or low-

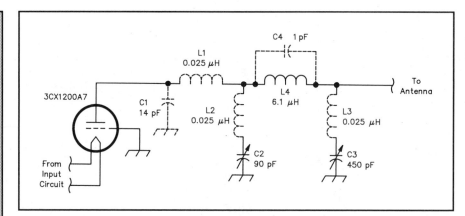

Fig 3—Schematic of a typical pi-output matching network in an HF vacuum-tube amplifier. C2 and C3 are the tuning and loading capacitors; L4 is the tank coil. C1, L1, L2, L3 and C4 are not actual components, but represent small stray reactances that are important at VHF frequencies. (The plate choke and coupling capacitor have been omitted for clarity.)

pass-filter components that cause excessive harmonic generation. This problem typically causes low power output as well. Check for signs of overheating, mechanical damage, or poor solder connections.

VHF Parasitic Oscillations In HF Vacuum-Tube Amplifiers

Although it doesn't show on the schematic diagram, every capacitor has some series self inductance from the connecting leads. Similarly, every inductor has some (parallel) capacitance distributed between the turns of wire in the coil. These stray (or parasitic) reactances become important at high frequencies. Fig 3 illustrates a typical kilowatt amplifier tuned for the 40-meter band. Parasitic inductances L1 and L2 form a high-Q resonance with C1 and C2 around 200 MHz, which causes the tube to have high voltage gain at this frequency. If enough feedback exists between output and input, the amplifier may oscillate.

How can you spot parasitic oscillations? They often reveal themselves through erratic plate-current fluctuations when tuning the amplifier output or input tuned circuits. Also try reducing the drive level to zero. If there is *any* change in grid or plate current when tuning with no drive, it is a sure sign of spurious oscillations.

It is helpful to determine the approxi-

mate frequency of the oscillation. The best way is to feed the transmitter output (through a high-power attenuator) into a spectrum analyzer. The traditional alternative (for those of us who don't have several thousand dollars to invest in test equipment) is a tuned wavemeter or a grid-dip oscillator in its wavemeter mode. Use the wavemeter coil to "sniff" near the amplifier tank coil while transmitting, and tune the wavemeter dial for maximum meter indication. Be *very* careful if you try this—there is high voltage nearby! A safer alternative is to listen for the oscillation on a nearby receiver. A general-coverage

shortwave receiver can detect oscillations in the HF region, and a TV set is a good detector of VHF parasitics.

Once the approximate frequency of the parasitic oscillation is determined, try to pin down which part of the tank circuit exhibits the spurious resonance. With the amplifier tuned for maximum interference, turn off the power, remove the covers and temporarily short the high-voltage capacitors to ground with an insulated screwdriver. Place the dip meter in its oscillator mode, and slowly tune it around the parasitic frequency while holding the dip-meter coil close to various circuit

"Don't overload the transmitter"

components. A sharp dip in meter current indicates the circuit with unwanted resonance.

VHF parasitic oscillations can often be reduced or eliminated by moving the spurious resonances higher in frequency. This is done by reducing the stray inductance of all final-amplifier RF connections: Make leads as short and fat as possible. Pay special attention to screen-bypass capacitors in tetrode and pentode tube circuits. Sockets with built-in low-inductance screen bypass capacitors are available for some high-power tubes. In grounded-grid amplifiers, connect the grid pins directly to the chassis with a short, wide copper strap. If the grid is not grounded directly, use the shortest possible leads on the grid bypass capacitor(s). Also, several capacitors in parallel have less inductance than one.

The plate-tuning capacitor should be bolted directly to the grounded chassis. The connection from tube plate through the coupling capacitor to the tuning capacitor should be as short and fat as possible. It is especially important to use low-inductance connections between tubes connected in parallel—otherwise they can act as a push-pull VHF oscillator.

Don't overlook sources of stray feedback from amplifier output to input. Make sure all power-supply leads are well bypassed and filtered, including the filament supply. The filament choke and input matching network should be well shielded from the output circuitry.

Parasitic Suppressors

You can "kill" a VHF parasitic resonance by installing a resistor in series with either the inductance or capacitance of the undesired VHF tuned circuit.[1] Unfortunately, this tends to "kill" the efficiency of the desired HF tuned circuit as well. The answer is to connect a small coil in parallel with the resistor, forming a circuit known as a *parasitic suppressor*. Since inductive reactance is proportional to frequency, the coil shorts out the resistor at HF frequencies, while presenting a high reactance at VHF.

If the VHF parasitic is too low in frequency, it may be difficult to design a parasitic suppressor capable of suppressing the oscillation without also degrading efficiency at the highest desired operating frequency. This is one reason to first move all parasitic resonances as high in frequency as possible.

Traditionally, parasitic suppressors are used in series with the plate, but they may also be useful in the control- and screen-grid circuits of grounded-cathode amplifiers. The resistance and inductance values are found by experimentation. If the inductance is too high, the resistor gets hot and efficiency at the high end of the HF range suffers. If the inductance is too small, the oscillation may not be suppressed (which may also cause resistor overheating!). A typical suppressor for the plate circuit of a 150-W transmitter consists of a half-dozen turns of wire in parallel with a 50- to 100-Ω, 1-W carbon or metal-film resistor. For a kilowatt amplifier, a good starting point would be three or four 220-Ω, 2-W resistors in parallel with 1 or 2 turns of no. 12 or 14 wire.

Parasitic suppressor resistors often overheat, not only from the RF current through them, but also because of the nearby hot final-amplifier tube. Replace any resistors that have become discolored.

Parasitic Oscillations In Solid-State Amplifiers

VHF parasitic oscillations are not normally a problem in HF solid-state amplifiers. Since the high-frequency current gain of a bipolar transistor drops 6 dB every time the frequency doubles, there is usually not enough gain at VHF to sustain an oscillation.

Low-frequency oscillations are another matter. Since transistors have so much gain at low frequencies, they oscillate readily if unwanted feedback is present. Detect low-frequency spurs by listening to the transmitted signal with a nearby receiver. Slowly tune back and forth several hundred kilohertz either side of the carrier frequency, while the transmitter is transmitting a series of CW dits. Try several different power levels. Alternatively, a modulation monitor or oscilloscope capable of displaying the RF signal will show any "fuzz" on the CW carrier.

A common low-frequency feedback path extends from the transistor collector, through the dc power supply and into the base bias network. Power supply decoupling must be effective not only at the RF frequencies the amplifier is designed for, but at low frequencies as well. A 0.05-μF

ceramic capacitor in parallel with a 220-μF electrolytic should serve to bypass the +12-V power supply to the final amplifier of a 100-200 W HF transmitter. The base bias network should be either supplied from an IC voltage regulator or decoupled with an RC low-pass filter.

Remember that every amplifier stage in the transmitter RF chain needs power supply decoupling. Failure of any decoupling components can cause low-frequency oscillations. Solder connections can degrade over time, especially in high-temperature environments like around high-power RF amplifiers. Resolder any suspicious solder joints.

Many amplifiers include negative-feedback networks to stabilize gain. The network often consists of a capacitor and a resistor (and sometimes an inductor) series-connected from the collector to the base. Often a resistor is connected from the base to ground in order to lower the low-frequency input impedance. If low-frequency oscillations develop in a solid-state amplifier, check to see that all parts still function and that the solder joints are good.

SHIELDS—PLUGGING THE LEAKY SIEVE

A low-pass filter can only reject harmonics that are exiting through the coaxial cable.

---

[1]R. Measures, "Parasitics Revisited," Sep 1990 QST, pp 15-18 and Oct 1990 QST, pp 32-35.

50-KHz CW SIDEBANDS

I once received an ARRL Official Observer report for transmitting a T-6 (hum-modulated) note just outside the bottom of the 20-meter band. My logbook revealed that I was operating about 50 kHz higher at the time. The transmitter I was using included a transistor amplifier stage with 0.01-μF coupling capacitors and 1-mH shunt-feed inductors at the input and output. A quick calculation revealed that these values resonate at about 50 kHz. The engineers who designed this rig had accidentally designed a tuned-collector, tuned-base oscillator! The 50-kHz oscillation amplitude modulated the transmitted signal, which caused sidebands to appear 50 kHz above and below the carrier frequency.—*Alan Bloom, N1AL, San Francisco Section Technical Coordinator, Santa Rosa, California*

"No ground problems here!"

increases. Ferromagnetic materials, such as steel, have a much shallower *skin depth*, and thus a much higher RF resistance than copper or aluminum. See Fig 5.

It is said that magnetic fields (low-impedance sources) are best shielded by ferromagnetic materials, such as steel. This is true only below a few hundred kilohertz. At those frequencies, the primary shielding mechanism is absorption. Absorptive loss is easily calculated: it is 8.7 (dB) times the thickness in skin depths. Since ferromagnetic materials have shallower skin depths, attenuation is greater.

At HF, however, reflection is more important than absorption, so once again, copper and aluminum make better shields. To summarize: use steel shielding only for low-frequency magnetic fields (such as stray radiation from a 60-Hz power transformer.) Aluminum or copper is better in all other cases.

Slots and Holes

Actually, the above discussion is incomplete. While it is true that steel is a poorer RF shield than aluminum, it doesn't matter in most practical cases. Most RF leakage is not through the shielding material itself, but through openings in the shield.

The trouble caused by a particular hole is determined more by its maximum dimension than by its area. A round hole passes less radiation than a long, thin slot

A filter won't help if interference radiates from the chassis. Just as water passes through a sieve, RF passes through a leaky chassis. The object is to plug the holes.

How important is grounding to shield effectiveness? If "ground" means "earth ground," the answer is not at all. A well shielded transmitter causes no RF currents to flow on the outside of the shield, so the presence or absence of an earth ground can have no effect. If poor shielding does allow RF currents to flow on the outside of the chassis, a ground wire will not cure the condition.

The FCC amateur rules do not regulate incidental radiation. However paragraph 97.307(c) states in part: "If any spurious emission, *including chassis or power-line radiation*, causes harmful interference to the reception of another radio station, the licensee of the interfering amateur station is required to take steps to eliminate the interference, in accordance with good engineering practice." (Emphasis added.)

The "Sniffer"

It is fortunate that the FCC doesn't require us to measure chassis radiation levels accurately, as this requires sophisticated equipment. The home-built "sniffer" illustrated in Fig 4 can give a qualitative measure of chassis leakage.

The sniffer does not respond strongly to electric fields, but favors magnetic fields, the type mainly radiated by breaks in RF shields. Connect the sniffer to a

device capable of receiving the interfering signal (such as a TV set) and move the loop around the suspect chassis. Maximum pickup occurs with the loop wire at right angles to any slots or seams and parallel to any wires exiting the shield enclosure.

Electric and Magnetic Fields

The fields radiated by an antenna are fundamentally different far away from the antenna (*far field*) than they are close to it (*near field*). In the far field (greater than about $1/6\ \lambda$ from the source), the electric (E) and magnetic (H) components of the electromagnetic field always have the same relative amplitude. The ratio E/H is known as the *impedance of free space*, which is equal to 377 Ω.

In the near field, however, E/H depends on the radiator. High-impedance sources generate mostly E fields. An example is a short rod or unterminated wire. Low-impedance sources generate mostly magnetic (H) fields. An example is a small wire loop.

Shields work by either attenuating the signal as it tries to pass through or by reflecting it. For high-impedance sources, the primary shielding mechanism is reflection. This is why the best shielding materials are high-conductivity metals such as copper and aluminum.

At RF frequencies, most of the current induced by the impinging signal flows near the surface of the shield. In fact, 63% of the current flows within one skin depth of the surface, falling off rapidly as depth

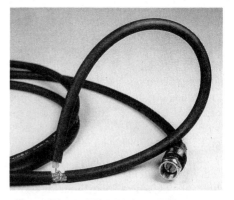

Fig 4—The "sniffer" is used to measure unwanted chassis leakage of harmonics and other spurious emissions. Remove 1-1/2 inches of outer insulation and shield from the end of the coaxial cable. Then remove 1 inch of inner insulation. About 12 inches from the end, score the outer insulation and carefully pull it back to expose 1/4 inch of shield. Now wrap and solder the center conductor to this exposed shield.

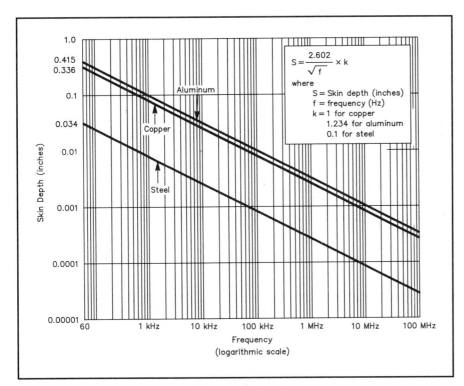

$$S = \frac{2.602}{\sqrt{f}} \times k$$

where
S = Skin depth (inches)
f = frequency (Hz)
k = 1 for copper
    1.234 for aluminum
    0.1 for steel

Fig 5—Skin depth of copper, aluminum and steel v frequency. Ideally, shield thickness should be at least 3-10 times the skin depth. The skin depth of a particular steel varies greatly depending on the alloy used.

of the same area. Pay special attention to seams around shield covers. It is amazing, but true: A long gap that is too tight to pass a piece of paper can radiate almost as though the shield were not present. As a "rule of thumb," space mounting screws no more than 1/20 λ (at the highest frequency of concern). In order to shield harmonics that fall within the North American VHF TV channels (up to 216 MHz), space shield screws no more than about 7 cm (2.75 inches) apart.

Improving Shields

In cases of stubborn leakage, shield seams can be sealed with "EMI tape." This can be especially effective in preventing microwave leakage. (EMI-tape adhesive is usually not conductive; capacitive coupling "bridges the gap" between the tape and the shield enclosure. The coupling works fine with microwaves, but not so well at lower frequencies.) 3M makes copper adhesive tape in several widths. Simply apply the tape across shield seams to effectively seal them. Holes in shield enclosures may be covered with conductive screen to prevent radiation. Copper screen can be attached to steel enclosures by soldering. Use aluminum window screen on aluminum

chassis, because any copper-to-aluminum contact is subject to corrosion.

Unshielded equipment may need an added shield. Despite its aesthetic disadvantages, most anyone can fabricate a suitable metal enclosure for equipment. Use copper or aluminum sheet. Make sure that the enclosure provides adequate ventilation via small holes, or large ones covered with screen. "Piano hinge" provides better shielding than small hinges where openings are needed. If the affected equipment is installed in a metal cabinet, such as an entertainment center, no fabrication is needed.

As a last resort, plastic enclosures can be shielded by the addition of metal foil or conductive paint (see GC-Thorsen in the Suppliers List) either inside or outside the cabinet. While internal shields usually look better, they have several disadvantages. Conductive material may short circuit the equipment and present a shock hazard. Internal shields should be installed only by qualfied technicians with the manufacturer's approval. Before painting a plastic enclosure, paint a small test area first to be sure the paint solvent does not damage the plastic. Also, check for good paint-to-plastic adhesion. Flaking conductive paint can cause shorts.

The orientation of interference-carrying wires inside the chassis is important. Obviously, it is best if they are far from any holes in the shield, but it's.even more important to orient them parallel to the seams. Wires at right angles to slots cause maximum leakage.

Paint is an insulator. Remove all paint from seam joints for the best connection. Where screw mounting is not practical (such as around an access door), use finger stock or one of the other commercial RF-gasket materials (Fig 6).

Meters are a source of shield leakage, because they usually require a large hole in the front panel. Shield round meter openings with part of a discarded tin can (an old trick, see Fig 7). Find a can slightly larger than the meter. Cut off a portion (including an unopened lid) somewhat deeper than the meter (leave enough extra depth to form tabs that can be bent and drilled for mounting screws). Bypass the meter leads to the can at the exit points. Fabricate shield enclosures for square or edgewise meters from PC-board material.

One final shielding tip: Two poor shields are usually better than one good one. A good shield might have 100 dB of isolation. Two 60-dB shields add up to 120 dB, a good 20 dB (100 times) better than the single "good" shield.

DECOUPLING AND BYPASSING

The Maginot line was France's defense against a German invasion before World War II. It was the strongest, most impregnable defensive structure ever built. The French high command was convinced that the Germans would never breach it. They didn't. They simply bypassed it by attacking around the end of the line, through Belgium and the Netherlands.

Similarly, the best shielding in the world won't work if there is a path through it. Any wire that pierces the shield wall acts as a receiving antenna inside the shield and a transmitting antenna outside. There are two ways to prevent this: by shielding the wire, or by filtering it at the point where it pierces the shield.

Shielded Wire

Shield the wire on the inside of the chassis, on the outside—or both. The most common method uses coaxial cable. Ground the coax shield right where it passes through the chassis. For best results, *don't* twist the braid into a pigtail

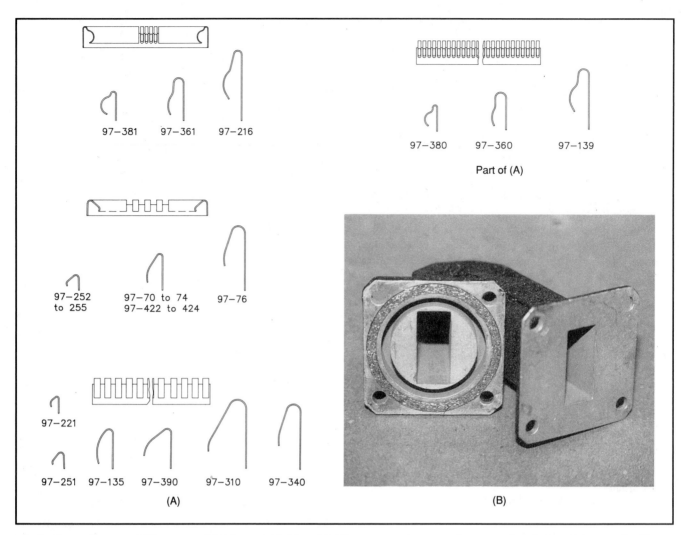

97–381   97–361   97–216

97–380   97–360   97–139

Part of (A)

97–252
to 255   97–70 to 74
97–422 to 424   97–76

97–221

97–251 97–135 97–390 97–310 97–340

(A)   (B)

Fig 6—Several types of "finger stock" (A) are useful for shielding seams that must be opened and closed frequently. (See Instrument Specialties Co and Richardson Electronics in the Suppliers List.) Wire-mesh RF gaskets (B) can give an adequate RF-tight seal up into the UHF and microwave range.

and solder it to a ground lug. It is much better to *connect the braid around the entire circumference* of the hole or connector, so that the center conductor is completely surrounded.

If interference is generated by a current flowing in a wire, it is magnetic (low impedance) in nature, and the H field is stronger. In that case, ground the coax shield at both ends. If the interference results from capacitive (voltage) coupling, then the E field is stronger. It may be better to ground only one end of the coax shield. This avoids a *ground loop,* where magnetic fields can cause current to flow in the loop formed by the coax shield and chassis. If the cable length is greater than 1/20 λ, however, always ground both ends.

If shielding a wire is impractical, try the "poor man's coax"—a "twisted pair."

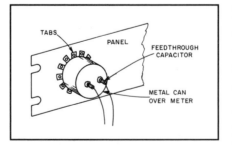

Fig 7—A shielding technique to prevent radiation from meter mounting holes.

Twisting the hot wire together with a ground-return wire reduces both electric and magnetic pickup. Ground the return wire according to the rules for coax shields, above.

Try to route the wire tight against the

chassis for its entire length—in the angle between bottom and sides, if possible. The shielding effect is greater when the wire is close to the chassis. This technique is useful even for coax, to prevent ground loops.

For the same reason, any PC board with potential radiation or susceptibility problems should have a ground plane on at least one side. To reduce radiated emissions, computer PC boards often use four or more layers: one or two for ground plane(s) and the others for circuitry.

Filtering Leads

If interference does manage to find its way onto a wire, the wire must be filtered before it exits the shield enclosure. Usually the desired signal is lower in frequency than the undesired signal, so a low-pass filter is called for. (A low-pass filter at a transmitter RF output connector is a special case that is covered later.)

Low-Pass Filters

The simplest low-pass filter is a *bypass capacitor* connected between the wire and ground. Install it with shortest possible leads directly where the wire exits the shield. Use additional bypass capacitors inside the shield, as close to the interference source(s) as possible. In a computer or other digital device, include a bypass capacitor from the power-supply lead to ground near each IC. In a transmitter, bypass the power-supply lead near each amplifier stage. Placing the capacitor close to the RF source minimizes the wire length carrying (and radiating) unwanted RF energy.

If a simple bypass capacitor doesn't work, add one or more resistors or inductors in series with the signal lead. Resistors are cheaper than inductors, and they work over a broader frequency range. They are not suitable in high-current applications, however, because of the IR (voltage) drop. Shunt capacitors should have low impedance, and the series resistors or inductors should have high impedance at the lowest frequency to be filtered. (An R:X ratio of 5:1 or 10:1 should be sufficient.) Use the standard reactance formulas:

$$X_L = 2 \pi f L \qquad \text{(Eq 1)}$$

$$X_C = \frac{1}{2 \pi f C} \qquad \text{(Eq 2)}$$

where

  f = frequency (MHz)
  L = inductance (μH)
  C = capacitance (μF)
  X_L = inductive reactance (Ω)
  X_C = capacitive reactance (Ω).

Typical values for power-supply decoupling of RF signals in the 3.5- to 200-MHz range would be 1 μH and 0.05 μF.

Be careful with LC filters as they suffer from a resonance effect. At frequencies near

$$f = \frac{1}{2 \pi \sqrt{LC}} \qquad \text{(Eq 3)}$$

they can *increase* the circuit response over that with no filter at all! One solution is to make L and C large enough that the reso-

nant frequency falls well below the lowest frequency of interest. Another solution is to load the resonant circuit, by adding a resistor in series with the coil or capacitor. The resistance should equal twice the reactance of either L or C at resonance.

Try "Brute Force"

As an example of interference filtering, consider the "brute-force" ac-line filter in Fig 8. Safety is an important consideration with ac-line filters. Three major organizations that promote safety standards are UL (Underwriters Laboratories) in the USA, CSA (Canadian Standards Association) in Canada, and IEC (International Electrotechnical Commission) in Europe. All capacitors connected to the ac line should be certified by one of these organizations for the line voltage in use. Certified parts are labeled with an agency symbol or *mark*. UL uses a backwards "R" attached to an "L." The CSA mark is "SA" inscribed in a large "C." Each European country has its own standards agency. For example, the mark of Germany's Verband Deutscher Elektrotechniker is "VDE" inscribed in a triangle.

Some references recommend using 1000-V dc ceramic-disc capacitors. Unfortunately, these parts are intended for dc bypass applications, and their performance with 60-Hz ac is unspecified. UL-recognized parts are inexpensive and readily available in mail-order parts catalogs, so there is no reason not to use them.

In an ac-line filter, capacitor size is limited by allowable 60-Hz leakage current. A 0.1-μF bypass capacitor connected from the 120-V ac line to a chassis would let up to 4.5 mA of 60-Hz current flow in anyone unlucky enough to touch both the chassis and ground at the same time. The total of all capacitors from each side of the ac line to ground should be less than 0.01 μF to keep leakage current within the UL limit of 0.5 mA. (Be cautious about installing ac-line filters in any home entertainment equipment, because the same 0.5-mA limit applies.)

The inductor value is limited by physical size (bigger is better). The coils described in Fig 8 are about 9 μH each, which results in a self-resonant frequency near 770 kHz. Install the filter inside the transmitter chassis if there is room, or in a small metal box bolted to the rear panel.

Ac-line filters are also available commercially, both as external "add ons" and for inclusion inside equipment. The inter-

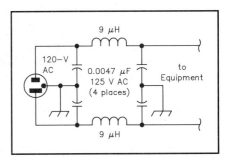

Fig 8—A "brute-force" ac-line filter. The inductors are 23 turns of no. 18 enameled wire on a 1-inch diameter form (9 μH). The 0.0047-μF, UL-recognized capacitors are available from Digi-Key Corporation, part no. P4600.

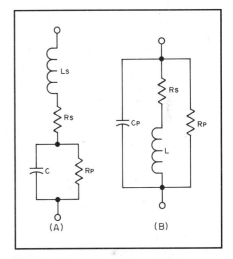

Fig 9—A, every capacitor includes stray series inductance and resistance. B, every inductor has stray shunt capacitance and series resistance.

nal units often include a line cord receptacle and a fuse holder. Manufacturers include Corcom, Cornell-Dubilier (CDE), Schaffner and Sprague. (See the Suppliers List at the end of this book.)

Filter Capacitors

Every capacitor you buy comes with an inductor and resistor included free of charge. The "equivalent series resistance" (ESR, R_S in Fig 9A) results primarily from losses in the dielectric. The "equivalent series inductance" (ESL, L_S in Fig 9A) results from self-inductance in the lead wires. As frequency increases, the ESL dominates; the capacitor acts like an inductor, which means the impedance increases with frequency. Hence it is important to keep bypass capacitor leads short.

Electrolytic capacitors, both the aluminum and tantalum versions, pack a lot of capacitance into a small space, but relatively high ESR and ESL limit their usefulness at high frequencies. Broadband bypassing requires a large-value electrolytic (for low frequencies) in parallel with a ceramic capacitor (to handle RF).

Ceramic capacitors have low ESL, and they are widely used in RF-bypass and filter applications. Those with larger values use high dielectric-constant materials, with high losses (high ESR). This can cause overheating in high-power RF circuits. The capacitance tends to be unstable as well. Thus, high-dielectric-constant ceramic capacitors are primarily used as power-supply bypass capacitors, where stability is not as important. The dielectric type is indicated by the temperature-coefficient code printed on the side of the capacitor body. Lossy dielectrics are indicated by TC codes like Z5U, Y5V and Z5P. Low-loss styles list the temperature coefficient (TC) in parts per million (PPM): like N750 (–750 PPM /°C), P150 (+150 PPM/°C) or NP0 (zero nominal TC).

Mica-dielectric capacitors have little loss, and they are stable, but they are not available in values greater than about 0.01 μF. Poly-film capacitors including polyester (mylar), polypropylene, polystyrene and polycarbonate are good low-ESR, medium-ESL styles with excellent stability. They are available in values up to a few μF, but they are physically large.

Fig 10 illustrates a special capacitor called a *feedthrough*. Feedthrough capacitors are useful for filtering leads that exit a shield enclosure. Both leads connect to one side of the capacitor, with the other side connected to the grounded shell. When the feedthrough capacitor is installed in the enclosure wall, there is almost no inductance from the "hot" side to ground. The result is excellent high-frequency performance.

A capacitor and inductor in series form a series-tuned circuit, which has very low impedance at its resonant frequency. This fact can be used to advantage in a trick called *series-resonant bypassing*. To bypass a single band of frequencies, choose a capacitor value that self-resonates with its own ESL. A typical ceramic capacitor mounted with short (1/16 inch) leads has around 3-4 nH (0.003-0.004 μH) of ESL. Table 1 lists the optimum series-resonant-bypass capacitor

values for various amateur bands.

For broadband bypassing, however, use the largest ceramic bypass capacitor that fits. It does a better job at low frequencies, and all ceramic capacitors have about the same reactance (ESL) at high frequencies anyway.

This Mortal Coil

Just as every capacitor includes stray inductance and resistance, so every inductor has stray capacitance and resistance (Fig 9B). The shunt capacitance results from distributed capacitance between turns, and the ESR is from the RF resistance of the wire.

Air-wound coils are used in high-power RF circuits or where good stability is required (as in VFOs). Their disadvantage is large physical size.

Coils wound on ferromagnetic forms pack higher inductance in less space. One disadvantage is core saturation: at some current level, the magnetic flux reaches the maximum density that the core can support. Inductance drops rapidly above this level, and the nonlinear response may cause distortion and harmonics.

Two magnetic materials are often used

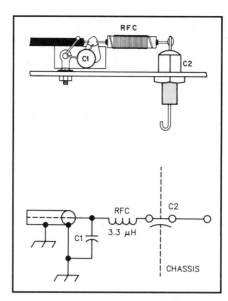

Fig 10—A feedthrough capacitor filtering a lead where it exits a shield enclosure. This filter rejects frequencies in the high-VHF TV band (174-216 MHz).

C1—0.001 μF disc ceramic.
C2—500 or 1000 pF feedthrough capacitor.
RFC—14 inches, no. 26 enameled wire close-wound on a 3/16-inch form (3.3 μH).

Table 1

Series-Resonant Bypass Capacitors for Various Amateur Bands

Frequency (MHz)	Capacitance
7	0.15 μF
14	0.039 μF
28	0.01 μF
50	0.0033 μF
144	390 pF
222	150 pF
440	39 pF

Values are based on measurements made using ceramic-disc capacitors mounted with minimum lead length on a 0.060-inch-thick PC board with zero effective trace length. Above 300 MHz or so, the resonance bandwidth becomes so narrow that the technique is not very useful. At such frequencies, it is better to use feedthrough or chip capacitors to reduce lead inductance.

in RF inductors: ferrite and powdered iron. Ferrite usually yields greater inductance, but it saturates more easily, and the inductance value is less stable. Both ferrite and powdered iron come in various mixes that are optimized for different frequency ranges and power levels.

Molded chokes look like resistors: they have a cylindrical body with pigtail leads on each end, and they are often color coded. Small-value chokes are usually wound on a simple insulating form, while the larger values use a ferromagnetic core to increase the inductance. Be sure to observe the maximum dc current rating, which can be limited either by core saturation or by wire heating.

Toroid coils nearly always use a magnetic core. The "doughnut" shape confines the lines of flux within the coil, so toroids are often used to reduce unwanted radiation (or susceptibility to incoming radiation).

A *ferrite bead*[2] is a small cylinder of ferrite with one or more holes through the

[2]Ferrite beads, rods and toroid cores are available from several sources: Amidon Associates, Palomar Engineers, RADIOKIT and others. See the Suppliers List at the end of this book.

center for wire. Fig 11 charts impedance v frequency for several ferrite mixes. For more impedance, use a bigger bead, string several beads in series on the same wire, or use a multihole bead with several turns of wire.

Split beads are available for wires or cables with connectors that would not fit through the hole. For best results, be sure that the mating bead surfaces make good solid contact.

DIFFERENTIAL VERSUS COMMON-MODE CURRENTS

Many cables used in electronic equipment include two wires. Examples are coaxial cable, ac-line cords and microphone wires. Interfering signals can flow in these cables in two ways: *Differential-mode* currents have equal strength and opposite phase in the wires. This is the mode used by the desired signal on a

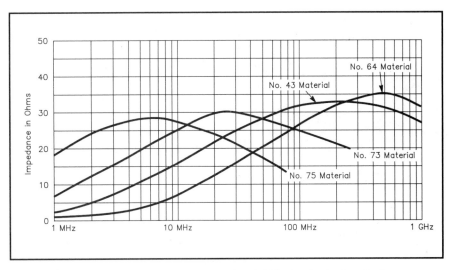

Fig 11—A plot of impedance v frequency for "101" size ferrite beads. Larger beads have higher impedance. Adding additional beads on a wire increases the impedance proportional to the number of beads. The impedance of different beads can be approximated by multiplying the impedance shown here by the appropriate "impedance factor" from Table 3 in the Stereos chapter.

HOW TO INSTALL A PL-259 CONNECTOR

Install PL-259s on RG-8 or RG-213 coax like this, and you will have good RF-tight connections that should not work loose. First, remove the coupling ring from the connector and slide it over the cable. (If you forget this step, you'll be sorry!) Now *carefully* score the outer insulation (jacket) around its circumference 1-1/8 inches from the end (A). If you nick the braid, cut the cable and start over. Remove the jacket without disturbing the braid. Tin the braid with rosin-core solder using a high-wattage soldering iron or gun (B). (Try not to melt the center insulation, and don't use too much solder.) Use a tubing cutter or sharp knife to cut through the tinned braid (and about half-way into the center insulator) 3/4 inches from the end (C). Twist off the cut braid and insulation. Clean and tin the exposed center conductor. Smooth any bumps or rough edges on the tinned braid with a file. Screw the coax into the connector body (D). (You may need to hold the connector body with pliers.) Solder the braid to the connector body through each access hole, and solder the center conductor to the pin (E). Trim the center conductor even with the end of the pin and file off any solder build-up. Finally, scrape away any solder flux on the outer surface of the pin.—*Alan Bloom, N1AL, San Francisco Section Technical Coordinator, Santa Rosa, California*

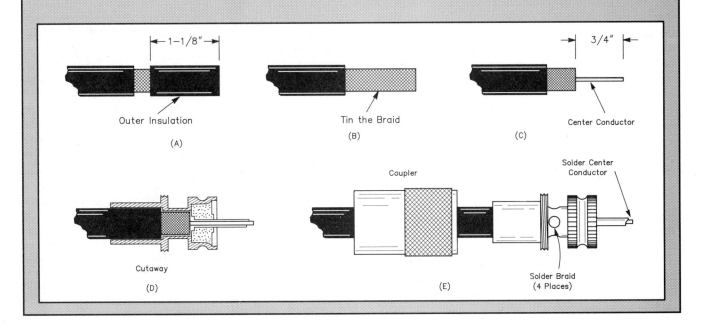

Fig 12—A common-mode choke to suppress shield current in coaxial cable.

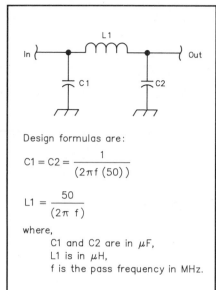

Design formulas are:

$$C1 = C2 = \frac{1}{(2\pi f (50))}$$

$$L1 = \frac{50}{(2\pi\ f)}$$

where,

C1 and C2 are in μF,
L1 is in μH,
f is the pass frequency in MHz.

Fig 13—The "quarter-wave" low-pass filter uses elements of 50-Ω reactance. Cascade as many sections as desired.

transmission line. *Common-mode* currents are equal in strength and phase in the wires. Common-mode currents on a transmission line cause feed-line radiation. On cables with more than two wires, common-mode current exists if the currents on all the wires do not sum to zero.

The cure for differential-mode interference is a filter that passes the desired signal but blocks the unwanted signal. Examples are: a low-pass filter on an HF transmitter and a high-pass filter on a TV set. These filters work only for differential-mode signals. If there is feed-line radiation from a transmitter, or large common-mode signals entering a TV tuner, a different technique is required.

The cure for unwanted common-mode signals must somehow present a high impedance in the common mode, without obstructing differential-mode signals. This can be done by inserting a transformer (with isolated windings) in series with the cable. (This will not work for dc power cables because a transformer cannot pass dc.)

A common-mode choke is an alternative. Such a choke is formed by wrapping all conductors around a ferrite core (toroid or cylinder) or placing them through one or more ferrite beads. (All conductors must pass through the same beads, not a bead for each conductor.) Since differential-mode signals have equal-strength and opposite-phase currents in the wires, their fluxes cancel within the core, and no inductance results. Fig 12 illustrates a toroid used as a common-mode choke for coaxial cable.

LOW-PASS TVI FILTERS

Before we get to the "nitty-gritty" of low-pass filter design, let's get one thing straight: A low-pass filter only acts upon signals coming through the feed line. If any VHF harmonics escape before reaching the filter, it cannot do its job.

The connection between the transmitter and filter is critical! Use only good-quality coax with at least 95% braid coverage. Be warned that "RG-8" is no longer a valid military specification for coaxial cable. Coax bearing that label may not be of high quality. RG-213 is the same size and impedance as RG-8, but it has guaranteed specifications. Pay special attention to the connection between the shield and the connectors at each end of the cable. The braid should be soldered around the entire connector-shell circumference. PL-259 "UHF" connectors have several solder-access holes: make sure that the braid is well-soldered through each hole. See the sidebar "How to Install a PL-259 Connector."

The coaxial cable between a transmitter and low-pass filter should be as short as possible. This not only reduces the chances of stray radiation, but also ensures that the transmitter "sees" the proper load impedance at VHF frequencies (above the filter cut-off frequency). Even better: use a double-male connector between the filter and transmitter.

Antenna SWR should be low at the operating frequency. High SWR can cause overheating and arcing in a low-pass filter. If necessary, install a Transmatch between the filter and antenna.

A Transmatch can also act as a filter, especially for low-order harmonics. Attenuation of high-order harmonics in the VHF TV range is limited, however: perhaps 5 to 20 dB, compared to 50-80 dB for a good low-pass filter. HF attenuation is best with tuners that use a low-pass circuit structure (with coils in series and capacitors shunting the load).

Designs

Low-pass filter designs appear in *The ARRL Handbook. Chebyshev, Butterworth* and *elliptic* filters are covered. Simply select a suitable design, look up the component values in a table, and scale the values for the desired impedance and frequency. The details are well covered in any recent *Handbook* as well as many other references.[3] Computer programs are also available to generate the component values.

Fig 13 illustrates a *quarter-wave* filter, so called because it electrically resembles a 1/4-λ transmission line below the cutoff frequency. This filter is easy to design because each of the three elements (C1, C2 and L1) have 50 Ω of reactance at the design frequency. Cascade as many

[3]Filter-design references, arranged in order of ease of use:

The ARRL Handbook for Radio Amateurs (published by ARRL) contains filter design information and construction projects (look for "filters" in the index). The element-value tables are the easiest to use, and some are set up for standard capacitor values. The selection of filter types is limited, however, and only low-pass and high-pass filters are covered.

E. Jordan, *Reference Data for Engineers* (Indianapolis, IN: Howard W. Sams & Co, Seventh Edition, 1985), Chapter 9. This resource is more difficult to use, but more inclusive than the *Handbook*. It includes bandpass and bandstop filter transformations.

A. Zverev, *Handbook of Filter Synthesis* (New York, NY: John Wiley & Sons, 1967). This is "everything you ever wanted to know about filter design." It contains 203 pages of design tables, plus charts and graphs and extensive mathematical theory. For the engineer.

A LOW-PASS TVI FILTER

Here's a low-pass filter that you can build. It's almost as good as commercial products. It should be adequate for all but the most severe TVI caused by transmitter harmonics.

Construction

The filter is constructed in an aluminum box measuring 3-1/2 × 2-1/8 × 1-5/8 inches. Input and output connectors are mounted at the center of each end. Use 5%-tolerance (2% would be better) capacitors. The 500-V capacitors specified should be more than adequate for a 200-W transmitter.

The coils are space wound from no. 18 enameled copper wire. Wind L1 and L3 using a 1/4-inch drill bit as a form, use a 1/8-inch bit for L2. To space the windings, wind two pieces of wire in parallel. To remove the extra winding, just grasp one wire and pull (with the windings still on the drill bit). Solder all parts to a two-terminal-with-ground solder strip as illustrated.

Tuning

Adjust the coils by spreading or squeezing the turns. Set L2 first, for maximum rejection of TV channel 2 (55.25 MHz). Then tune L1 and L3 for lowest SWR, or minimum insertion loss, at 28.5 MHz.

Here are two ways to adjust L2: Method 1 requires a grid-dip oscillator or solid-state dip meter. Temporarily short L2 (the end that connects to L1 and L3) to the grounded end of C2 with a VERY SHORT, FAT conductor. (A piece of tinned coax braid should do nicely.) Set the dip meter frequency by placing it near a TV set tuned to channel 2. Tune the dip meter to produce broad horizontal interference bars on the TV. Loosely couple the dip meter to L2, and adjust the coil for a dip at 55.25 MHz.

Method 2 requires a strong channel-2 signal and a TV fed with coax. CATV service should suffice. Connect the filter between the TV antenna connector and the feed line. Adjust L2 for maximum "snow" in the channel-2 picture. It may be necessary to temporarily short L1 and L3 to yield a strong enough signal.

Performance

The filter attenuates all VHF TV frequencies a minimum of about 48 dB. It has about 70 dB of rejection at the channel-2 carrier frequency (often one of the worst trouble spots). The worst-case insertion loss is about 0.3 dB, and the SWR is less than 1.3:1 below 29.0 MHz. The loss rises to over 0.4 dB at 29.7 MHz. If you operate at the high end of 10 meters, you may want to peak L1 and L3 at 29.0 MHz instead of 28.5 MHz.

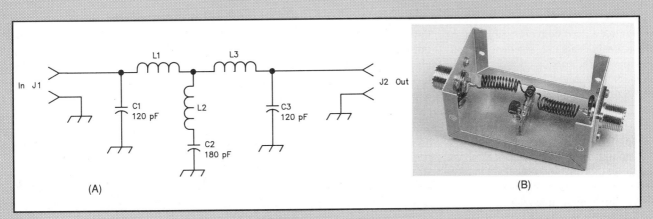

(A)   (B)

A low-pass filter for amateur transmitting use. The circuit (A) is constructed on a two-terminal-with-ground terminal strip (15TS003), which is fastened to the chassis with no. 4 hardware. (Part numbers in parentheses are catalog numbers from Mouser Electronics, see Suppliers List for address.) B is a photo of the completed filter.

C1, C3—120 pF, 5% tolerance, 500 V silver mica (232-1500-120).
C2—180 pF, 5% tolerance, 500 V silver mica (232-1500-180).

J1, J2—SO-239 "UHF" connectors (16SO239) or RF connectors of the builder's choice.
L1, L3—13 turns no. 18 wire, space wound 1/8-inch ID.

L2—4 turns no. 18 wire, space wound 1/8-inch ID.
Aluminum chassis—3-1/2 × 2-1/8 × 1-5/8 inches (537-TF-773).

sections as needed to get additional high-frequency rejection. Unfortunately, the low-frequency ripple is not well controlled, especially for multisection designs (Fig 14). This kind of filter is only good for single-band applications.

Absorptive Filters

Most filters are designed for a 50-Ω load impedance at both terminations. In a typical amateur installation, this is only true at the transmit frequency. Most antennas are not 50 Ω outside the band. The same is true for the transmitter output circuit.

Unfortunately, filter response into a highly reactive load is unspecified. Harmonic attenuation may be degraded at some frequencies (and improved at others).

Take the 50-Ω frequency-response plots as rough approximations of performance in actual installations.

An *absorptive filter* (Fig 15) is one partial solution. In addition to a low-pass filter, this circuit includes a matching high-pass filter connected to a 50-Ω load, which provides a matched, 50-Ω resistive load at VHF frequencies. Connecting this

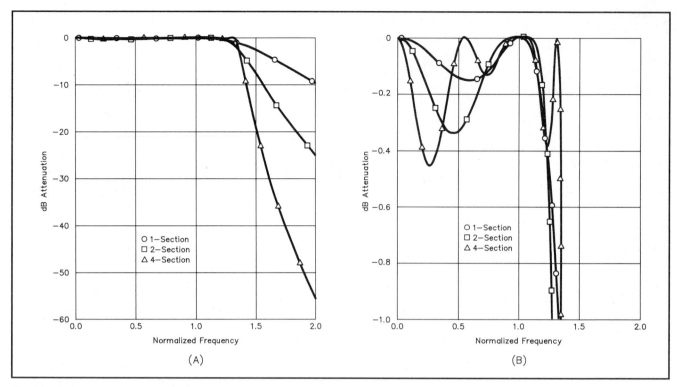

Fig 14—A, frequency response of 1, 2 and 4-section quarter-wave filters. The pass frequency has been normalized. B shows the same curves with the attenuation scale expanded to illustrate passband ripple.

filter after a good conventional low-pass filter should guarantee excellent harmonic suppression.

BANDSTOP (TRAP) FILTERS

For interference problems resulting from a single spurious frequency, a series- or parallel-tuned *wave trap* within the transmitter, or near the antenna terminal, can help. (See Fig 16.) Each LC circuit is resonant at the interfering frequency.

With vacuum-tube transmitters, install a parallel-tuned trap in the plate lead of the final amplifier. This might cure mild TVI cases. The final amplifier output impedance is too low for this technique in solid-state amplifiers, but a series-tuned trap in parallel with the antenna connector may help. For either type of trap, the inductor and capacitor should each have a reactance in the neighborhood of 200 Ω. To trap out TV channel 2, for example, suitable values would be about 0.56 μH and 15 pF (use a 20-pF variable capacitor). Tune the trap to the frequency of the affected TV channel. *Beware the high voltage present on the tube plate!* Use a plastic alignment tool to adjust the capacitor, or use a dip meter to align the trap with the transmitter switched off.

A tuned *transmission-line stub* is sim-ilar to an LC trap. A 1/4-λ transmission line (open at the far end) acts as a series-tuned circuit. Place such a 1/4-λ stub (cut for the interfering frequency) in parallel with the transmitter output; it acts like the circuit in Fig 16B.

Parallel-conductor transmission lines present a special problem: balanced low-pass filters are not readily available. A shunt trap could be used, either an LC circuit or a tuned stub. However, the best solution is a standard coaxial low-pass filter at the transmitter output followed by either a balun or a balanced-output Transmatch.

INTERFERENCE FROM VHF TRANSMITTERS

Some amateur VHF and UHF transmitters use a lower-frequency oscillator that is multiplied to the transmit frequency. In the process, unwanted oscillator harmonics can leak through, causing interference to other services. Of course, VHF and HF transmitters are equally likely to have output-frequency harmonics and unwanted mixer products. For any of these problems, a bandpass filter between the transmitter output and the antenna can help.

VHF Bandpass Filter Designs

A *cavity* or *stripline* filter is a tuned transmission line. Instead of using coaxial cable, a high-Q line is built out of wide strap or tubing for the center conductor

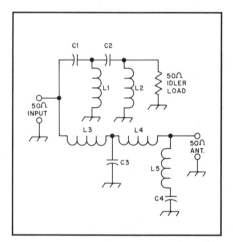

Fig 15—An absorptive low-pass TVI filter. The following values are for a cutoff frequency of 40 MHz and a rejection peak in TV channel 2:

C1—52 pF.
C2—73 pF.
C3—126 pF.
C4—15 pF.
L1—0.125 μH.
L2—0.52 μH.
L3—0.3 μH.
L4—0.212 μH.
L5—0.55 μH.

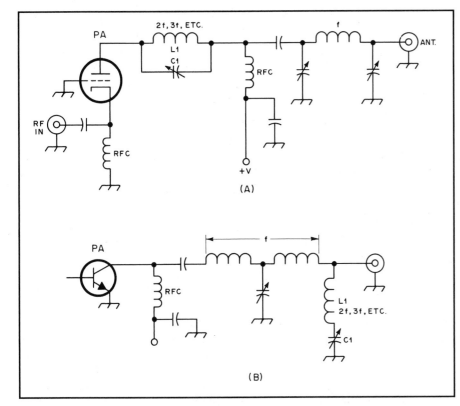

Fig 16—A parallel-resonant harmonic trap is shown in the plate lead of the circuit at A. The example at B uses a shunt connected series-resonant trap.

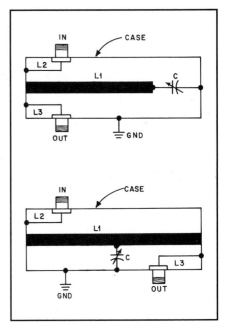

Fig 17—Equivalent circuits for the stripline filters. The circuit at A applies to the 6- and 2-meter designs. L2 and L3 are the input and output links. At B is the circuit for the 220- and 432-MHz filters. All of the filters are bilateral (the input and output terminals may be interchanged).

and some kind of chassis or enclosure for the outer conductor.

Figs 17 through 20 illustrate stripline filters for the amateur VHF and UHF bands from 50 through 450 MHz. Construction is easy, and the cost is low. Standard boxes are used for ease of duplication.

The filter of Fig 18 is selective enough to pass 50-MHz energy and attenuate the seventh harmonic of an 8-MHz oscillator that falls in TV channel 2. With an insertion loss (at 50 MHz) of about 1 dB, it can provide up to 40 dB of attenuation to energy at 58 MHz.

The filter uses a folded transmission line to keep it within the confines of a standard 6 × 17 × 3-inch chassis. The aluminum partition down the middle of the assembly is 14 inches long. The inner conductor is 32 inches long and 13/16 inch wide, made of 1/16-inch brass, copper or aluminum. In the prototype, two pieces of aluminum were spliced together to provide the 32-inch length. The sides of the "U" are 2-7/8 inches apart, with the partition at the center. The line is supported on ceramic standoffs shimmed with sections of hardwood or Bakelite rod to give the required 1-1/2 inch height.

The tuning capacitor is a double-spaced variable (Hammarlund HF-30-X, 4.9-30 pF, 1600 V) mounted 1-1/2 inches from the right end of the chassis. The input- and output-coupling loops are made from no. 10 or 12 wire, 10 inches long. They are spaced about 1/4 inch from the line.

The 144-MHz model shown in Fig 19 is housed in a 2-1/4 × 2-1/2 × 12-inch box. One end of the tubing is slotted (1/4-inch deep) with a hacksaw. This slot takes a brass angle bracket 1-1/2 inches wide and 1/4-inch high, with a 1/2-inch mounting lip. The 1/4-inch lip is soldered into the tubing slot, and the bracket is bolted to the end of the box to position the tubing at the center of the end plate. The tuning capacitor (Hammarlund HF-15-X, 3.6-15 pF, 1600 V) is mounted 1-1/4 inches from the other end of the box with the two stator bars soldered to the inner conductor.

The two SO-239 coaxial connectors are 11/16 inch in from each side of the box, 3-1/2 inches from the left end. The coupling loops are no. 12 wire, bent so that each is parallel to the center line of the inner conductor and about 1/8 inch from its surface. Their "cold" ends are

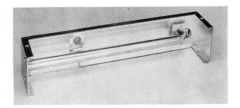

Fig 18—Interior of the 50-MHz stripline filter. The aluminum inner conductor is bent into a "U" shape to fit inside a standard 17-inch chassis.

Fig 19—The 144-MHz filter has an inner conductor of 1/2-inch (5/8 inch OD) copper tubing, 10 inches long. It is grounded to the left end of the case and supported at the right end by the tuning capacitor.

soldered to the brass mounting bracket.

The 220-MHz filter (Fig 20) uses the same size box as the 144-MHz model. The

Fig 20—The 220-MHz filter uses a 1/2-λ stripline. It is grounded at both ends and tuned at the center.

inner conductor is 1/16-inch brass or copper, 5/8-inch wide, just long enough to fold over at each end and bolt to the box. It is positioned so that there is a 1/8-inch clearance between it and the rotor plates of the tuning capacitor. The latter is a Hammarlund HF-15-X (3.9-15 pF, 1600 V), mounted slightly off-center in the box, so that its stator plates connect to the exact mid-point of the line. The 5/16-inch mounting hole in the case is 5-1/2 inches from one end. The SO-239 coaxial fittings are 1 inch in from the opposite sides of the box, 2 inches from the ends. The coupling links are no. 14 wire, spaced 1/8 inch from the inner conductor.

The 420-MHz filter is similar in design, using a 1-5/8 × 2 × 10-inch box. A half-wave line is used, with a tuning capacitor (fabricated from brass discs) at the center. The two discs are each 1/16 thick and 1-1/4 inch in diameter. The fixed disc is centered on the inner conductor, the other is mounted on a no. 6 brass lead-screw. This passes through a threaded bushing (which can be taken from the end of a discarded slug-tuned coil form). Such bushings usually include a tension device for the screw. If there is none, use a lock nut.

The 420-MHz model uses N connectors. They are located 5/8 inch from each side of the box, and 1-3/8 inches from the ends. The coupling links are no. 14 wire, spaced 1/16 inch from the inner conductor.

To adjust the filters, simply connect them between the station SWR meter and the antenna; tune for minimum SWR. The SWR should be close to 1:1 into a VHF dummy load or a well-matched antenna. All coaxial cable between the transmitter and filter should be good quality (95% braid coverage), with the shield properly soldered to the connector shells.

Parasitics in VHF/UHF Transmitters

Parasitic oscillations far above the operating frequency should not be a problem in VHF or UHF vacuum-tube transmitters. If short, low-inductance connections are used throughout, the frequency of any parasitic resonances in the final amplifier tank circuit are so high that the tube won't have enough gain to sustain oscillation.

Oscillation near the operating frequency can be a problem, however. Grounded-grid amplifiers should be stable if the input and output circuits are well isolated from each other. Grounded-cathode triode amplifiers usually need to be neutralized. Some tetrode or pentode circuits require neutralization as well. Series-resonant screen bypassing is a possibility here: Above the tube self-resonant frequency, a certain bypass capacitor value will tune out the self-inductance of the screen.

Solid-state VHF amplifiers can oscillate at either low or RF frequencies. For low-frequency oscillations, apply the same solutions previously mentioned for HF solid-state amplifiers.

To determine whether high-frequency oscillations are present, retune the final amplifier while watching its collector current (or total power-supply current). Any erratic current fluctuations during tuning indicate unwanted oscillation, probably in the VHF/UHF range. The problem may be caused by simple misadjustment. If, after proper alignment, each tuning control can be slightly detuned (near the optimum setting) without causing oscillations, the transmitter is probably operating correctly.

If there is no stable tuning condition, check all circuitry around the affected stage for poor solder joints and damaged or overheated components. If there are no defects, circuit modifications may be in order.

One quick-and-dirty solution: Add a low-value resistor (5-50 Ω) in series with a 0.001-µF ceramic disc capacitor from the base of the affected stage to ground. With luck, a resistor value that is low enough to kill the oscillation will not reduce output power excessively. If there is a coil or RF choke from the base to ground, try adding a series ferrite bead. The same trick sometimes works with a choke in the collector power-supply lead.

BE AN OPTIMIST

When troubleshooting interference problems in a transmitter, keep in mind that there is almost always a technical solution. After all, there are literally millions of transmitters around the world that work, day in and day out, without interference-causing spurious emissions. With proper shields, filters and bypassing, any transmitter can be clean.

Hints & Kinks

---

from August 1979 *QST*, p 50:
ELIMINATING HW-101 TVI

An aggravating problem of TVI caused by my HW-101 transceiver ended after I corrected the poor continuity between the chassis and the top and bottom metal covers of the rig. The liberal amount of paint applied during the manufacture of the HW-101 prevented the housing cover and bottom from serving effectively as a shield.

I removed a small area of paint around one of the screw holes on the back of both the upper and lower covers. From each of these holes I connected a wire to the ground terminal on the transceiver. To ensure, furthermore, that the HW-101 has a good earth connection, I installed a ground lead from the set to the cold-water-system plumbing. Now peace has been restored to the VE3HBP household.—*Doug McLennon, VE3HBP*

from December 1985 *QST*, p 50:
TV AND THE HEATH SB-230

I experienced TVI trouble with my SB-230 amplifier. The TVI was not appreciably reduced by improving the PA-compartment shield. For significant TVI reduction, the open-frame antenna relay must be enclosed (with copper-clad circuit board or equivalent), the outer conductors of all coaxial cable leads must be soldered to the relay enclosure, and both coil leads must be brought into the enclosure via feedthrough capacitors. Solidly connect this new relay enclosure to the chassis of the SB-230.—*Bob Loving, Jr, K9JU*

from June 1979 *QST*, p 41:
DRAKE TR-4 TVI PROBLEM SOLVED

I've experienced a TVI problem with my Drake TR-4. For the benefit of others who may have this difficulty, I wish to offer my explanation of the cause and cure.

Problem: Harmonic interference appeared on TV channel 2 (when my TR-4 was operating on 14 MHz) and on channel 4 (when the transceiver was on 21 MHz). Use of a low-pass filter at the output of the transmitter and high-pass filters on the TV sets partially eliminated the TVI. Further steps had to be taken for a complete cure.

Cause: Strong radiation on the 2nd, 3rd and 4th harmonics was noted. A low-pass filter effectively reduced interference received via the transmitting antenna. With the help of a search loop and spectrum analyzer, I established that undesirable harmonic radiation was coming from the chassis. Further investigation disclosed the presence of relatively strong harmonics at the grids of the 6JB6 final amplifiers. These unwanted signals passed right through [or perhaps *around*—Ed.] the 9.0-MHz sideband crystal filter. Correcting the condition and eliminating harmonic content in the final amplifier by adjusting the amplifier for proper linearity brought the matter to a successful conclusion.

Cure: Oscillator distortion was corrected by installing a 9.0-MHz parallel-resonant circuit from the wiper of the carrier-balance potentiometer to the cathode circuit of V16. This circuit is tuned to approximately 9.0 MHz with the aid of a dip oscillator. Final adjustment is accomplished by observing an oscilloscope connected to the grids of the 6JB6 amplifiers while tuning for maximum amplitude. The transmit-gain potentiometer should be set in the linear region (near the nine o'clock position) for this adjustment. Realignment of the 9.0-MHz oscillator according to instructions in the Drake manual is also required.

Linearizing the final amplifiers is done simply by providing for additional cathode degeneration. To do this merely remove the 470-pF capacitors from pin 3 of the 6JB6 amplifiers. The power reduction is not noticeable.

Results: Harmonic radiation from my TR-4 was reduced on 14 MHz by at least 17 dB at the 2nd, 3rd and 4th harmonics. Measurements were made at the output connector.—*J. H. Mehaffey*

from December 1985 *QST*, p 50:
RFI TO THE KENWOOD TS-430S

It was much to my pleasure that my friend, AF1U, decided to replace his old rig with a new Kenwood TS-430S. (I picked up the old rig as a standby.) After several weeks, George called to confess that he thought the change was a mistake. His new '430 would not operate with his SB-220 amplifier on 20-meter CW. He had never before encountered this problem.

When I visited him to see if I could help solve the problem, George had set up the '430 and '220 on the 20-meter band. The combination locked in a key-down condition when the first code element was sent. We first checked the keyer for a malfunction, but found no problem. RF was affecting the sensitive circuitry of the '430—but how? Next, I eliminated all leads exiting the radio except for the antenna lead and key line, but the problem remained. Although the key line was a shielded wire, I suspected it as an RF path into the '430. To solve this problem, I wound five turns of the key line on a toroid core, very close to the key jack. This solved the problem nicely. My friend George is, once again, happily operating on 20-meter CW.—*Louis Parascondola, WA1GSO*

TVI EXAMPLES

Here are photos of typical television interference patterns. The photos and descriptions are taken from the FCC *Interference Handbook* (1990 edition). That booklet is available from the Superintendent of Documents, US Government Printing Office, Washington, DC 20402.

is probably caused by something within your home. The operation of electric razors, hair dryers, electric drills and saws also cause temporary interference problems. You may choose to tolerate these types of interference since they are temporary and often expensive to eliminate.

Two-Way Radio Transmitter Interference

This pattern may appear on your television screen when your set is receiving signals from a CB, amateur, police or other two-way radio transmitter. The pattern will only appear when the operator transmits. It may vary according to the type of signal being received. You may also hear the operator's voice.

Electrical Interference

Pattern A above may appear on your television screen when your set is reacting to an electrical device operating in or near your home. Pattern B can be caused by radio frequency energy generated by power lines or power-line equipment. One way to determine if the interference is confined to your home is to check whether your neighbors are also experiencing the problem. If not, the interference

FM Radio Transmitter Interference

This pattern may appear on your TV screen when interference is caused by signals from an FM radio transmitter. The Commission encourages new FM station applicants to consider potential wide-spread interference problems and minimize them by selecting a proper transmitter location. FM station operators are required to respond to all reasonable interference complaints once the station begins operating. Therefore, you should first contact the FM Station Manager, Chief Operator or Chief Engineer for assistance. If only TV channel 6 is affected, you should contact the TV station.

There are two common interference problems that involve FM stations. The first problem may arise when you are receiving a distant TV signal and a new FM station begins operating in your area. Your TV receiver may respond to the nearby FM signal, which may overpower your TV signals. It is important to recognize that TV signals are only designed to be received within a limited number of miles (or radius) of the transmitter. Any TV beyond this is called "fringe area" reception. Although fringe-area reception may be satisfactory to you, you are only receiving a weak signal. That signal is highly susceptible to interference. Sometimes, installation of an FM band rejection filter and/or a highly directional antenna may reduce your interference problem. The Chief Engineer of the FM station is a good source of information about interference-resolution techniques. If you are not satisfied with the station's response, contact your local FCC office.

Chapter 6

Televisions

By Art Block, W3YK   **and**   **Ed Hare, KA1CV**
ARRL Official Observer   ARRL Senior Lab Engineer
PO Box 15185
Las Cruces, NM 88004

Television Interference (TVI) is not a thing of the past. Although the number of cases has dropped in recent years, it can be difficult to determine the cause when TVI does occur. Amateur transmitters must meet stringent FCC requirements for spectral purity, and most TV manufacturers design TVs to meet the ANSI standards for rejection of unwanted signals. These two factors make it less likely that the TVI is caused directly by the amateur transmitter or TV receiver.

The FCC does not provide protection to fringe-area TV signals. The FCC *Interference Handbook* says, "Although fringe area signals may be satisfactory, they are weak signals that are highly susceptible to interference and are not protected by the Commission." You may want to help your neighbors if you live in a weak-signal area, but if you are outside the television service area (grade-B contour) this is a voluntary action on your part.

PRELIMINARY STEPS

Amateurs involved with or assisting with a TVI problem should take a few preliminary steps before meeting with the TV owner. Many of these first steps are discussed in earlier chapters of this book, but major points are repeated here for those who may have skipped over those important chapters.

Education

Electromagnetic Interference (EMI) problems can be complex. Stock answers do not always result in solutions. It is important for all concerned parties—the amateur, Technical Coordinator, Technical Specialist, TVI Committee members and the neighbor—to learn as much as possible about the subject of EMI. Books about EMI, Amateur Radio magazine articles, local clubs and other hams are all possible sources of information and education.

Help

The amateur community has programs to help hams and consumers with EMI problems. Start with the ARRL Section Technical Coordinator (TC). The TC knows the best help available in your area. He or she may refer you to an assistant (Technical Specialist), a local TVI Committee, the Amateur Auxiliary or a local club that has an EMI expert or two. The technical needs and resources of each ARRL Section are different, and each Section Manager and TC have determined what works best for their own Section. If you don't know the name of your Section TC, contact your Section Mana-

"Here's a choke that will fix the whole problem!"

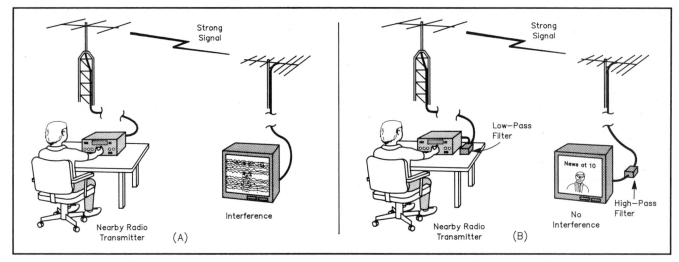

Fig 1—(A) This TV doesn't have enough rejection to keep out an amateur fundamental signal: The TV set is rather unhappy about the whole thing. (B) The appropriate cures have been installed: Everyone is happy again.

ger (SM). Page 8 of any recent *QST* gives you the name, address and usually the telephone number of your SM. ARRL HQ can also supply the name and address of your TC. This book refers to such help collectively as the Technical Coordinator or TC (to remind you that the TC is the first place to turn for help with EMI problems).

The consumer-equipment manufacturer should offer some assistance. Some manufacturers have EMI or RFI modifications available for their equipment. Sometimes the parts for the modifications are furnished free of charge, and a few progressive companies install the parts (at no charge) as well. The Electronic Industries Association (EIA), Director of Consumer Affairs, 2001 Pennsyvania Ave NW, Washington, DC 20006, 202-457-4977, can supply the address of the best contact at any of the companies who are members of the EIA.

Personal Diplomacy

Personal diplomacy is another important subject. An argument *never* solves anything, so keep your cool when dealing with people involved in an EMI problem. This advice is not just for the amateur operator: it applies to the ham, the neighbor and any others helping solve the problem on behalf of these two key parties in the EMI problem. (In keeping with this spirit, the complainant is named the neighbor in this chapter.)

As an amateur, remember that *you* have access to the *technical* resources to solve an EMI problem. Your neighbor

may not understand the technical issues or know about helpful ARRL programs. He or she usually is not sure how to approach you about the problem. A neighbor usually reacts positively if you express a willingness to help solve the problem *regardless of fault*. This entire issue is discussed in depth in the First Steps chapter. Read it!

The Dark Side of Diplomacy

There are problems you can encounter if you are too accommodating. A helpful attitude sometimes gives people the idea that you will do anything that they wish. They may wish that you stop operating permanently, or at least when they want to watch TV. You can best judge what constitutes the optimum balance between accommodation and cooperation. In cases where your transmitter is clean, you have a federally granted privilege to operate. If you decide to curtail operation until a TVI problem is solved, make certain it is clearly understood that you are taking a voluntary, temporary measure. It is a good idea to set a time limit for voluntary quiet hours, perhaps a few weeks. This ensures that the neighbor will extend timely cooperation.

Remember that many people take advantage of others in any way they can. We live in a litigation-prone society. There is an unknown degree of liability (and risk) when dealing with strangers. An amateur, TC or TVI Committee should *locate* solutions, not *implement* them. Even installing a high-pass filter in a neighbor's external feed line can create

Fig 2—Simplify the problem. This rat's nest of equipment and cables contains too many potential problems to allow for easy troubleshooting.
(photo by Karen Sullivan)

problems. Each amateur and TC must decide the best approach for each individual case.

It is a good idea to have the neighbor install any external filters used to cure TVI. This minimizes liability, and helps show the neighbor that fundamental overload and its cures are the problems of the equipment manufacturer and owner. Amateurs are there to assist and provide information about solutions.

Fig 3—This chart shows CATV and broadcast channels used in the United States and their relationship to the harmonics of MF, HF, VHF and UHF amateur bands.

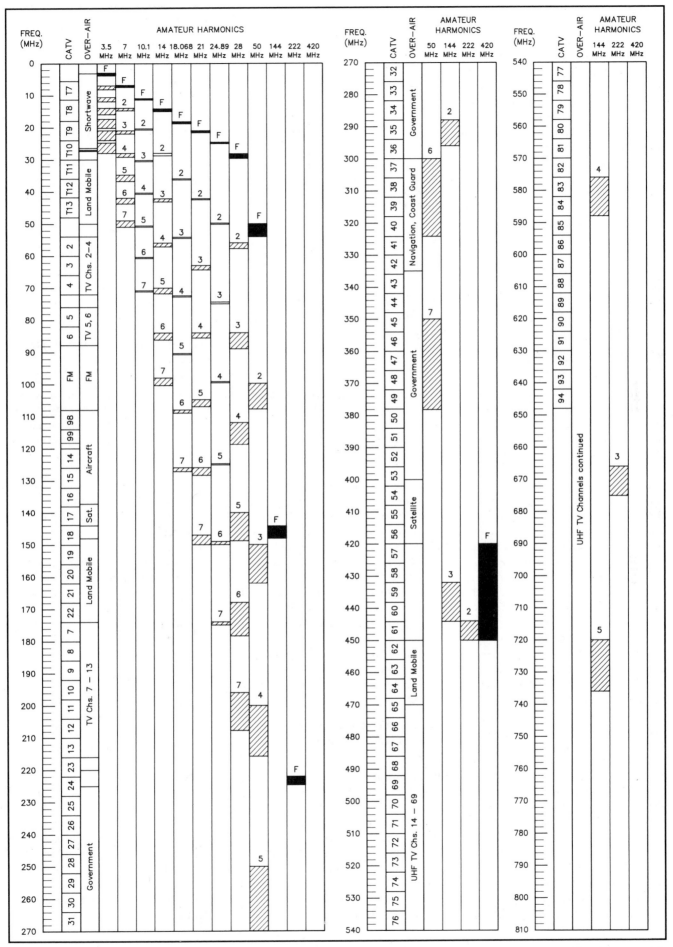

ARRL-affiliated clubs with ARRL club-liability insurance have the official activities of a club-authorized TVI committee covered if that activity has been properly described on the insurance application form. Contact ARRL HQ for details. A homeowner's or renter's insurance policy may also protect you from liability. Check with your insurance agent to find out if you have this coverage, or if a rider can be written.

Reporting

The Report Form chapter contains an EMI report form and instructions. Informal surveys indicate that most cases of EMI are not reported to the FCC, the EIA or ARRL HQ. These groups cannot assess the EMI situation without complete data. *Please* fill out and mail the EMI report form. ARRL HQ will use the information in its EMI programs and policies.

TROUBLESHOOTING TECHNIQUES FOR TVI

Complete elimination of TVI may not be a simple process. A single measure, such as a high-pass filter at the TV does not always cure the problem. Sometimes, a number of methods must be applied. An important factor in any TVI case is the ratio of TV signal strength to interference level (also called signal-to-noise ratio, S/N). This includes interference of all kinds, such as ignition noise, random or thermal noise (which sets the minimum signal required for snow-free reception) and unwanted signals that fall within the TV channel. An S/N ratio greater than approximately 35 to 40 dB is required for good picture quality (see Fig 1).

One thing is worth noting up front: For years amateur literature has not emphasized the difference between common-mode and differential-mode signals, nor has it fully explained the different cures needed for each mode. There are critical differences! It can't be said too many times—*a differential-mode high-pass filter is not enough! Common-mode cures are usually required as well.* Remember this as you read on. The EMI Fundamentals chapter contains an explanation of differential- and common-mode signals.

Many cases of TVI have one easily located cause and cure. When you are confronted with a case caused by multiple factors, however, much time can be wasted trying different things that appar-

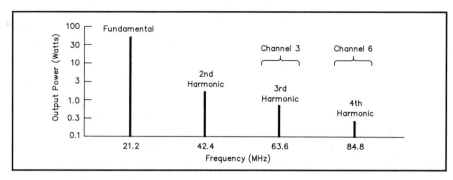

Fig 4—A 21-MHz signal and its harmonics. In actual practice, the harmonics might not have the amplitudes shown. It is important to remember there are always harmonics present.

ently don't work. The first important step in troubleshooting a case of TVI is to *simplify the problem* (see Fig 2). This principle applies to the amateur station as well as to the affected equipment, and especially to VCRs. They (especially the older ones) are notoriously susceptible to EMI problems.

Try one, and only one thing at a time. If too many variables change at once, you will never know which cure actually solved the problem.

The frequency relationship between amateur bands and TV channels is shown in Fig 3. Use it to determine if there is a harmonic relationship between operating frequency and any TV channels affected by station operation. Harmonics are likely in cases where TVI is experienced only when the transmitter is operated on specific bands (Fig 4). Harmonics can be generated by the transmitter or external devices, such as an unpowered VHF transmitter connected to a VHF antenna. They can also be generated inside the TV tuner, a form of fundamental overload.

The Troubleshooting chapter teaches how to identify EMI causes. Take the time to read it. Fig 5 is a TVI troubleshooting flow chart.

Ensure That The Amateur's Own House Is Clean!

One of the best ways to show that a station is not generating spurious emissions that cause TVI is to be sure that the home television equipment functions properly. If an antenna-connected television next to the station is not subject to interference on any channel during amateur transmissions, the amateur station is certainly not the source of the problem. The FCC sees it that way, too!

A high-pass filter, ac-line filter and a

common-mode choke should be installed on the amateur's TV sets, whether they need them or not. This demonstrates that filters are effective and do no harm.

THE AMATEUR STATION
Always Check The Amateur Station

If TVI is experienced on all channels (not only on those harmonically related to the operating frequency), it is likely that the interference is caused (at least in part) by TV susceptibility to fundamental overload. It is a good idea to proceed with the amateur station check-out anyway. It ensures that interference is not *also* caused by station problems. It is not uncommon to have multiple problems and sources associated with a single TVI case.

An amateur transmitter in perfect working condition (with no harmonics or other spurious radiation) can still interact with TVs. This occurs because the TV manufacturer did not build the product to reject high-level signals that are a part of its normal environment.

The Transmit Filter

Every amateur that expects to encounter interference problems, and this includes nearly every amateur, should have a properly installed transmit filter on the transmitter or transceiver antenna lead (see Figs 6 and 7). It is not a cure-all for interference, but it can be considered "good engineering practice," which is required by the FCC. When confronted with interference, an amateur can point to the filter with pride. For an HF transmitter this filter is usually a commercially available low-pass filter.

Commercial low-pass filters are very effective and relatively inexpensive.

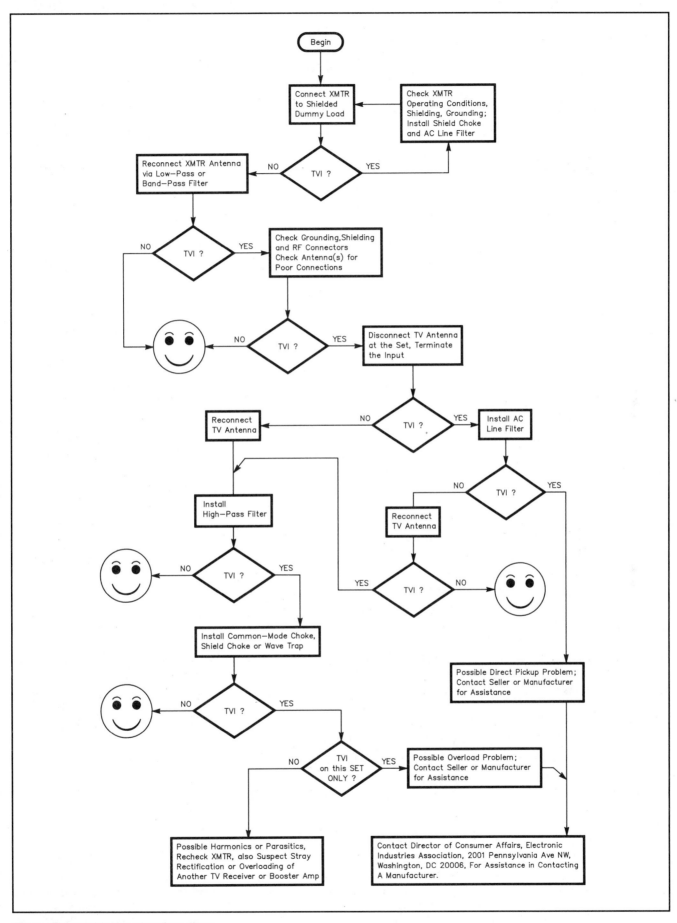

Fig 5—TVI troubleshooting flow chart.

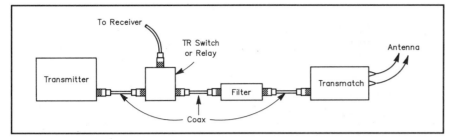

Fig 6—The proper method of installing a low-pass filter between an HF transmitter and Transmatch. The transmitter and filter must be well-shielded and grounded. If a TR switch is used, install it between the transmitter and low-pass filter. (TR switches can generate harmonics.) If the antenna is well matched (SWR = 1.5:1 or less), the Transmatch can be eliminated. A mismatch degrades filter performance.

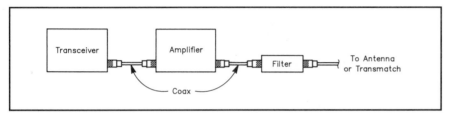

Fig 7—This figure shows the proper location for a low-pass filter in stations that use a linear amplifier. Some hams insert a low-pass filter between the transmitter and amplifier, but it isn't necessary in most cases.

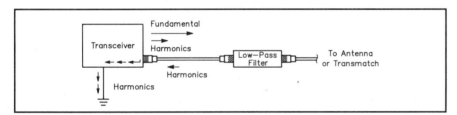

Fig 8—A good RF ground can help a low-pass filter reject unwanted VHF harmonics.

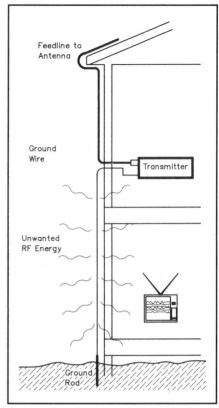

Fig 9—A second-floor apartment presents a nearly impossible grounding situation. The ground lead acts as a VHF long-wire antenna that can radiate unwanted energy. This station might be better with no ground connection. A lossy ground plane, described in "Eliminate TVI with Common-Mode Current Controls," May 1984 *QST*, pp 22-25, offers a better alternative in cases where a good RF ground is required.

Details of construction can be found in the Transmitters chapter of this book or the *ARRL Handbook.*

The Station Grounds

A transmit filter works better with a good RF ground. Equipment grounding has long been considered a first step in eliminating interference (see Fig 8). While the method is very effective in the MF range and below, it is less useful in suppressing VHF energy; even short lengths of wire have considerable reactance at VHF. The delay effects along the wire are similar to those on the surface of an antenna. As the electrical length approaches 1/4 λ, the wire appears as an open circuit, rather than a low-impedance

ground. It may also radiate like an antenna, as shown in Fig 9.

In most cases, this does not directly contribute to interference. When the ground lead is much closer to the affected equipment than the antenna, however, the susceptible equipment may receive more energy from the ground lead than from the antenna.

AC-Line Filters

An ac-line filter with an additional common-mode choke should be installed on the transmitter power cord. This reduces the possibility of RF energy entering the ac power system by that route. Examples of common-mode and ac-line filters that can be used on trans-

mitters or home-electronic equipment are shown in Figs 10 and 11. Fig 12 shows a "brute-force" ac-line filter.

Operating Habits

It is easy for even the best operator to develop habits that can contribute to TVI. The operator can become careless when tuning the rig for the millionth time. An overdriven final stage usually increases spurious output without any significant increase in output power. ALC and speech-compressor settings are important considerations that are frequently overlooked.

An overdriven transceiver or amplifier has a wider bandwidth than necessary and an increased potential for creating TVI. The station operator can demonstrate to the TC the procedure used during normal station operation. (Read the

Fig 10—An ac-line common-mode choke. This eliminates most forms of ac-line related EMI that results when power lines pick up a radio signal. Wrap about 10 turns of the ac-line cord through a ferrite toroid or around a ferrite rod. Use no. 43 (VHF) or no. 75 (HF) material.

Fig 11—A commercial ac-line filter installed with a common-mode choke. Note that the excess line cord has been coiled around the common-mode choke, minimizing the "capture area" of the ac-line cord "antenna."

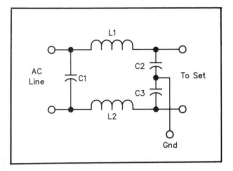

Fig 12—A "brute force" ac-line filter. C1, C2 and C3 can be any value from 0.001 to 0.01 µF, rated for ac-line service (1.4 kV dc). L1 and L2 are each 2-inch-long windings of no. 18 enameled wire on a 1/2-inch diameter form (7 µH). If installed outside the equipment cabinet, enclose the filter to eliminate shock hazard. This filter is similar to commercial ac-line filters.

manual the night before!) Correct any problems in the operating procedure.

VERIFY THAT YOUR OWN RECEPTION IS INTERFERENCE-FREE

Once it has been demonstrated that the amateur station is in perfect condition, verify that there is no TVI on the amateur's TV sets. An assistant can switch the TV through all channels, verifying that each channel is interference-free while the amateur operator transmits. The test frequencies, bands and modes should be those the operator uses most frequently.

MEET THE NEIGHBOR

When the TC is convinced that the amateur station is in perfect order, it is time to call on the TVI complainant. A phone call to schedule the visit serves as good introduction for the TC interface. The caller (amateur) should explain that the TC is a volunteer who functions to help identify TVI causes and suggest solutions.

The first encounter with the neighbor should build confidence and establish good communication. The TC team can do this through a professional approach. A TC team of more than three people can be intimidating to an apprehensive neighbor, so keep the group small. Ask the complainant to demonstrate the problem. If the TVI is absent, ask for a description of the TV picture and sound when interference is present. If possible, get descriptions from several household members.

In a recent case reported to ARRL

HQ, the husband said that they could hear voices but they were not intelligible. The wife said "Oh, we could hear a few words …and by the way what is this ten-four and good-buddy stuff they are always talking about?" The TC had been 98% convinced that the amateur was not part of the problem, and this conversation strengthened that conclusion.

It may be natural to laugh at this point, but tact is required. Give a short explanation of the difference between municipal communications, business communications, Citizens Band operations and Amateur Radio. Many citizens lump all of these activities into one category—"ham radio."

At this point, some TCs volunteer to identify the signals causing the problem, whether or not the source is an amateur station. There are *many* possible sources of interference other than Amateur Radio, some of which may be in the neighbor's own home. Do not overlook such things as electrical interference or Part 15 devices as sources.

Often the set owner can tape the set

behavior during a period of TVI. Too often, Murphy's Law takes over while the TC is present, and the set performs perfectly.

There is always the possibility that the TV set itself is generating internal signals that interfere with the desired signal. In fringe areas, the broadcast signals may not be strong enough for a good picture.

Figs 13 through 18 show different aspects of TV installations and problems. The captions explain each figure.

THE TV AND ITS INSTALLATION

After developing a friendly relationship, the TC should investigate the technical aspects of the problem. (Refer to the Troubleshooting chapter.) Start with the TV installation, antenna, feed line and lightning arrestor (if present).

The investigation usually indicates: (1) fundamental overload, (2) harmonics or (3) mixing products (IMD). Most likely, the TV set is suffering fundamental overload. Since the ham station was checked before contacting the neighbor, harmonics or IMD must be generated *outside* the ham station.

Externally generated harmonics are relatively rare, and they are discussed in detail in the External Rectification chapter. Let's concentrate on things that can be done at the TV to improve its rejection of strong unwanted signals.

Again (and again if necessary), *simplify the problem! Anything connected to the TV is a possible source of interference.* It is impossible to efficiently troubleshoot a system with multiple TVs,

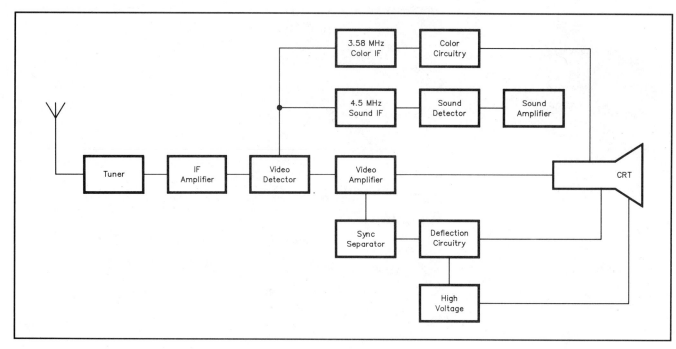

Fig 13—Block diagram of a typical color television receiver.

video games, VCRs, splitters, amplifiers, remote TV tuners, CATV converters, A/B switches and long cable runs. Simplify things to one TV and one short feed line (close to where the feed line enters the house). Once the bugs are out of the simple system, reconnect accessories one at a time. Eliminate problems as you go along.

The TV

Fig 13 shows a typical color-TV block diagram. The major systems are designed to work together. Study the way that the various circuits interact. This helps you understand how interfering signals can enter each circuit and affect the picture or sound.

The television set may have a defective component. This is not likely in a well-maintained set. If a defect exists in the TV set, however, the TC can spend a lot of time looking for problems outside the set when the actual problem is inside the set: Broken shields, defective AGC circuits or improper IF alignment can all contribute to TVI.

Leave TV repair to professionals. Safety, licensing and liability issues make it inappropriate for amateurs to repair a neighbor's equipment!

The TV Antenna and Feed Line

Check the TV and antenna installation. The TV antenna is usually strapped to a chimney, where it has been subjected

to years of rain, wind, and contaminants. The twin-lead feed line has alternately rattled in the icy wind of winter and deteriorated in the summer sun. When the TC finds defective twin lead and corroded antennas, recommend that the owner have them replaced. The corroded connections of an old antenna system are just waiting for a transmitter to induce a little current that they can rectify and radiate again as harmonics or IMD.

Perform a quick check on the overall condition of the television antenna system by temporarily substituting an indoor antenna and selecting a weak station. If the signal is not stronger on the outside antenna, there is clearly something wrong with the outdoor TV antenna system.

Sometimes TVI susceptibility is reduced by replacing normal twin-lead feed line with coaxial cable or shielded twin lead. Fig 14 shows the basic hookup for using 75-Ω coaxial cable with a 300-Ω

Fig 14—This is how to use 75-Ω coax with a 300-Ω TV antenna and television.

Fig 15—A TV preamp. These devices are not usually necessary, and in many cases are the *cause* of TVI.

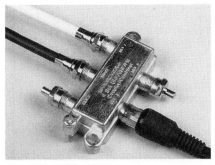

Fig 16—All multi-tap splitters should be properly terminated. 75-Ω resistors terminate unused taps of this multi-tap splitter.

Fig 17—All connections in a CATV system should be tight and secure. This shield crimp connection has loosened, resulting in poor shield integrity, possible leaks and resultant TVI problems. Call CATV service personnel to repair such problems.

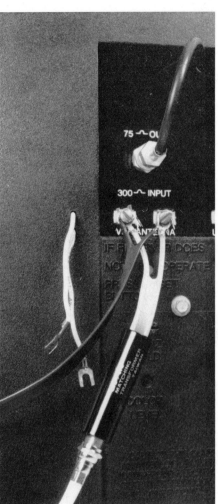

Fig 18—The 300-Ω twin lead used in this illegal CATV hookup certainly causes CATV leaks and results in interference to and from amateur stations. All CATV installation work should be done by CATV personnel.

antenna and TV tuner input. Of course, if CATV service is available in your area, you can always hope that your neighbor will consider subscribing. A properly maintained CATV system is well shielded and delivers a strong signal to the TV. Both factors significantly lessen the potential for interference.

The TV Preamp

TV preamplifiers are notoriously susceptible to strong RF fields. They usually lack front-end selectivity and are easily overloaded. They can also oscillate; when they do, they may radiate an interfering signal over many blocks. Many times, TVI lessens when the preamplifier is removed. Check the TV installation carefully for a preamp; sometimes they have been installed in the attic or a closet and forgotten.

Proper Terminations

All connections on a coaxial system should be properly terminated. If multi-tap splitters or distribution amplifiers are used, *all inputs or outputs should be connected through quality feed line to matched loads*. Unused inputs or outputs should have a suitable resistive termination. Radio Shack sells 75-Ω terminating resistors for this purpose.

IF ALL ELSE FAILS— TRY A TVI LOG

If after verifying that everything you have inspected is in order, and the interference cannot be duplicated when the amateur transmits, ask the owner to watch for a few days and keep a TVI log of any interference that occurs. The TVI log should contain the date and time, the channel selected and a description of the interference. If the TV system includes accessories (other TVs, VCRs or FM receivers), ask the neighbor to note what

other equipment was in use at the time.

Perhaps a VCR recording another channel, or maybe a video game, is generating interference to the whole system. Before leaving, set up a convenient time (to all parties) for the next visit.

Hopefully, the TVI log will give some clues to the nature of the problem. If the problem persists until the next visit, compare the TVI log with those of amateurs in the area.

The TVI log is especially important when you are not sure of the interference cause. If, for example, the interference is caused by harmonics of a nearby 11-meter (CB) transmitter, much time could be wasted on various fundamental-overload cures.

CATVI
CATV Installation

CATV investigation is a little simpler. In the ideal CATV installation, the cable runs directly to the house, vertically to the ground, then into the house through the basement or crawl space. It should be

connected to an earth ground where it enters the house. This may shunt most common-mode signals harmlessly to earth ground.

In a typical installation, the CATV coax comes to the outside of the house, drops vertically to a grounding block and enters the house through a wall. The grounding block connection to ground is governed by local electrical codes, which are usually consistent with the *National Electrical Code*. Connections through the ground block, therefore, are made in compliance with these codes. If there is latitude in the applicable code that allows for a separate ground rod, this may be a better solution. Any changes to the CATV ground connection should be performed by CATV personnel—who would be expected to know and obey the applicable electrical codes.

Many CATV installations do not minimize EMI. In a large apartment building, there are too many mechanical and constructional constraints for EMI to be a main consideration during installation. High-pass filters and common-mode chokes remain the most effective tools against CATVI.

The CATV installer usually uses crimp-on coax connectors throughout the installation. If the shield connections are not tightly crimped, the shield can float and act as a receiving antenna. Wear and tear can result in shield breaks at cable crimps. People who try to repair the damage themselves often do not crimp connectors properly, and a leak results.

Illegal hookups are another major source of CATVI problems. People do the strangest things. Spliced-on 300-Ω twin-lead runs to another set (or another house!) are not inconceivable. Some people hook an outdoor TV antenna in parallel with the incoming cable line! That makes a real mess—radiated CATV throughout the neighborhood! If a neighbor is reluctant to consult with the CATV company about TVI problems, he or she may have something to hide.

Let CATV repair personnel perform *any and all* repairs to the CATV installation. Under FCC regulations, they are responsible for any leaks. If they are responsible for leaks, it's only fair that the entire system remain under their direct control!

The CATV-Connected TV

After making sure the antenna and feed line (or CATV cable) are in good order, look at the TV set. The only parts you should check are the external connections on the back of the set. *Do not open the case of equipment that you do not own!* Internal problems (such as broken connections, dirty tuners and so on) should be left to qualified service personnel. Most states permit only state-licensed persons to repair electronic equipment. The license is required even if you do not charge for the service. A neighbor is more likely to hold you liable for future TV problems if you have opened the set. The ARRL Lab staff tell some real horror tales about hams who have not heeded this advice. Don't become part of the next story!

CATV Leakage

In cases of interference from an HF or VHF transmitter, CATV leakage may be a factor. If the cures outlined do not work for a TV connected to a CATV system, refer to the CATV section later in this chapter.

CATV service technicians should determine if there is a leak in the CATV system. Leaks may be caused by broken shields, improperly terminated splitters, illegal hookups or improperly installed TVs, VCRs or video games. The latter problems may be uncovered when the problem is simplified by temporarily removing these devices.

THEORY
Review the Fundamentals

The processes of understanding, troubleshooting and solving EMI problems are all related; a solution cannot be found without troubleshooting, and you can't troubleshoot what you don't understand. In order to fully understand the technical solutions to TVI, first review the EMI Fundamentals chapter. Nonetheless some important EMI fundamentals are summarized here for your convenience.

There are three ways that undesired signals can couple into a TV set. Each way has several variations. These ways are:

Radiation

The undesired signal is radiated directly from the amateur station to the TV-set circuits. The undesired signal is usually the fundamental signal of a transmitting station, which is radiated by the antenna (see Fig 19). The term may, however, include spurious emissions and signals radiated by the feed line, ground lead or station equipment. It can also include any noise that is radiated by any source, such as a neon sign or an electrical storm.

Conduction

The undesired signal is conducted by wires between the source and the victim (see Fig 19). This includes ac wiring and shared ground leads. An induced signal (see the next paragraph) is also a conducted signal by the time it reaches the victim.

There are two modes of conduction (see Fig 20). A differential-mode signal arrives on one wire of a two-wire transmission line, and returns via the other wire. A common-mode signal arrives, in phase, on one or more wires of a system and returns via ground.

It is important that you fully understand the difference between differential-

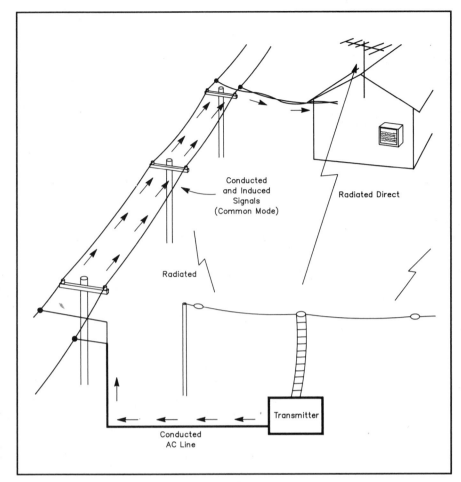

Fig 19—Conducted and radiated interference.

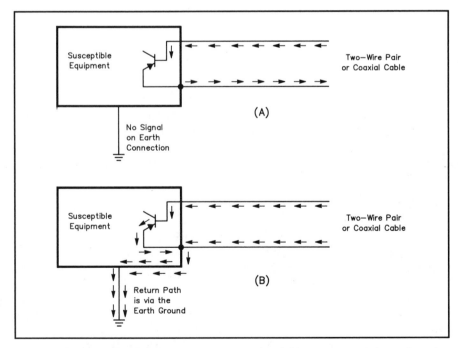

Fig 20—(A) Differential-mode signals are conducted between two wires of a pair. This signal is independent of earth ground. (B) A common-mode signal is in phase on all wires that form the conductor (this includes coaxial cable). All wires act as if they are one wire. The ground forms the return path, as with a long-wire antenna.

mode and common-mode signals because *the cures are different.*

Induction

Induction is a combination of radiation and conduction. An induced signal is a radiated signal that is picked up by wiring and conducted to the victim. Most induced interference occurs in the common mode.

Basic Cures

Table 1 summarizes the cures that are used to fix these major paths of interference propagation. The EMI Fundamentals chapter describes the different interference paths in detail.

The feed line between the TV receiver and the antenna usually picks up a great deal more energy from nearby HF transmitters than does the TV antenna. These induced feed-line currents are common-mode signals. (The line acts like two wires connected together to operate as one.) If the TV input circuit is perfectly balanced, it rejects common-mode signals and responds only to true transmission-line (differential-mode) currents. That is, only signals from the TV antenna (and not the feed line) would cause a receiver response.

No receiver is perfect in this respect, however, and many TV receivers respond *strongly* to such common-mode currents. The result is that signals from nearby transmitters are much more intense at the TV front end than they would be if the TV was well designed.

The feed line is much larger than the antenna in most TV receiving installations. Thus, the feed line picks up more interfering signals than does the antenna. To reduce this problem: (1) shield the feed lines, (2) use a common-mode filter at the TV antenna terminals and (3) relocate and/or reposition the transmitting and/or receiving antenna to take advantage of directive effects.

Intermodulation (IMD)

Under some circumstances, fundamental overload results in IMD, or mixing of the overloading signal with others (such as those from local FM or television stations). The most common place for this to happen is in the front end of a TV set or VCR.

For example, a 14-MHz signal can mix with a 92-MHz FM station to produce a beat at 78 MHz, the difference between the two frequencies (92 – 14 =

Table 1
TVI CURE SUMMARY

TVI Problem: action

HF or VHF radiated: contact set manufacturer through the EIA

HF differential mode: install high-pass filter

HF common mode: filter from Fig 22 or common-mode choke from Fig 23

HF cure-all: Filter from Fig 22, or high-pass filter in conjunction with common-mode choke shown in Fig 23, and the cures from electrical cure-all

VHF differential mode: Notch filter or tuned stub. See Figs 27-29

VHF common mode: common-mode choke from Fig 23

VHF cure-all: Combine notch filter or tuned stub with the VHF common-mode choke, plus the cures from electrical cure-all

Electrical differential mode: ac-line filter (Fig 12)

Electrical common mode: common-mode choke from Fig 11

Electrical cure-all: ac-line filter (Fig 12) with common-mode choke from Fig 11

DISCUSSION
HF or VHF Radiated

This implies that the interfering signal is being picked up directly by the victim circuitry, wiring or printed-circuit traces. This type of interference is rare in modern TVs, but it may still be a problem with older equipment. External shielding may help, but it is not always practical. In cases of direct radiation pickup, contact the set manufacturer through the Electronic Industries Association.

HF Differential Mode

A differential-mode high-pass filter (see Figs 30 and 31) or the unusual filter shown in Fig 22, will usually help with this type of interference. Keep in mind that a differential-mode high-pass filter will not work against a common-mode signal that may also be present.

HF Common Mode

The unusual filter arrangement shown in Fig 22 works, as will the common-mode choke shown in Fig 23. The common-mode choke is not effective against any differential-mode signals that are also present.

HF "Cure-All"

Many times the interfering signal will have both common- and differential-mode components. This means that the filter must be able to eliminate both types of interference, or two filters must be used. Either use the filter shown in Fig 22, or use a commercially available high-pass filter along with the common-mode choke shown in Fig 23. The cures described under electrical cure-all should also be applied.

---

78). Since 78 MHz is within TV channel 5, TVI would occur on that channel. The affected TV channel has no harmonic relationship to 14 MHz. Both signals must be on the air for interference to occur. Eliminating either at the receiver will eliminate the interference.

There are many possible IMD combinations, depending on the band in use and the frequency assignments of local stations. The interfering frequency is equal to the sum or difference of the stations involved. Their harmonics may enter into the mixing process as well. Whenever interference occurs at a frequency that is not harmonically related to the transmitter frequency, investigate IMD possibilities.

IF Interference

Some TV receivers do not have sufficient selectivity to reject strong signals at the TV intermediate-frequency (IF). The third harmonic of 14 MHz and second harmonic of 21 MHz fall at the television IF, as do some local-oscillator frequencies used in heterodyne transmitters or transceivers. If these frequencies break through the TV tuner, a high-pass (cutoff above 42 MHz) filter and common-mode choke can significantly improve the situation.

There is a form of IF interference that is peculiar to 50-MHz operation near the low edge of the band. Some TVs with the standard 41-MHz IF (sound carrier = 41.25 MHz; picture carrier = 45.75 MHz) pass a 50-MHz signal into the IF stages. The 50-MHz signal beats with the IF picture carrier to give a spurious response at (or near) the IF sound carrier, even though the interfering signal is not actually in the normal passband of the IF amplifier.

IF interference is easily identified because it affects all channels (although the intensity sometimes varies from channel to channel) and the cross-hatch pattern it causes rotates as the TV fine-tuning control is moved. With harmonic, fundamental-overload, and IMD interference, the orientation of the interference pattern does not change (its intensity may change) as the fine-tuning control is varied.

VHF TVI CAUSES AND CURES

Table 2 lists common causes of VHF TVI. There are other possibilities. Nearly all can be corrected completely; the rest can be substantially reduced.

Items 1, 4 and 5 are receiver faults; no amount of filtering at the transmitter can eliminate them. (A reduction of transmit power can help, but that is a volun-

VHF/UHF Differential Mode

A commercially available high-pass filter will not work to eliminate interference from a VHF transmitter because the VHF signal falls within the filter passband. The same problem applies to the filter shown in Fig 22. It would be necessary to use a filter with a sharp notch tuned to the fundamental, or a tuned stub. Examples of these types of filtering techniques are shown in Figs 27-29.

VHF/UHF Common Mode

The Fig 22 filter is effective only against HF signals. It does not offer significant attenuation to VHF common-mode signals. The common-mode choke shown in Fig 23 rejects VHF common-mode interference.

VHF "Cure-All"

A VHF notch filter used in conjunction with a common-mode choke will sometimes help against interference from a VHF signal. For a number of reasons interference from VHF transmitters is not as common as interference from HF transmitters. When it does occur, however, it is more difficult to eliminate because TV tuners don't offer any attenuation to VHF signals. CATV systems use several amateur bands to carry TV signals. The cures described under Electrical Cure-All should also be applied.

Electrical Differential Mode

This is the kind of interference generated by electric motors. A commercial ac-line filter will usually reduce this kind of interference.

Electrical Common Mode

Many noise sources place a common-mode signal on the ac power line. When ac wiring picks up an amateur fundamental signal, that signal is nearly always carried in the common mode. On an ac line, the common-mode signal is carried on the hot, neutral *and ground* leads. Most ac-line filters do not filter the ground lead, so they are *ineffective* against the most common forms of ac-line interference. The common-mode choke shown in Fig 10 does a good job with ac-line interference. It is not effective against ac-line differential-mode interference.

Electrical "Cure All"

Most noise sources, and some RF pickup, on the ac line contains both differential- and common-mode signals. Use a commercial ac-line filter in conjunction with the common-mode choke shown in Fig 23 to ensure that all interfering ac-line signals are properly filtered. This book has a chapter on Power Lines and Electrical Devices.

---

tary concession, not a cure.) The cure for these problems must be applied to the TV receiver. In mild cases, increasing the antenna separation may reduce or eliminate the problem. Item 6 is also a receiver fault; it may be minimized by using FM or CW instead of SSB.

Treat harmonic troubles (items 2 and 3) with standard methods given in the Transmitters chapter. Builders of VHF equipment should become familiar with TVI-prevention techniques and incorporate them in projects.

Locate the radiating portion of antennas as far as possible from TV receivers and their antenna systems. Use vertically polarized transmitting antennas for a 10-30 dB improvement (from antenna cross-polarization effects).

These steps should eliminate interference from VHF transmitters. If TVI is still present after these steps have been taken, follow the procedure outlined in the Troubleshooting chapter to determine whether the TV is subject to direct radiation pick-up. If the test indicates that the TV is okay, CATV leakage is a strong possibility. If the cause is not readily apparent (illegal hookup, broken shield and so on), contact the CATV repair personnel.

UHF TELEVISION

Harmonic TVI in the UHF-TV band is far less troublesome than in the VHF band. Harmonics from transmitters operating below 30 MHz are normally weak. In addition, the components, circuit conditions and construction of HF transmitters tend to prevent strong UHF harmonics. This is not true of amateur VHF transmitters, particularly those working in the 144-MHz and higher bands. There

the problem is similar to that of the low-VHF TV channels with respect to HF transmitters.

THE SUSCEPTIBLE TV

By now, you have determined that the TVI is caused by deficiencies in the TV set filters or shields: a genuine case of fundamental overload. Hopefully, the neighbor understands the difference between interference caused by transmitters and interference caused by poor TV design (inability to reject strong out-of-band signals). ARRL HQ has prepared an RFI Consumer Information package that helps with this explanation. It is available for an SASE from ARRL HQ, Box RFI, 225 Main St, Newington, CT 06111.

Fundamental overload (which can also affect FM receivers) is cured by preventing undesired signals from entering the affected receiver. The remedies are simple and well-established. Use a combination of high-pass filters, ac-line filters and common-mode chokes on the feed line and ac power leads.

TV sets receive their signals from direct broadcast, CATV, or a satellite dish. Direct-broadcast TV (because of the TV antenna system) is more susceptible to TVI than are CATV and satellite TV. Direct-broadcast TV is frequently the only TV service available in remote areas, and TV signals there are weak. When the TV signal is weak, the chance for TVI is greatly enhanced.

If only one neighbor has TVI problems, the TV set and its installation are suspect. If the external solutions described in the rest of this chapter do not effect a cure, contact the TV manufacturer (through the EIA) to obtain additional assistance.

Most set manufacturers produce service bulletins that describe TVI cures for their models. If it is not practical to contact the manufacturer or dealer, the amateur and set owner can work with a TV-service technician who can install the appropriate cures given in this chapter. These can be effective for fundamental-overload interference.

When a television receiver is located close to a non-TV transmitter, the intense RF signal from the transmitter's fundamental may overload one or more of the receiver circuits to produce spurious responses or blocking that cause interference.

When overload is moderate, the interference is much like harmonic interference (the harmonics are generated in the early stages of the receiver). Since it occurs only on channels harmonically related to the transmit frequency, it is difficult to distinguish light overload from transmitter harmonics. In such cases, additional harmonic suppression (transmit filters) at the transmitter does no good, but measures to reduce amateur signal strength at the TV improve performance. When fundamental overload is severe, interference affects all channels.

TV Filtering

Many cases of TVI are caused by faulty equipment or installation, and the interference is eliminated during the inspections described thus far. When TVI

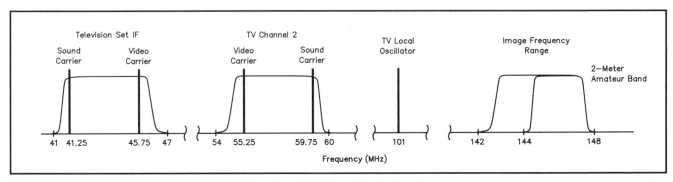

Fig 21—A receiver tuned to channel 2 but with inadequate selectivity in the input stages will pass enough strong 144-MHz band signal to cause image interference with reception. Adding a 144-MHz trap in the antenna feed line improves the selectivity of the TV receiver and prevents image interference from 144-MHz band signals.

Table 2
The principal causes of TVI from VHF transmitters are:

1. Interference in channels 2 and 3 from 50 MHz
2. Fourth harmonic of 50 MHz in channels 11, 12 or 13, depending on the operating frequency
3. Radiation of local-oscillator or multiplier stage harmonics
4. 50-MHz fundamental-overload (blocking) effects, including modulation bars, usually found only in the lower channels
5. IF-image interference in channel 2 from 144-MHz transmitters (in TVs with a 45-MHz IF—see Fig 21)
6. Sound interference (picture is clear in some cases) resulting from RF pickup by the TV audio circuits

"FB, but I still don't see how polarizing my antennas will fix my TVI problem"

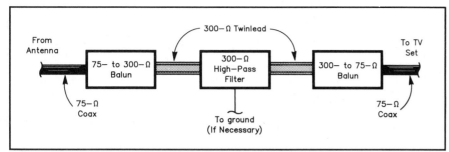

Fig 22—Insert two transformers and a 300-Ω filter in a 75-Ω TV antenna system to quash RFI caused by shield-borne HF energy.

Fig 24—A 75-Ω high-pass filter has been installed with a common-mode choke. This combination will eliminate nearly all TVI from signals that are entering the set by way of the feed line.

Fig 23—This is how to wind an antenna-lead common-mode choke. Wrap ten to fifteen turns of feed line through the core. Use no. 75 core material for interference from HF signals. Number 43 core material works well for common-mode interference from VHF signals. In some installations there is not enough slack feed line to allow the required number of turns, so it is a good idea to have a common-mode choke prepared, using either 75-Ω cable with the appropriate connectors (two crimp-on male connectors, along with a double female "barrel" connector work well) or 300-Ω twin lead, as appropriate.

remains, consider adding filters or shields.

Let's begin with proper filter application. TVI from an HF transmitter is often cured with a commercially available, differential-mode high-pass filter. The equivalent for VHF-transmitter interference is a band-reject (or notch) filter, but these are not the first solution to consider for a VHF-transmitter interference problem. Common-mode chokes and ac-line filters may be needed as well in both cases.

Filters for HF Signals

The following techniques should eliminate over 90% of TVI caused by HF

fundamental overload:

Differential-Mode Filters

Install a commercially made differential-mode high-pass filter in the TV feed line, as close to the TV as practical. Such filters are sold by Radio Shack and many department stores that sell TV accessories. High-pass filters are available for 300-Ω twin lead or 75-Ω coax.

The filter may not work well with some newer TVs. Sets with a 75-Ω coaxial input and an "ac/dc" chassis (no isolation transformer) use a capacitor in the shield lead to isolate the coaxial shield from the ac line. This provides a low-impedance path for VHF energy, while presenting a high impedance at 60 Hz, and some moderate impedance at HF. Because of this poor HF ground, high-pass filters at the antenna input terminals may not be as effective as those at the TV tuner.

Common-Mode Filters

If a stock high-pass filter doesn't cure the problem, suspect common-mode interference. If the TV is fed by a 75-Ω cable, ensure that the shield of the cable is properly grounded where it enters the house. This will shunt most of the common-mode signal harmlessly to ground. CATV cable grounds should only be installed by CATV service personnel.

It is not always practical to install a ground. In apartment buildings it is impossible to install a separate ground for a 10th-floor apartment. At that height, there is just no such thing as a good RF ground! This concept is mentioned for those cases where it may provide an easy cure.

A common-mode choke is another possible cure for common-mode interference. The filter arrangement shown in Fig 22 is effective against both differential- and common-mode signals. As an alter-

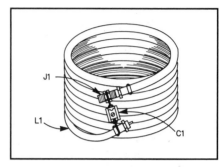

Fig 25—A compact shield-breaker assembly. This is another way to eliminate common-mode interference. It forms a high-impedance, parallel-resonant circuit at the operating frequency. It is effective, but only for interference resulting from one HF band.
C1—150 pF ceramic trimmer.
J1—F-81 bushing (Radio Shack 278-213).
L1—8 feet (2.4 meters) 75-Ω coaxial cable with an F-59 connector on each end (Radio Shack 15-1530).

native, the appropriate common-mode choke shown in Fig 23 can be used. A photograph of a common-mode choke and a differential-mode high-pass filter installed on a 75-Ω cable is seen in Fig 24. Additional cures for common-mode problems appear in Figs 25 and 26. Any of these methods should effectively eliminate the common-mode signal picked up by twin-lead or coaxial cable.

AC-Line Filters

An ac-line filter and common-mode ac-line choke eliminate the possibility that the interfering signal enters through the ac line. Most ac-line filters are effective only against differential-mode signals present between the hot and neutral leads; they do not filter common-mode signals that may also be present. In most cases, an ac-line common-mode choke is

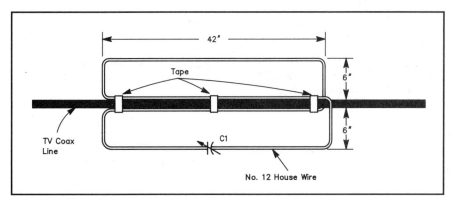

Fig 26—A resonance breaker loop for 80 and 40 meters. The loops should be taped securely to the coaxial cable at several locations. Gaps between the loops and the outer surface of the coaxial cable greatly lessen the coupling effectiveness of the assembly. C1 should be adjustable and have a maximum capacitance of 150 pF or more. Either air-variable or compression-trimmer capacitors are satisfactory.

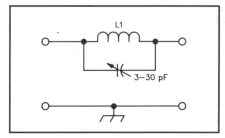

Fig 27—A tunable VHF notch filter. For 50-MHz use, L1 is 9 turns of no. 16 enameled wire close wound on a 1/2-inch-diam form (air core). For 144 MHz, L1 is 6 turns of no. 16 enameled wire close wound on a 1/4-inch-diameter form (air core).

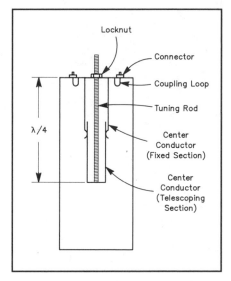

Fig 28—A cutaway view of a typical cavity. Further discussion of cavity filters and construction details for a 144-MHz duplexer (an assembly of cavities) appear in *The ARRL Antenna Book.*

needed as well.

These steps should eliminate fundamental-overload interference resulting from HF transmitters. If TVI is still present after these steps have been taken, follow the steps outlined in the Troubleshooting chapter to determine if the TV is subject to direct radiation pickup. If the test indicates that the TV is clean, there is a strong possibility that there is a leak in the cable. If the cause is not readily apparent (illegal hookup, broken shield and so on), contact the CATV repair personnel.

Filters for VHF Signals

Try common-mode cures first in VHF TVI cases because VHF-notch (band-reject) filters are not easy to find. If the TV is fed by a 75-Ω cable, ensure that the cable shield is properly grounded where it enters the house. This will shunt most of the common-mode signal harmlessly to ground. CATV cable grounds should only be installed by CATV service personnel.

Another possible cure for VHF common-mode interference is the installation of a common-mode choke. A common-mode choke that is effective for VHF common-mode signals is shown in Fig 23.

AC-Line Filters

An ac-line filter and common-mode ac-line choke eliminate the possibility that the interfering signal enters through the ac line. Most ac-line filters are effective only against differential-mode signals present between the hot and neutral leads; they do not filter common-mode signals that may also be present. In most cases, an ac-line common-mode choke is needed as well.

Band-Reject Filters

VHF notch band-reject filters are not as readily available to amateurs as TV high-pass filters. Generally, they must be home built (see Figs 27 through 29). For this reason, the installation of this kind of filter is considered last for VHF TVI problems. Notch filters should not be used in a CATV installation; most CATV systems use the VHF and UHF ham bands for active channels (see Fig 3); a notch filter would remove one or more of these channels.

These steps should eliminate TVI resulting from VHF transmitters. If TVI is still present after these steps have been taken,

---

6-METER TVI

Very little information has been written lately about TVI caused by operation in the 6-meter band, where frequencies are in close proximity to TV channels 2 and 3. Six-meter signals are some of the most difficult to isolate from these two TV channels. Normal high-pass filters do not effectively filter out strong VHF amateur signals. Other steps must be taken.

To reduce, or preferably eliminate, TVI caused by a 50-MHz signal, a resonant circuit of some sort is needed at the TV receiver. One such circuit is a resonant stub installed at the TV antenna terminals and tuned to the 6-meter operating frequency. There are two drawbacks to the use of a stub: It only notches out one operating frequency, and it also notches out all odd harmonics of the operating frequency (it's probably unsuitable for CATV use). If the amateur station operates on a frequency much removed from that of the stub, interference returns. The amateur can only operate over a limited frequency range, for which the stub is tuned.—*Roger Laroche, N6FOU, Santa Barbara Section Emergency Coordinator, Arroyo Grande, California*

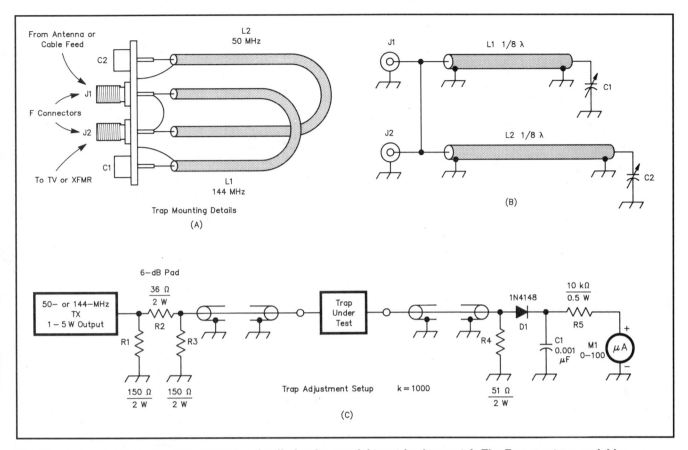

Fig 29—A pictorial diagram of the mounting details for the coaxial traps is shown at A. The F connectors, variable capacitors and cable loops are mounted on a scrap of double-sided PC-board material. At B, the schematic diagram of the trap circuit is shown. The filter is bilateral; either connector can be used for input or output. C shows a test setup to pretune the coaxial traps. Adjust the capacitors for minimum meter deflection at the desired frequency. Some retuning is often necessary after the traps are connected to the TV.

C1—3-12 pF ceramic trimmer.
C2—7-45 pF ceramic trimmer.
L1 (144 MHz)—6.7 inches for cables

with 0.66 velocity factor (VF) (RG-8, RG-58); 8.2 inches for cables with 0.80 VF (semirigid and foam-dielectric cables).

L2 (50 MHz)—18.7 inches for cables with 0.66 VF; 22.7 inches for cables with 0.80 VF.

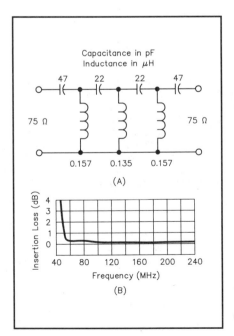

(A)

(B)

Fig 30—The schematic diagram of a 75-Ω Chebyshev filter assembled on PC board is shown at A. At B, the passband response of the 75-Ω filter. Design inductances: 0.157 μH: 12 turns no. 24 wire on T-44-0 core. 0.135 μH: 11 turns no. 24 wire on T-44-0 core. Turns should be evenly spaced, with approximately 1/4 inch between the ends of the winding. If T-37-0 cores are used, wind 14 and 12 turns, respectively.

follow the steps outlined in the Troubleshooting chapter to determine if the TV is subject to direct radiation pickup. If the test indicates that the TV is clean, there is a strong possibility that there is a leak in the cable. If the cause is not readily apparent (illegal hookup, broken shield and so on), contact the CATV repair personnel.

Filter Construction
TV High-Pass Filters

In most TVI cases, interference can be eliminated if the fundamental signal strength is reduced to a level that the receiver can reject. To accomplish this with signals below 30 MHz, the most satisfactory device is a high-pass filter (with a cutoff frequency just below 54 MHz) installed at the tuner-input terminals of the TV.

Refer to Figs 30 and 31. Slice off the extruded insulation around the solder pins on two type-F coaxial connectors (Radio Shack 278-212). Butt the connectors directly against the PC board. Solder the connector shells to the bottom ground plane and the center pins to the microstrip line. Cut the micro-strip line in four equally spaced places. The capacitors should be mounted across the spaces;

Fig 31—Photo showing construction of the 75-Ω unbalanced filter. Double-sided 1/16-inch FR-4 epoxy-glass PC board is used as a base for the filter components. A section of copper on the top is stripped away on both sides of center to approximate a 75-Ω microstrip line about 3/32-inch wide. Both sides of the top copper foil (at the edges) are connected to the ground plane foil underneath.

inductors can be connected between the capacitor junctions and the ground plane on the top of the board. Use NP0-ceramic or silver-mica capacitors. Inductors are wound on toroidal powdered-iron cores; winding details are given with the schematic.

Fig 32 shows the schematic and pictorial diagrams of a 300-Ω balanced elliptical high-pass filter that uses PC-board capacitors. Use double-sided 1/32-inch FR-4 glass-epoxy PC board. Thicker board will require more area for the desired capacitances. C2, C4 and C6 should be NP0-ceramic or silver-mica capacitors. It is easier to strip away rather than etch copper to form the series of capacitive elements. Mark the edges by cutting with a sharp knife; soldering-iron heat helps lift the strips more easily. Top and bottom views of the filter are shown in Fig 33.

Neither of the high-pass filters described requires a shielded enclosure. For mounting outside a receiver, some kind of protective housing is desirable, however. These filters were first presented in "Practical 75- and 300-Ohm High-Pass Filters," *QST*, February 1982, pp 30-34, by Ed Wetherhold, W3NQN.

VHF Filters

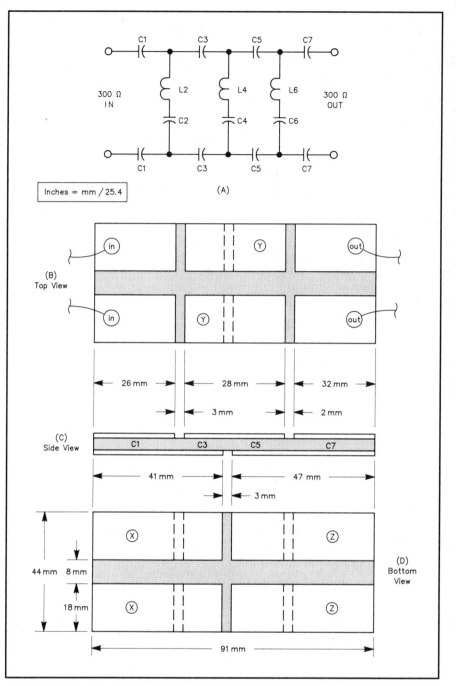

Fig 32—Schematic and pictorial diagrams of the 300-Ω balanced elliptical high-pass filter with PC-board capacitors. Shaded areas indicate where copper has been removed. Dimensions are given in millimeters for ease of measurement. L2-C2 connects between the points marked "X," L4-C4 connects between the points marked "Y" and L6-C6 connects between the points marked "Z" on the pictorial. C1 = 28.0 pF, C3 = 14.0 pF, C5 = 14.8 pF, C7 = 34.2 pF, C2 = 162 pF, C4 = 36.0 pF, and C6 = 46.5 pF. Design inductances: L2 = 0.721 μH: 14 turns no. 26 wire evenly wound on a T-44-10 core. L4 = 0.766 μH: 14 turns no. 26 wire bunched as required on a T-44-10 core. L6 = 0.855 μH: 15 turns evenly wound on a T-44-10 core. These coils should be adjusted for resonance at 14.7, 30.3 and 25.2 MHz.

Simple high-pass filters cannot always be applied successfully in the case of 50-MHz transmissions, because most filters do not have sufficiently sharp cut-off characteristics to give both good attenuation at 50-54 MHz and no attenuation above 54 MHz. A more elaborate design capable of giving the required sharp cutoff has been described (F. Ladd, "50-MHz TVI—Its Causes and Cures,"

(A)　　　　　　　　　　　　　　　　　(B)

Fig 33—A top view of the 300-Ω elliptic filter using PC-board capacitors is at A. Twin-lead should be soldered to the left and right ends of the board. At B, bottom view of the filter.

MOST TVI PER DOLLAR!—WAR STORIES FROM W6KFV

Accessory equipment is often overlooked in TVI cases. This includes things like unshielded A/B switches, remote TV tuners (with and without built-in CATV converters), video games, preamps and VCRs.

The Electronic Industries Association, in their document "Consumers Should Know Something About Interference" states,

"Anything connected to your TV is a suspected source of interference."

A recent RFI case involved a Magnavox VCR that was rated by *Consumer Reports* to have the most features per dollar. This VCR caused severe audio interference to the TV set even when the TV set was turned off! No amount of external filtering on the CATV leads and ac-line reduced the interference. The manufacturer said that it could not be fixed.

Another case involved an automatic remote TV tuner (with a built-in CATV converter). No degree of external filtering would solve the audio-interference problem. The manufacturer said, "What can I do that you have not already done? It cannot be fixed."

In another case involving an unshielded A/B switch the switch had to be removed from the circuit to solve the associated interference problem.—*John Norback, W6KFV, SB Section Technical Coordinator*

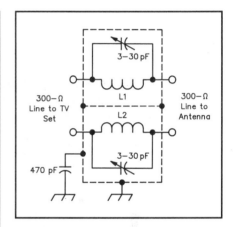

Fig 34—Parallel-tuned traps for installation in the 300-Ω line to a TV set. Mount the traps in a shielded enclosure with a shield partition as indicated. For 50-MHz use, L1 and L2 are each 9 turns of no. 16 enameled wire close wound on a 1/2-inch-diam form (air core). For 144-MHz use, L1 and L2 are each 6 turns of no. 16 enameled wire close wound on a 1/4-inch-diameter form (air core). This trap technique can be used to overcome HF fundamental overload as well.

QST, June 1954, pp 21-23, 114, 116 and July 1954, pp 32-33, 124 and 126.) This article also contains other information useful in coping with the TVI problems peculiar to 50-MHz operation.

A High-Q Wave Trap

As an alternative to such a filter, a high-Q wave trap (tuned to the transmitting frequency) may be used. It suffers only the disadvantage that it is quite selective and therefore protects a receiver from overloading over only a small range of transmitting frequencies. A trap of this kind is shown in Fig 34. These "suck-out" traps absorb energy at the trap frequency, but do not otherwise affect receiver operation. The assembly should be mounted near the input terminals of the TV tuner and its case should be RF grounded to the TV set chassis by means of a small capacitor. The traps should be tuned for minimum TVI at the transmitter operating frequency. An insulated tuning tool should be used for adjustment of the trimmer capacitors, since they are at a hot point and will show considerable body-capacitance effect.

High-pass filters are available commercially at moderate prices. All should understand however, that while an amateur is responsible for harmonic radiation from the station transmitter, the amateur is not responsible for filters, wave traps or other devices needed to protect the receiver from interference caused by the *fundamental frequency*. The question of cost should be settled between the set owner and the manufacturer or seller. Don't overlook the possibility that the TV manufacturer may supply a high-pass filter free of charge.

Shielding

Additional shielding may help in a few cases of direct radiation pickup, when the manufacturer can't supply a cure. This is not always a practical solution: Envision telling an irate, non-technical neighbor that a brand-new, expensive, rather attractive VCR must be placed inside a sheet-metal box. Just the thought of the resultant conversation would give most hams nightmares for weeks!

There are sometimes less-objectionable ways to improve shielding. Sometimes it is possible to add shielding to a home-entertainment-center cabinet. It may also be possible to relocate the TV so that there are more walls between the TV and the amateur antenna. A qualified service technician can sometimes add shielding inside the TV, either metal or a spray-on shielding material.[1] Added shielding can cause short circuits if not done properly, and it may upset the air flow around heat-generating components. Internal

[1] A spray EMI-RFI coating is available from GC-Thorsen (see the Supplier's List at the end of this book). It is catalog no. 10-4807. Beware: Such sprays do not adhere well to all plastics. Some cases may require special primers so that the conductive coating does not flake off and cause short circuits.

shielding is a task best left to the manufacturer or qualified service personnel.

SUMMARY

TVI can be cured! It takes a combination of technical knowledge, troubleshooting skills, tried-and-true cures and a good dose of personal diplomacy. Most difficult cases of TVI have involved problems with common-mode signals on a coaxial or twin-lead feed line. Those cases were only difficult because cures for common-mode signals were not well known. Armed with cures for most types of TVI, amateurs should be able to operate in peace.

CATV

By Robert V. C. Dickinson, W2CCE
President, Dovetail Systems Corp
125 Goodman Dr
Bethlehem, PA 18015

A properly maintained CATV (cable-television) system can be a real blessing to the amateur. A cable system is (ideally) self-contained: Signals inside the cable cannot (should not) be radiated to the outside world, and over-the-air signals should not be able to get inside the cable. The CATV system delivers a strong TV signal to each subscriber. These factors combine to make CATV systems relatively free from interference.

CATV is a relatively new phenomenon (about 40 years old). While not nearly as ancient as broadcast radio or even television, it has been estimated that CATV serves 54 million homes throughout the United States and multimillions more overseas. While CATV is an outgrowth of modern technologies, the leakage problems that we encounter stem from well-understood phenomena and are addressed by modern forms of traditional remedies.

TVI began early in the history of broadcast television. Amateurs were particularly concerned with TVI prevention in the 1950s and early 1960s, but the problem has decreased as a result of improved transmitter and TV performance. With the arrival of CATV, instances of TVI have reduced considerably because television signals are delivered to viewers via coaxial cable at a much greater signal strength than provided by broadcast TV. This method of transmission is designed to be a "closed system," and it does provide sufficient shielding to reduce the occurrence and severity of TVI caused by radio transmissions. CATV systems, however, are often only partially closed. To some extent, they are susceptible to radio transmissions and can cause interference to other communications services, including Amateur Radio.

CATV HISTORY

In order to gain a proper perspective on the subject, let's briefly review the history of CATV. In the late 1940s there were the early signs of what is now a pervasive technology. With the advent of television broadcasting, many viewers were far from the television transmitters and shielded by terrain. A number of innovators began businesses providing television signals to such remote viewers (and to satisfy their own desires for TV entertainment). The normal procedure was to locate a high vantage point with reasonable reception and then carry the signals by wire to viewers in a community where direct reception was poor or impossible.

Early CATV systems varied in their approaches. It is said that some used barbed-wire fences to carry the signal, although more classical methods were required for reasonable signal quality. These methods include balanced open-wire line, twin lead, G-line (an approach where the signal was launched down a single wire) and, of course, coaxial cable. Coaxial cable won out and is currently the method of choice.

By design, a CATV system is a unity-gain network, where all users can receive approximately the same TV signal levels over a broad range of frequencies. Since CATV is considered to be a closed system, any frequency within the transmission capabilities of the network can be used inside the cable. CATV can, therefore, use frequencies allocated to other services. Present CATV frequencies range from 5 MHz through 648 MHz and beyond. This is the fundamental problem in CATV interference: CATV systems use frequencies allocated to other services. In a case of CATV interference, a legal over-the-air fundamental signal may fall within a CATV channel. Once inside the CATV cable, such signals cannot be removed by filtering. Obviously, if CATV is not a closed system, signals can enter the network and degrade TV reception, while CATV signals can leave the system and interfere with over-the-air services.

By the early 1970s CATV was becoming more sophisticated. Rather than the original 12 over-the-air channels, CATV systems were beginning to carry 26, 30, 36 channels and even more. In those days, little formal attention was paid to CATV signal leakage, except when: (1) leaky systems began interfering with over-the-air signals (which gave some people free CATV service by way of local leaks); or (2) a radio transmitter interfered with CATV service.

NCTA Sees The CATVI Problem

In the early 1970s the National Cable Television Association (NCTA, the largest CATV-industry organization) became aware of numerous complaints of signals leaking from CATV facilities.[2] NCTA felt the growing concern that this leakage might interfere with other services, particularly aircraft-navigation and radio-communication (we'll call them NAVCOM) circuits.

Department of Commerce (DoC) Tests CATV v NAVCOM

At the same time, the DoC ran susceptibility tests on typical navigation receivers and pointed out the possible consequences of excess CATV leakage. Other work explored the modes of leakage from coaxial cable, including probable levels, radiation patterns and so on.

The results of the navigation receiver testing showed that it is virtually impossible for a random signal(s) to yield improper navigational information that

[2]National Cable Television Association (NCTA), 1724 Massachusetts Ave NW, Washington DC 20036, 202-775-3550.

would lead a pilot (unknowingly) in the wrong direction. It was demonstrated, however, that a significant interfering signal could make a navigational service unusable. Such a situation would, fortunately, notify a pilot that the navigational system was inoperative or unreliable. This took away the fear that leakage would cause someone to fly the wrong course. Nonetheless, it left the possibility that CATV leakage could obstruct instrument approaches in bad weather.

A NAVCOM Crisis— FAA and FCC Step In

A situation in Harrisburg, Pennsylvania, was probably the most significant instance of real-life CATV leakage. After the FAA made a new air-traffic-control communications-frequency assignment, pilots in the area heard various tones on the FAA channel. Although the tones did not cover the controller's voice, they were annoying and sometimes held the receiver squelch open. The FAA was unable to locate the cause of the problem over a period of weeks and called the FCC. The FCC Field Operations Bureau decided that the problem had something to do with the local CATV system.

The CATV operator was cooperative and the final explanation was rather surprising: The CATV system had four separate "headends." The AGC for the cable amplifiers was derived from a pilot signal on the same frequency as the FAA. The four headends had separate pilot generators, on slightly differing frequencies. Although a single interfering frequency would simply cause a squelch break and a little background noise, leakage from two (or more) CATV sectors was heard in some areas, and AF beats between the pilot carriers appeared as audio tones. Once the cause was found, the CATV system was shut down, and the AGC system was reconfigured to operate on TV-carrier frequencies. (This was an old and large system that was preparing for rebuild.)

The Harrisburg situation was the proverbial "straw that broke the camel's back." A huge uproar in aviation circles followed, and great efforts were made to prevent CATV systems from using any frequencies not used for standard TV broadcasting. Had those efforts been successful, the industry and the public interest would have been severely affected. As it worked out, the FCC threatened operators of leaky

CATV systems with severe consequences. This edict got the attention of some, who began to seriously address the problem.

The Advisory Committee on Cable Signal Leakage was formed by the FCC and the FAA. It included industry participants who studied the situation and made recommendations to the FCC for further rule making. The committee reported in 1979 and recommended measures which ultimately became part of the FCC Rules and Regulations Part 76. One of the most significant FCC actions (even before the report of the committee) requires continuous monitoring of CATV systems for leaks and timely repair of any leaks. A few CATV operators took the situation very seriously and began to clean up their act. Others virtually ignored the problem.

CATV Leakage Controls

Cable-industry signal-leakage control programs have a long history, with a great deal of success and some notable failures. The FCC rules enacted in 1985 not only require an ongoing monitoring program to find and fix leaks, but also annual qualification of all CATV systems to demonstrate compliance by actual measurements. (This part of the rules did not take effect until July 1, 1990.) At present, all CATV systems using aeronautical frequencies (108-137 MHz and 225-400 MHz) must qualify under §76.611 annually. A brief description of the qualification requirements is given in the "Cable Leakage" sidebar.

We Must Work Together

What does all of this mean to the Amateur Radio community? Primarily, it means that CATV is here to stay, and we amateurs must be cognizant of the facts and diligent to avoid or reduce the inherent problems of coexistence.

CATV Operator Responsibilities

The CATV industry has some comprehensive legal responsibilities under the FCC regulations. The permitted signal leakage is limited according to frequency range as shown in Table 3. Note that 20 μV/m at three meters is a small, but not insignificant, signal. A 20-μV/m field indicates over S9 on an average amateur two-meter transceiver. This means that (under these FCC rules) occasional squelch breaks while riding through town are possible and interference with weak signals is acceptable.

Table 3

CATV Leakage Limits

Frequency (MHz)	Field Strength (μV/m)	Distance (ft)
0-54	15	100
54-216	20	10
216+	15	100

Present and Future CATVI Hot Spots

Fortunately there are usually only a few CATV signals in a given area that might cause trouble. One is a video-carrier frequency that falls in the amateur two-meter band (probably 145.25 MHz, and possibly the first 15.75 kHz of sidebands). In some CATV systems, the TV carrier is at 144.000 MHz, where it can interfere with amateur EME work. There are equivalent signals in the 222-MHz and the 70-cm bands. If and when CATV moves up toward 1 GHz, other frequencies may be involved.

A Mutual Responsibility

There is a second responsibility of the CATV operator (as well as anyone using FCC frequency assignments): systems may not cause "harmful interference." Harmful interference is defined under FCC Rules and Regulations Part 76, paragraph 76.613(a) as "any emission, radiation or induction which endangers the functioning of a radio navigation service or other safety service or seriously degrades, obstructs or repeatedly interrupts a radio communications service operating in accordance with this chapter." "This chapter" covers essentially all radio frequency use and therefore includes Amateur Radio operation.

The scope of §76.613 is broad, and the rule can be difficult to interpret. Statements from high-ranking FCC personnel indicate that resolution of interference under §76.613 is a *cooperative venture* between the interfered and interfering parties. Cases referred to the FCC may well be resolved in ways not favorable to either party. The "bottom line" is: Work *with* the CATV company to find an acceptable solution, rather than calling in the FCC. Although the FCC decides exactly what constitutes interference, amateurs should apply common sense to any potential interference situation. It is unreasonable to interpret merely break-

ing squelch as interference.

CATV FUNDAMENTALS— A TECHNICAL DESCRIPTION

Let's look at some technical fundamentals of CATV. (See Fig 35.) A CATV system is a very unusual network, and it is quite interesting from a technical point of view. Its signals are transmitted from a central point normally called a headend. There, over-the-air signals are collected and combined with others (public access, satellite feeds and so on) to form a broad spectrum of television programming. This often includes FM radio stations and occasionally includes data transmissions.

From the headend, signals are routed throughout the system over a network of

WHO IS RESPONSIBLE FOR CATV INTERFERENCE PROBLEMS?

The CATV Operator

It would be easy (for amateurs) if we could blame all CATV interference problems on the CATV system. This is not always a correct assessment of responsibility, however. Many CATV-interference problems are the direct result of a leak or other defect in the CATV system. Although no regulation directs a CATV operator to correct leaks that let signals *into* a CATV system, leaks work both ways. The CATV operator must fix the leak to prevent signals from leaking out of the system. In this sense, interference caused by leaks is the responsibility of the CATV operator.

The Amateur Operator

Unfortunately (for hams) leaks are not the only cause of interference to a TV connected to a CATV system. There are even ways that the amateur may be responsible. If, for example, a spurious emission generated by the amateur station causes the interference, the amateur should eliminate the problem. This is true even when a CATV leak contributes to the problem, because the spurious emission may interfere with antenna-connected TVs or disrupt communication in other services.

The TV Owner Or Manufacturer

Some TVs do not have adequate shielding and filtering. A strong radio signal, especially at VHF, may be picked up directly by the TV circuitry, resulting in interference. There is little that a CATV company can do when interference results from direct radiation pickup in TV or VCR circuitry.

The shield of the CATV system forms a large long-wire antenna. This antenna can pick up large amounts of RF energy from nearby radio transmitters. (This does not indicate a leak or defect in the CATV system.) Most of the time, the RF present on the CATV shield does not cause a problem. Some TVs and VCRs are quite susceptible to this common-mode signal, however. The CATV company is not responsible for such common-mode interference.

CATV repair personnel should understand the various ways that interference can occur in their CATV system. In order to determine that interference is not caused by a leak, it is usually necessary to diagnose how interference does occur. A common-mode choke should be installed in all cases to determine whether interference is caused by common-mode signals on the CATV shield. The proper construction and use of a common-mode shield are discussed in earlier sections of this chapter.—*Ed Hare, KA1CV, ARRL Senior Lab Engineer*

coaxial cables. Amplifiers in the system replace cable losses. As you may know, cable losses increase with frequency. Equalizers compensate for the cable frequency response. In fact, signals are usually compensated at each cable-amplifier output such that the levels are relatively flat by the end of the cable run (at the next amplifier).

Transmission quality is governed by two general factors: Noise contributed by the amplifiers tends to degrade picture quality. Distortion products generated by the amplifiers cause interference that gradually increases toward the system end.

Since noise is generated in each amplifier, the system noise is kept low by keeping a good signal-to-noise ratio at

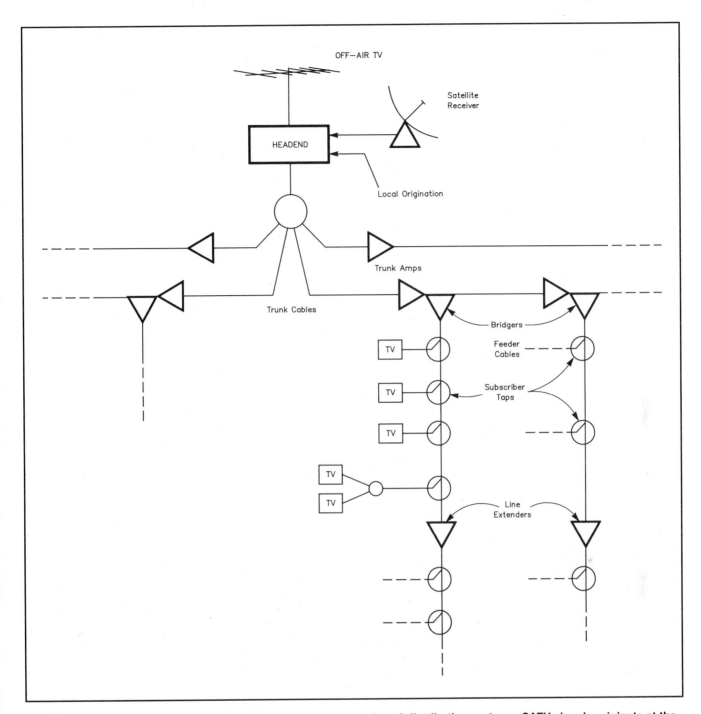

Fig 35—A typical CATV installation consists of the headend, trunk and distribution systems. CATV signals originate at the headend. The trunk system carries the signals to the various parts of the service area. Individual subscribers receive their signals from the distribution system.

each amplifier input (so that the amplifier noise contribution is relatively insignificant). The signal level must not be too high, however, or distortion may reduce picture quality. This combination of effects defines a window of acceptable operation. The window gets smaller as more amplifiers are cascaded. There is a finite limit to the number of cascaded amplifiers, which is known as the "depth" of the CATV system.

From the headend, CATV signals enter the "trunk" system. Generally, CATV systems are limited to a few tens of trunk amplifiers in any one run. Trunk signal levels are relatively low, as are the noise and distortion degradation. In a normal CATV system, subscribers are connected to distribution legs or "feeders." Signals are extracted from the trunk, amplified and sent down the feeders. Signal levels in the distribution leg are considerably

higher than on the trunk, but the depth of distribution amplifiers is usually limited to two or three. Thus, the distortion products remain insignificant.

The TV-signal level at any point in the CATV network seldom exceeds 1 mW (0 dBm), and the nominal level at a TV set may be on the order of –45 dBm. The CATV system is designed to maintain a constant impedance (nominally 75 Ω). Hence, 75-Ω cable is used throughout the system. Most modern TVs have a 75-Ω coaxial input that allows direct connection to the CATV system.

Channelization Schemes

CATV channelization has never been officially standardized. There are three general schemes, which are known as standard, HRC and IRC. Fig 3 shows nominal channel frequency assignments to 648 MHz. Channels are often identified by numbers only, but there is no universal standard. Standard channelization begins with the normal frequency assignments of the VHF-television bands and the extension of these channels (6n + 1.25 MHz) into the CATV-only frequencies (the spectrum above FM and below channel 7, 108-174 MHz, and all channels above 13, 216 MHz).

Harmonically Related Carriers (HRC)

During the development of CATV-system amplifiers, amplifier-semiconductor distortion limited the number of channels and signal levels that could be carried. Efforts were made to "hide" distortion products by placing them on the frequencies of other-channel carriers, where they would be the less visible. The Harmonically Related Carriers (HRC) approach resulted. In this system, all channel frequencies are multiples of 6 MHz and locked to a precise 6.000-MHz comb. (A graph of an array of equally spaced channels resembles a comb. See Fig 36.) Therefore, IMD products between channels also fall on a 6-MHz comb, and the distortion products are somewhat hidden by the carriers of other channels. This is a successful approach that is still used in some CATV systems (but not without its own problems).

Incrementally Related Carriers (IRC)

In order to achieve some of the HRC benefits without using nonstandard TV-channel frequencies, the Incrementally Related Carriers (IRC) system was developed. It uses the basic VHF comb based on (6n + 1.25) MHz for all channels. In this system, all carriers are locked to a coherently generated comb of frequencies to stabilize the CATV channels. IRC has some of the IMD-hiding effects of HRC while not displacing TV-carrier frequencies significantly from those of over-the-air channels. Obviously, CATV channel frequencies vary with the system used.

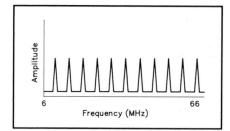

Fig 36—Equal channel spacing is called a "comb" because of a graph of amplitude v frequency resembles a comb.

Two-Way CATV

CATV systems are capable of two-way operation. Two-way signals originate at the user and travel back to the headend (upstream: opposite to the entertainment-signal flow). In a normal CATV system, the frequency range from 5-30 MHz is available for upstream operations. Unfortunately, few CATV companies use these facilities. (It is unfortunate because a wide range of services can be implemented via this return path.)

Some parts of the CATV network are often used for non-entertainment purposes. These are generally included in what is known as an institutional network (I-Net). In such systems the number of upstream channels is expanded, while the number of downstream channels is reduced to provide approximately equal capacity in both directions.

The upstream channels (5 to 30 MHz) may be somewhat more susceptible to interference than those in the VHF range (susceptibility depends on a number of factors). It may be more important that amateur power levels in the HF range are typically higher than at VHF: More serious interference can be expected. When the CATV system is known to have reverse transmissions, the energy from these signals can be monitored for leakage with HF equipment. This, however, involves the use of unwieldy or inefficient antennas and has no particular advantage over VHF monitoring. Things to remember: In general, a leak leaks at all frequencies; where energy exits, energy can enter.

FCC Requires Offsets To Protect NAVCOM Frequencies

In the process of the FCC rulemaking about CATV signal leakage, some extra caution was applied. Based on the possi-

bility of a catastrophic leak (such as a severed cable) FCC established a system of frequency offsets in §76.612. The offsets protect aircraft NAVCOM channels from strong carriers in CATV systems by assigning CATV frequencies midway between the NAVCOM channels. The offsets are odd multiples of 12.5 kHz in the 118-137, 225-328.6, and 335.4-400 MHz ranges and odd multiples of 25 kHz in the 108-118 and 328.6-335.4 MHz bands.

This is important to amateurs because interference from CATV leakage may not appear exactly where indicated by channel plans. Note that offsets are not required in the amateur bands, but CATV operators may choose to use them for consistency with the aeronautical channels.

For instance, 145.25 MHz is the nominal carrier frequency of a channel in the standard channelization plan (144.00 MHz in an HRC system). If this channel were offset like those in the 118-137 MHz band, 145.25 would not be allowable. The CATV signal might be found 12.5 kHz higher or lower. (Actually, the offset can be any odd multiple of 12.5 kHz, but converter and receiver problems usually determine the use of plus or minus 12.5 kHz.)

HRC Offsets

An HRC system doesn't use offsets in the way described above. Under the offset rules, the 6-MHz base frequency must be set to 6.0003 MHz ±1 Hz. The HRC system provides sufficient offset in most of the aeronautical frequency bands. This may not be so good for operators near 144 MHz, but the CATV frequency there is known to be 144.0072 MHz ± 24 Hz.

Bottom Line: A Leaky System

When all is said and done, our attention focuses on one fact: the shield integrity of a CATV system not perfect. Where CATV-system energy can leak out, external RF energy can enter. Energy that leaks out can interfere with amateur communication, while energy that leaks in can interfere with viewer reception (which often results in unpleasantness for the radio amateur).

WHY CATV SYSTEMS LEAK

CATV systems leak for several reasons. The seamless aluminum jacket (the cable shield) has a relatively high coefficient of expansion: As air temperature rises, the shield lengthens. When the air temperature lowers (after sundown for

Fig 37—(A) A Hardline connector as used on CATV trunk lines. (B) A ring crack breaks the shield connection completely.

example), the shield shortens. This constant cycling develops small cracks in the shield (usually behind the connector compression fitting (see Fig 37A). The cracks present effective escape routes for signal leakage. Left untreated, cracks can develop into complete shield breaks, called "ring cracks," which expose the dielectric and center conductor (see Fig 37B). If this happens, ground continuity along the shield can be lost!

(This same condition can occur anywhere Hardline is used. For amateurs, a worsening SWR is usually the first clue of a mechanical fault such as a crack. This is an excellent reason why all amateurs should keep detailed antenna-system records and check system operation several times each year. Measurement comparisons "flag" developing defects in the system. They may also be of assistance in convincing the FCC that your neighbor's nasty RFI problem isn't due to technical deficiencies at your station!)

Connector Continuity

CATV systems often leak because technicians haven't properly tightened the compression fittings used on Hardline and subscriber-drop cables. As with ring cracks, connectors with poor ground continuity present excellent paths for signal leakage.

Illegal Connections

Not all leaks result from maintenance problems. Illegal connections to the CATV system (made by technically inept people without the proper tools and parts) are a major source of leaks. It is not uncommon for an illegal hookup to use 300-Ω twin-lead cable for runs to additional outlets, or even to a neighboring home. Needless to

---

---

say, such hookups are strong leaks.

Inferior Cable From 1960s And '70s

The "drop" cables used to wire homes during the 1960s and '70s are another major source of CATV leaks. Before CATV operators started programming on frequencies used primarily by other services, it was common practice to use drop cables with 67% braided shields. While such cable does provide ground continuity, it does not always reduce leakage to levels acceptable to amateurs with 0.1-µV-sensitivity receivers! Modern drop cable is

much better. It may have from one to four shield layers of braid and/or foil. Quad-shield cable is so different from standard coax that it requires different connectors; see Fig 38. Modern cables efficiently keep signals within the CATV system. (If you experience interference on 144.0 MHz or 145.25 MHz, ask your CATV operator to determine if your home/station is wired with the old 67%-braid drop cable.)

Leakage At The TV And Accessories

Even if the Hardline is free of cracks, the connectors are properly installed and

Fig 38—CATV connectors for two different contemporary cables. Cable differences can be so slight that inappropriate connectors appear okay (another reason to leave CATV installation and maintenance to professionals). All connectors should have integral sleeves, as shown, rather than separate crimp rings.

the drop coax is modern quad-shield, there may still be leakage interference from a TV set! Many television receivers with 75-Ω "F" connectors on the back use a balun to convert the 75-Ω incoming signal to 300 Ω for the tuner. The short section of 300-Ω twin lead can act as an antenna for CATV system leakage. Your CATV operator can't do anything about this leakage, except to disconnect the service until the leaky TV

set is corrected.

Some other leaks originate in the subscriber's home. These may result from fittings or extemporaneous connections and electronic devices. Electronic devices include poorly shielded TV sets and FM Tuners, switches for video games and the like, high output levels from distribution amplifiers used to feed extra drops within the dwelling and other similar sources. It is obvious that these are difficult for the CATV operator to address because they are inside the subscriber's home. Note that the FCC allows CATV operators to disconnect subscribers who have excessive leakage within the dwelling. The normal scenario is: Notify the subscriber and ask for entrance to find and repair the problem. If such is not granted, the drop may be disconnected.

HOW CATV OPERATORS FIND LEAKS
FCC Requires Vigilance

The FCC requires all but the smallest CATV operators to have and follow a signal-leakage patrol program. That requirement is part of the FCC rules (see "CATV Leakage"). The FCC requires that a

CATV operator continuously patrol the system looking for leaks, logging them and repairing them.

Most CATV operators use truck-mounted commercial antennas and special commercial receivers to patrol the system for leaks. The receivers are usually meter-calibrated so that the technician can read or compute leak strength. When a leak is heard, the technician narrows down and logs its general location and strength. In large systems, those logs are given to other technicians who pin-point and repair the leak. In small systems, the same technician often patrols, finds and fixes leaks.

Locating Leaks

Leak location takes many of the same skills that amateur transmitter hunters use. The technician localizes the area of the leak by monitoring signal strength while driving. As signal strength increases, the technician may use an attenuator to avoid overloading the receiver. When the apparent maximum is found, the technician usually uses a hand-held receiver with a small antenna to "walk the area." When the leak is determined within several feet, the technician may use a "near-

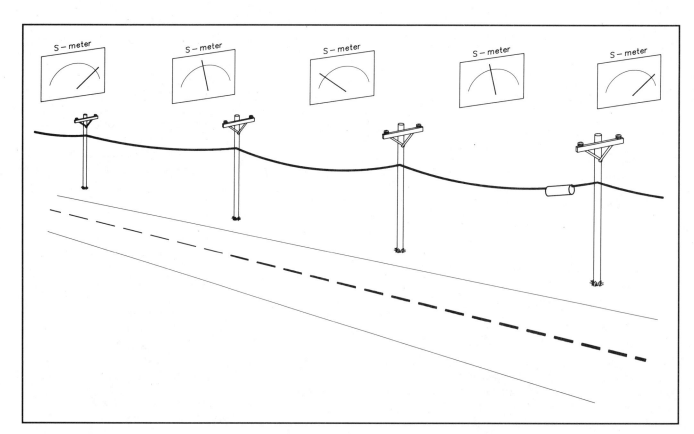

Fig 39—A CATV leak may set up a pattern of RF peaks and nulls at 1/2-λ intervals.

UNDERSTANDING AND IDENTIFYING CATVI

CATV interference to Amateur Radio typically occurs on 2 meters and 222 MHz. The usual frequencies are at or near 144.00 MHz or 145.25 MHz (CATV channel 18[E]) and 223.25 MHz (CATV channel 24[K]). The interference is a constant "BUZZ" (the picture-sync buzz), or a steady carrier that gets stronger as you get closer to the CATV leak.

Causes

Such interference happens when CATV carriers leak out of the system. The reasons for leakage usually fall into several general categories:

Poor CATV Cable Shielding

Until the early 1980s, coaxial drop cables (which connect the TV to the CATV feeder line) employed a single layer of braided wire as the shield. Some drop cables had less than 70% of the center conductor covered. This allows much more leakage than contemporary quad-shield cables.

Cracked CATV Cable Shields

The main trunk and distribution coaxial cables used in CATV are constructed with a copper-clad aluminum center conductor surrounded by hardened foam. An aluminum tube forms the electrical shield. A plastic jacket usually provides an environmental barrier. Unfortunately, aluminum expands and contracts with heat and cold. Circumferential ("ring") cracks can develop at the point of maximum stress, usually at the back nut of a Hardline connector. Leakage then occurs. Leakage can also occur if "smog" erodes unjacketed cables.

Loose Connectors and Hardware

Another major source of signal leakage is loose connectors (commonly known as "F" connectors) on drop cables. Newer connectors have an integral sleeve to improve shielding, but millions of older "two-piece" connectors are still in use, and are still sold at neighborhood electronics stores. Leaks also occur when amplifier housings are not properly closed or signal-tap face plates are not tightened properly.

Law Of Reciprocity

If you can hear them, they can hear you. Not only do cracked or poorly shielded cables emit potential interference to your operations, leaks also permit amateur signals to enter the CATV system. Ever wonder why your neighbors complain about TVI on their channels 18 and/or 24? Now you know!

Depending on the ingress point and the strength of your signal, you could cause herringbone (beat) patterns for hundreds (or thousands) of your neighbors! Whose responsibility is it? As long as you are "...a radio-communication service operating in accordance with [the FCC rules] ..." it's the CATV operator's responsibility. But for the sake of your neighborly relations and general sanity, work with your CATV operator to help identify the leak! (See the FCC CATV rules §76.613.)

Use Transmitter-Hunt Skills To Find CATVI

Brush up on your transmitter-hunting skills and dial up 145.25 MHz (or 144.0 MHz for "HRC" systems) on your hand-held radio. Set your squelch to just close. As you get closer to the leak, the squelch will break, the signal will get louder, and the background noise will quiet down. (You may hear the buzz of the television sync pulse.) Keep tightening the squelch as you get closer. When you can no longer squelch the radio, remove the antenna and then use the radio as a "near field" detector. When you've located the leak, or the approximate location, call your CATV operator.

Don't try to repair CATV leaks yourself. The lines are the property of the CATV company. They are responsible for system condition, and they must ensure that leakage is within FCC specifications. *Also, some trunk or distribution CATV lines carry dangerous voltages.*

While you can try to describe the signal leakage problem to the customer service representative who answers the phone, it's better to speak directly with the "Technical Manager" or "Chief Technician." Always log the date, time, and names when you refer CATVI complaints or inquiries to your CATV operator! You'll never know when that log will be needed to substantiate your claim of interference.

Finally

CATV operators do want to resolve service problems, especially leakage. The FCC requires every CATV system to file an annual report of leakage control. When you call about a CATVI problem, it's usually just a matter of getting through to the "right person," who can understand you. Local radio clubs should initiate and maintain a liaison with local CATV operators before problems occur.

In the event that you are unable to resolve a problem at the local level, report (in writing, see the Report Form chapter) your complaint to ARRL HQ. Mark it "Attention: Box RFI." For many years, the League and the NCTA (the CATV-industry trade association based in Washington, DC) have formally cooperated to resolve problems that couldn't be resolved on the local level.—*Jonathan Kramer, KD6MR, President, Communications Support Corporation*

field" antenna with the hand-held receiver to find the exact point of leakage. (Because many CATV systems use special heat-shrink boots over Hardline connectors, cracks may not be visible.)

Sometimes a leak is difficult to locate because energy couples into the cable shield or support "strand." There, the leak propagates as a series of peaks and nulls (see Fig 39). The largest peak occurs very near the leak. Sometimes, it is difficult to determine which is the largest peak. Careful measurement techniques will lead to the location of the actual leak.

LOCATING CATV INTERFERENCE NEAR YOUR SHACK
Safety

First follow basic safety rules: Don't ever touch your antenna or antenna connector to any CATV system surface. You never know where you'll find ac power! Leave the actual repairs to trained CATV technicians who are equipped with the proper tools and parts.

Interference From CATV

If you believe that you are experiencing CATVI, you can perform your own CATVI-hunt with a hand-held 2-meter receiver, a detachable antenna, and some basic transmitter-hunting skills. First, tune your receiver to 144.00 MHz or 145.25 MHz. If you hear a loud buzzing (the TV sync pulse) or your receiver quiets, start sweeping the area looking for the strongest signal. (It helps if your receiver has an S-meter, but it is possible to gauge signal strength by ear alone.) When you locate the strongest signal, loosen the antenna connector (to reduce the received signal strength) and continue to hunt for the strongest signal. (It sometimes helps to use your body as a shield for increased directivity.) When the signal strength again reaches the point of receiver overload, remove the antenna entirely, and probe with the open receiver antenna connector for the strongest signal. (You'll probably locate the leak this way.)

Don't be surprised if you find yourself staring at a wall! It's not unusual for CATV coax inside walls to deteriorate and begin leaking. Also, a careless installer may have accidentally pierced the cable with a staple during installation. As the staple rusts, the cable deteriorates and begins to leak.

Interference To CATV

Sometimes CATV leakage is encountered by the fixed-station operator who uses a scanner to monitor the entire spectrum. The scanner may routinely stop at the CATV-leakage frequency. Conversely, the fixed-station operator may find that his or her transmissions interfere with CATV connected television sets. In such cases, follow the diagnostic and repair procedures set forth in the Troubleshooting chapter and in the first part of this chapter.

When all receivers in the same area are interfered with, it is possible that energy is entering the CATV cable and propagating to other receivers on the system. As you can see, there are few situations to be dealt with, and the solutions to the problems have very few options. When the problem is interference on the cable, it will disappear when the cable is disconnected and the TV properly terminated.

WHEN CATV LEAKAGE IS FOUND

It is clear that the amateur has rights in regard to CATV leakage. These rights come from the privileges and protections supported by FCC regulations. On the other hand, amateurs should be careful and thorough in working out solutions to specific problems and only involve the FCC as a last recourse.

The "political" procedure for gaining relief from CATV leakage problems is generally as follows:

1. Contact the CATV operator. It is best to contact a supervisor in the technical (preferably maintenance) side of the company. Normal approaches may not work efficiently here; if possible, find the name of the chief technician and give him a call. (Your ARRL Section Technical Coordinator, TC, may know the best CATV contact. If you don't know your TC, ask your Section Manager.)
2. Discuss the problem with the chief technician in a matter-of-fact way. Give him the details of your observations. (The focus on CATV leakage over the past several years has made most CATV operators cooperative. In fact, good amateur relations can help with their leakage program.)

ARRL + NCTA—A Strong Hand

If there is no positive result from your initial and follow-up contact with a CATV operator, the next step should involve the ARRL. The ARRL, in conjunction with the NCTA, has developed a program of complaint processing designed to assist peaceful and constructive resolution of CATV problems (and avoid involving the FCC). Begin by providing complete information on the incident as described in the chapter on EMI reporting.

If necessary, the HQ staff may become involved in some CATVI cases. In cases where the amateur and CATV operator are unable to agree on the technical or regulatory issues, the League can call on the NCTA liaison committee. The committee exercises their contacts with the particular CATV operator in an effort to induce cooperation. Give the committee a chance to operate, and help them wherever possible.

Complaints should only go to the FCC as a very last resort. When all else fails, do contact the FCC, but do so in cooperation with the ARRL. This procedure ensures that all possibilities for a peaceful solution have been exhausted and that the complaint is properly addressed at the Commission level.

This procedure has been in effect for a number of years, and it has been a substantial help in resolving leakage problems. The CATV operator has real incentives to cooperate; resolution of amateur leakage complaints: (1) improves the CATV system integrity, (2) reduces complaints from other services and (3) improves customer signal quality. The amateur incentive is that a cooperative CATV operator better controls leakage. There will be fewer future problems.

The procedure outlined above has worked in many cases. Certain cases (where complaints were made directly to the Commission) have been rejected by the FCC and sent through the above channels— and usually resolved. A sensible, cooperative approach is best for all. It leaves a better image in the mind of the Commission, who licenses both amateurs and CATV operators. It leads toward better performance in both the CATV and Amateur Radio communities.

VCRs

By John Frank, WB9TQG
Wisconsin Section Technical Specialist
POB 5113
Madison, WI 53705

Try to visualize the following scenario: your brother-in-law has been awarded a Nobel prize, and the presentation will be on TV the same night that you hope to work a DXpedition to Outer Elbonia. If you watch the awards presentation on TV, you miss working Outer Elbonia on 80 meters. If you work Outer Elbonia, you incur the wrath of your family for decades to come.

Wait, a VCR can solve this problem for you! You can automatically record the historic event (the award, not the DX contact) while you are in the ham shack making a once-in-a-lifetime DX contact.

After working Outer Elbonia, however, you rewind the videotape and play it back. Your brother-in-law is barely recognizable: The interference is so bad you can't tell the Nobel-prize presentation from an old rerun. Now you incur your family's wrath because you missed your brother-in-law's moment of glory and you interfered with the VCR.

Fortunately, most VCR owners don't experience any interference from radio transmitters. The potential for interference is clearly present, however, because of the design and construction of VCRs. VCRs are highly susceptible to interference because their circuitry uses the frequencies of (and near) several heavily populated radio services. Additionally, older VCRs lack shielding needed to keep unwanted signals out of their video modulators and demodulators.

Is the situation hopeless? Must VCR users endure interference? The answer to both questions is "No, not necessarily." Newer VCRs are better shielded than their predecessors. Interference to VCRs can be reduced or eliminated with some relatively simple procedures.

Although some of the techniques used to curb interference to VCRs are similar to those used to eliminate interference to other home entertainment devices, others are quite different because of the frequencies involved in the video recording and playback process. Also, as the ham who worked Outer Elbonia discovered: Once interference has been recorded, it's on the tape as long as the program is there.

WHY ARE VCRs PRONE TO INTERFERENCE?

In order to understand why VCRs are susceptible to interference, it is important to understand how video is recorded and how the process differs from audio recording. The audio spectrum is generally regarded as frequencies from 20 Hz to 20 kHz. It is relatively easy and inexpensive to obtain good frequency response by recording the audio signal directly onto the tape. Video signals, however, contain frequencies from 30 Hz to about 4.5 MHz.[3]

Although it is possible to obtain good frequency response over such a wide bandwidth with direct recording, the cost would be prohibitive for home entertainment equipment. The simplest way to record this bandwidth on magnetic tape (at a reasonable price) is to frequency modulate a sine wave with the baseband video signal. At first this may seem like technological overkill, but it works and offers good immunity to noise.

Recording The Video Signal

The problem for amateurs is that most VCRs use frequencies in or near amateur bands to record the FM video signal. For example, VHS video recorders use the frequencies from 3.4 to 4.4 MHz to record the luminance (brightness) portion of the video signal.[4] This frequency choice makes VCRs especially prone to interference from amateur stations operating in the 80-meter band.[5] It also makes harmonic suppression an absolute must

for amateur stations using the 160-meter band. Remember that VCRs are high-gain, broadband devices that can suffer interference from other HF amateur bands.

The Super VHS system uses different frequencies and a wider bandwidth to deliver better video resolution. The FM video signal is recorded from 5.4 to 7.0 MHz.[6] This higher frequency and greater bandwidth provides almost twice the resolution of conventional VHS recording. The video signal, however, is adjacent to the 40-meter amateur band.

Beta video recorders use 3.5 to 4.8 MHz to record luminance. Although the frequencies are slightly different from VHS, beta machines are also prone to interference from amateur stations in the 80-meter band.

Recording The Audio Signal

The audio portion of a television program can be recorded onto tape as an audio track. Several other techniques are used as well.

In the case of stereo audio on videotapes, the audio is used to frequency modulate a carrier, which is then recorded on the tape. In Beta stereo, the left and right audio channels frequency modulate carriers between 1 and 2 MHz. The frequency difference between the carriers helps minimize crosstalk between audio channels. (This range is also occupied by AM-broadcast stations, however.)

VHS stereo also uses frequency modulated audio. In this system the audio-channel carriers are at 1.3 and 1.7 MHz.[7] Unfortunately, the lower-channel carrier is in the AM-broadcast band, and the upper-channel carrier is perilously close to the 160-meter amateur band. Even though FM is less prone to interference than some other modes of recording

[3]R. Goodman, *Maintaining & Repairing Videocassette Recorders*, Tab Books, Inc © 1983, p 1.

[4]*Consumer Guide, Video Buying Guide*, Publications International Limited, © 1985, p 23.

[5]W. Nelson, *Interference Handbook*, Radio Publications Inc, © 1981, second edition, fourth printing, 1990, p 247 (available from ARRL).

[6]J. Meigs, "Super Home Video," Jan 1988 *Popular Mechanics*, p 59.

[7]*Consumer Guide, Video Buying Guide*, Publications International Limited, © 1985, p 23.

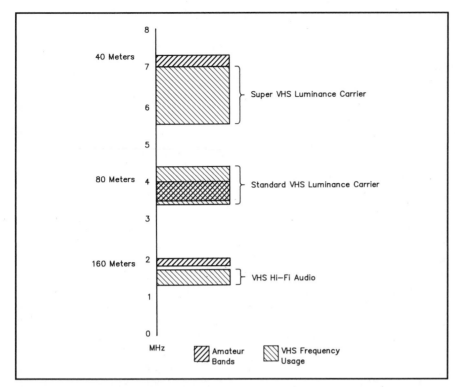

Fig 40—The VHS recording format uses frequencies in and near the US 160, 80 and 40-meter amateur bands. Beta recorders use similar, but not identical, frequencies.

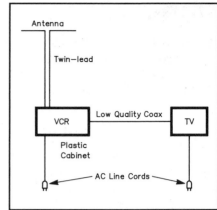

Fig 41—This is a typical home VCR installation. It has no protection against EMI.

might be, it is not immune to all interference. VCRs that record audio directly onto tape (as baseband audio rather than as FM signals) are susceptible to interference in much the same way as are audio tape recorders.

The relationships between frequencies used by VCRs and the amateur bands are shown in Fig 40.

CURBING INTERFERENCE TO VCRs

Although the technology of video recording may make interference suppression seem a hopeless task, the key to curbing VCR interference is isolation. Isolation is achieved through the proper use of shields, filters and grounds. The VCR installation shown in Fig 41 is probably susceptible to interference from the transmitters of several different radio services. The proper use of filters (shown in Fig 42) usually eliminates VCR interference.

Some Stock Cures Are Unsuitable For VCRs

Although interference to many home entertainment devices can be cured with the proper value of bypass capacitor or RF choke in the right place, interference

to VCRs is a bit more complex. If the frequency of the interfering signal is within the bandwidth of the desired signal, conventional bypass and filter techniques can have disastrous effects on VCR performance.

Suppose that a CW signal on the 80-meter amateur band is interfering with a VCR. Although installing capacitors in

the VCR might bypass the intruding signal to ground, the sync pulses of the video signal would be bypassed as well. *Bypass capacitors* in the rotating drum that contains the video heads might cause even greater problems if the balance of the drum were upset. Because the video signal occupies the same frequencies as some amateur HF bands, unwanted signals must be removed in ways that do not affect the desired signal: It is more appropriate to remove the 80-meter signal with a common-mode choke on the feed line or a filter on the power cord, or place the VCR in a shielded enclosure (as determined by the interference path).

Do not install *metal foil or screen shields* inside a VCR cabinet. While this technique is common with audio ampli-

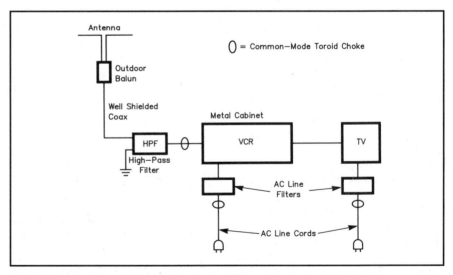

Fig 42—Proper filters and shields make VCRs more immune to interference. See Fig 44 for details of the common-mode chokes.

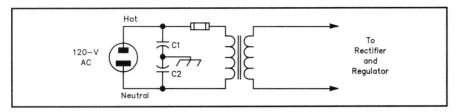

Fig 43—C1 and C2 form a voltage divider, which can present a shock hazard in equipment that does not have a three-wire line cord.

Fig 45—A common-mode choke using a snap-on core installed on twin-lead feed line. Install the filter as close to the VCR terminals as possible.

fiers and stereo receivers, they don't contain many moving parts. Since VCRs contain a multitude of moving parts, aftermarket shielding inside a VCR cabinet is extremely hazardous. If a wayward piece of foil or screen got tangled in the videotape, it could cause considerable damage. The EMI-RFI coatings mentioned elsewhere in this chapter are preferable shielding methods. The spray must be used with care, however, to reduce the likelihood of damage to the environment, individual or equipment.

Avoid the use of *capacitors across the ac line* as shown in Fig 43. Although this RFI cure might actually work on a VCR, it has several disadvantages. C1 and C2 form an ac voltage divider that can hold the VCR chassis above ac ground. Since home VCRs usually do not have three-wire line cords, there is a shock hazard should C1 short out. Furthermore, if the VCR is connected to a properly grounded CATV system, ac could show up on the coax shield. An ac-line common-mode choke (see Fig 44) is a much safer alternative.

It is common practice to connect a VCR to the antenna with twin lead when CATV is not used. There is nothing wrong with this practice so long as interference cures are not improperly applied. Use a common-mode choke as shown in Fig 45.[8] *Do not place a ferrite bead over each wire of the twin lead as shown in Fig 46:* the TV signal will be suppressed along with the interference.

Filter Shopping

When shopping for a high-pass filter, try to find a unit with appropriate connectors for the suffering VCR. This eliminates the need for additional adapters or balun transformers. Additional compo-

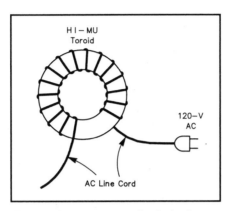

Fig 44—A common-mode choke for line cords. Wind 10-15 turns of the appropriate cord around an FT-240-43 ferrite toroid core.

Fig 46—*THE WRONG WAY!* Ferrite beads installed this way suppress the desired (differential-mode) TV signals.

nents may degrade performance because they do not present a proper termination to the high-pass filter. Also, use filters with metal cases. The metal is an extra shield against unwanted signals. Finally, install the filter as close to the VCR input as possible (as shown in Fig 47). High-pass filters are compact and easy to install. They are relatively inexpensive insurance against interference.

TCE Labs makes a filter (model #BX-#2S) that is effective against common-mode interference on coaxial cable.[9]

To Find A Cure, Understand The Interference Mechanism

Even though the design and construction of VCRs makes them susceptible to interference from radio transmitters, it is generally possible to reduce or eliminate the effects of the interfering signal. The first thing to do is to determine how the offending signal is entering the VCR.

Antenna/Feed Line

The TV antenna system is one of the most common VCR interference paths.

Fig 47—Commercial high-pass filters are compact and easy to install. Place them as close to the VCR terminals as possible.

[8]Palomar Engineers, *RFI Tip Sheet,* Palomar Engineers PO Box 455, Escondido, CA 92033.

[9]TCE Labs, see Supplier's List.

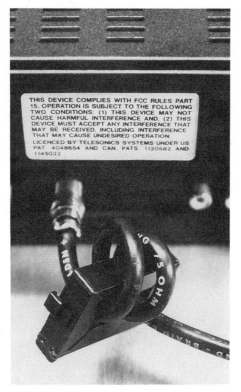

Fig 48—A common-mode choke using a snap-on core installed on coaxial cable. Install the filter as close to the VCR terminals as possible.

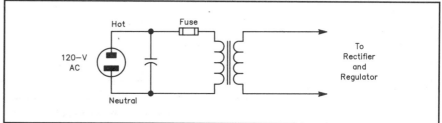

Fig 49—Most consumer-grade VCRs use a power supply with a two-wire line cord and a polarized plug. There is no on-off switch in the primary circuit of the power transformer.

VCRs connected to CATV systems can also accept interference from the cable, either as a differential-mode signal or a common-mode signal. You can determine whether interference is entering through the antenna connection by disconnecting the feed line and viewing a prerecorded tape while the amateur transmits. The accepted cure for this interference path is a quality high-pass filter installed at the VCR input. A common-mode choke is often necessary as well.

In many interference cases the unwanted signal enters the VCR via the coax shield.[10] This is called common-mode interference, and its cure is not the same as the cures for fundamental or harmonic interference. You can construct a ferrite toroid common-mode choke as shown in Fig 48; more turns of cable through the core give more suppression. In some stubborn interference cases, it

[10]ARRL Handbook for Radio Amateurs (Newington: ARRL, 1991) p 39-14.

may be necessary to use more than one choke. In some cases, it may be necessary to use a normal (differential-mode) high-pass filter as well.

Common-mode interference can also occur in systems using twin-lead feed line. When using twin lead, remember that both conductors must go through the core.

Fig 46 shows the wrong way to suppress common-mode signals on twin lead. Fig 45 shows the right way to suppress common-mode signals on twin lead.

AC Line

In some cases, interference reaches VCRs through the ac line. Although most VCRs have some filtering in their power supply circuitry, there is considerable room for improvement. Fig 49 is a simplified representation of a VCR power-transformer filter circuit. If you suspect the ac line as an interference path, wrap the ac-line cord around a ferrite rod as shown in Fig 50 (a common-mode choke).

To make the most of this simple filtering method, wrap the line cord around the ferrite rod in one smooth layer with no gaps between turns of wire. Use nylon cable ties to hold the line cord in place, and the filter is complete. When using this technique, place the ferrite rod close to the VCR as shown. This helps minimize line-cord RF pickup between the choke and the VCR.

Stubborn cases of ac-line interference may require a "brute force" ac-line filter (Fig 44). Only qualified electronics technicians should install components inside a VCR. Most states have licensing and certification requirements for home-electronic service personnel.

Direct Radiation Pickup

Direct penetration of the cabinet by

unwanted signals is not nearly the problem it once was. Many early top-loading VCRs were built in plastic cabinets. Newer video recorders are certified to comply with FCC Rules and Regulations, Part 15 (as a result of excessive radiation from RF modulators). When VCR manufacturers installed the shielding necessary to suppress modulator radiation, the shielding reduced direct radiation pickup.

Fig 51 shows the inside of a typical consumer-grade VCR. The metal box near the head drum contains much of the RF circuitry, and it is well shielded to prevent the exit and/or entrance of unwanted

Fig 50—A common-mode choke made by winding the ac-line cord around a ferrite rod. Toroids are better (higher permeability and self shielding) than rods, but rods are easier to wind. They are adequate for some EMI cases.

Fig 51—The inside of a typical consumer-grade VCR.

"MY COUSIN HOOKED IT UP THAT WAY"

Several years ago, a friend of mine received a VCR as a gift. Before long the neighborhood was abuzz with stories of a pirate TV broadcaster showing racy movies late at night. The neighbors were more amused and entertained than shocked, but my friend was embarrassed and confused as to how the movies she and her husband were watching could be seen on their neighbors' TV sets.

This happened because the VCR output and the outdoor TV antenna were parallel connected at the antenna terminals of the TV set. Whenever the VCR was used to watch a videotape, the signal was not only fed to the TV set, but also to the antenna—and radiated around the neighborhood.

The cure was very simple: Connect the antenna to the VCR input and the TV to the VCR output. This shows the importance of proper VCR connections. Follow the instructions in the owner's manual! Who connected the VCR to the TV the first time? The owner replied "My cousin hooked it up that way." The late night broadcasts stopped and none of the neighbors suspected a thing.—*John Frank, WB9TQG, Wisconsin Section Technical Specialist, Madison, Wisconsin*

signals. The Phillips-head screw just left of the large PC board holds down one end of a springy metal piece; the other end of that metal presses up against the bottom of the cover and grounds it to the chassis.

Even though newer VCRs have metal cabinets, many older plastic-cased units are still in use. If direct radiation pickup is the prime suspect in an interference complaint, there are two ways to RF proof a plastic cabinet. (Again, keep in mind that these internal cures should be performed only by qualified service personnel.)

The first involves careful use of an aerosol spray on the inside surfaces of the VCR cabinet. There are several antiRFI sprays on the market. One such spray advertises shielding ability of 35-50 dB depending on the frequency involved.[11] When using RFI shield sprays: (1) Exercise caution when removing the VCR from its cabinet, (2) Mask areas you don't want coated, and (3) Allow adequate dry-

[11]A spray EMI-RFI coating is available from GC-Thorsen (see the Supplier's List at the end of this book). It is catalog no. 10-4807. Beware: Such sprays do not adhere well to all plastics. Some cases may require special primers so that the conductive coating does not flake off and cause short circuits.

ing time before reassembly. Remember that the resulting coating is conductive. Make sure there are no component leads touching the coated areas (where they might cause a short circuit).

A second, much less attractive, way of shielding plastic cabinets involves constructing a metal shield around the outside of the VCR. Use perforated aluminum rather than solid sheets because VCRs need ventilation. (The small-diameter holes do not significantly affect the shield quality.) If a VCR shield enclosure is necessary, make it as RF tight as possible. Overlap the seams to help keep RF out, and use a "piano" hinge. Such exterior VCR shields are a last resort, but serious interference sometimes calls for serious shields.

Accessory Connections

Some, but not all, VCRs have input and output jacks for audio and video signals. These jacks are often on the back of the VCR, and videophiles sometimes leave patch cords attached when they are not being used. This practice is convenient, but it is also an invitation to interference. Cables can act as antennas and feed interference into the VCR.

Audio and video cables should be well shielded and equipped with proper connectors (adapters may add poor connections). Patch cords should also be unplugged from the VCR when they are not in use. A common-mode choke formed with about 10-15 turns of cable on a ferrite core will sometimes help. Audio cables can be bypassed as shown in the Stereos chapter.

INTERFERENCE TO VCR AUDIO

Interference to the audio portion of videotape recordings is often more complex than interference to audio tape recordings. VCR audio can be interfered with as part of the TV signal, after demodulation, during recording and during playback.

As discussed under Recording the Audio Signal, some VCR formats record separate audio tracks. Others use the audio to frequency modulate carriers that are then recorded via audio heads mounted in the same drum that carries the video heads. To cure audio interference in VCRs, determine how the offending signal enters the VCR and which part of the circuit is affected.

Audio Examples And Cures

If audio interference occurs when a VCR is used with a microphone and camera but not during playback of prerecorded tapes, the interference is probably affecting the audio circuitry or microphone as it might an audio tape recorder. Such interference to the audio circuitry of VCRs can usually be cured with the same techniques used for audio tape recorders: ferrite beads, RF chokes, and bypass capacitors (as described in the Stereos chapter).

On the other hand, VCRs that record audio as FM signals may also receive interference via record audio modulator or the playback audio demodulator: Suppose a VHS-stereo VCR picks up interference in one audio channel. Further suppose that there is a nearby AM-broadcast station at 1310 kHz. (VHS stereo uses FM audio with carriers at 1300 and 1700 kHz.) Even though FM is often considered immune to AM interference, the AM broadcast station is interfering with the VCR audio circuitry. To cure this kind of interference, use shields and filters to keep the unwanted signal from reaching the VCR by using techniques described elsewhere in this chapter.

Summary

When confronting audio interference in VCRs, determine which circuitry is affected and how the unwanted signal reaches that circuit. Remember that VCRs use frequencies in and near the AM broadcast band, international short-wave bands and amateur bands. Place filter and bypass components carefully so that they do not disrupt the normal operation of the VCR.

ACCESSORIES AND VCR INTERFERENCE

High-quality baluns, splitters, switches, combiners, and patch cords can make the difference between interference immunity and susceptibility. Some baluns are built in plastic boxes while others have metal enclosures. The same holds

true for splitters, combiners and switches. Most good-quality video accessories are built in metal enclosures, which provide much better shielding than plastic enclosures. Don't save money by using low-quality accessories; they are an open invitation to interference.

Patch Cords

Poorly shielded patch cords contribute to VCR interference by allowing unwanted signals to enter the system. All 75-Ω coaxial cable is not the same: Some cables (with a single braided shield) have as little as 70% shield coverage. Better cables (such as RG-6) use a foil shield together with a braided shield for 100% coverage.

CATV Converters

The converters and decoders used on some CATV systems are not very well shielded. Ideally, such devices should be built in RF-tight metal boxes. Unfortunately, many converters and decoders are built in plastic boxes, which offer no shielding. Although it might be possible to shield the inside of the plastic box with an antiRFI spray, this is not advisable if the device is sealed to prevent tampering. Do not attempt to shield or modify CATV-company owned converters or decoders to make them resistant to interference. That is the job of the CATV system operator.

If CATV equipment is the point of interference ingress, notify the CATV system operator of the problem. The responsibilities of system operators are defined in FCC regulations and explained in the CATV section of this chapter.

WIRELESS VCR REMOTE CONTROLS

The wireless remote controls that come with most VCRs (Fig 52) are not RF devices, but rather infrared devices. Hence, they are generally not affected by strong RF fields. In a few rare cases, however, the VCR-remote logic circuitry is adversely affected by strong RF signals. For example, if a VCR switches from "playback" to "pause" or from "record" to "rewind" whenever a nearby transmitter is keyed, it is reasonable to assume that RF is affecting the logic circuits that control the VCR mode.

Again, it is easier to keep RF out of the VCR than to cure the problem at the affected circuitry. Proper shields and a good ac-line filter should eliminate most cases of interference to VCR wireless

Fig 52—Typical VCR wireless remote controls operate with infrared, rather than RF. That's a blessing.

Fig 53—Will Eduardo, KA9YIT, experience interference to the camcorder if his neighbor (AC9J) calls CQ on 80-meter CW?

remote controls.

WIRED VCR REMOTE CONTROLS

Wired remote controls require a different approach than their wireless counterparts. The cable between the control and the VCR can act as an antenna and feed unwanted signals to the control circuitry. The simplest cure is to feed the cable through a ferrite bead, toroid, or ferrite core in much the same way that twin-lead feed line is treated to eliminate common-mode interference.

If the common-mode choke does not cure the problem, contact the VCR manufacturer for help. Other components should only be added as directed by VCR manufacturer service bulletins. Only qualified personnel should perform service on home-entertainment electronic equipment.

THE CAMCORDER DILEMMA

Regrettably, camcorders do not respond well to the interference cures described in this chapter. Since there is usually no antenna or ac power connected to a camcorder, most interference results from direct radiation pickup. Camcorder construction and use generally preclude home-installed shields (see Fig 53).

Can anything be done to eliminate interference to camcorders? Yes. Contact the manufacturer. When a camcorder functions as a shortwave receiver (as it does when receiving interference), it is defective or unsuitable for the use environment.

In desperate situations, an amateur can voluntarily alter amateur station operation. Careful selection of operating frequency and transmitter power can reduce or eliminate interference to camcorders.

AT THE AMATEUR STATION

Finally, the way in which an amateur station is assembled and operated can influence interference to VCRs. Proper RF grounding is important, as is an ac-line filter to reduce RF on the power line. Use the minimum power necessary for communication to reduce the possibility of interfering with VCRs.

TRANSMITTER TREATMENTS

Although the nature of the recording system used in VCRs makes them prone to interference, it is every amateur's duty to make sure that station transmitters don't radiate harmonics or spurious signals that could cause interference. Part 97 of the FCC rules and regulations states, "All spurious emissions from a station transmitter must be reduced to the greatest extent practicable. If any spurious emissions, including chassis or power line radiation, causes harmful interference to the reception of another radio, the licensee of the interfering amateur station is required to take steps to eliminate the interference, in accordance with good engineering practice."[12]

A Transmit Filter May Not Help

Many amateurs think that a low-pass filter between the transmitter and antenna will solve all interference problems. It may solve cases of broadcast television interference, but it probably won't reduce VCR interference. Since amateur 160- and 80-meter fundamental signals are within the VCR passband, no transmit filter can help.

For example, when an Amateur Radio station transmits in the 160-meter band, the second harmonic is likely to fall in the luminance portion of a VCR record or playback signal. Since the harmonic is within the passband of the low-pass filter, it is not attenuated. How can harmonics below the transmit filter cut-off frequency be eliminated? A properly designed and carefully tuned matching network (Transmatch) can sometimes suppress harmonics

[12]*FCC Rules and Regulations* Part 97.307(c).

by 20 dB or more. Information about the construction and use of matching networks can be found in *The ARRL Antenna Book*.

Transmitter AC-Line Filters

Install an ac-line filter as close to the transmitter as possible. If the filter is located any appreciable distance from the transmitter, the wires between the transmitter and the filter might still radiate RF. Remember, the object is to keep the RF off the power line and out of the air (except at the antenna). RF signals must be directed to the antenna and nowhere else.

SOME FINAL THOUGHTS

Some radio amateurs are under the impression that nothing can be done about interference to VCRs because of the frequencies they use to record the video. While it is true that VCRs use frequencies in and near amateur bands, interference to these devices can be controlled by the proper use of shielding and filtering to keep the unwanted signals out of the VCR.

Proper selection of VCR accessories can be a factor in VCR interference susceptibility. Top quality baluns, coax cable, splitters, combiners and switches help protect VCRs from unwanted signals.

The nature of FM video recording, and in some cases the FM audio recording, is such that interference control techniques used in other home entertainment devices like tape decks and turntables are not appropriate for VCRs. In some cases, improper interference control measures could actually degrade VCR performance or damage the unit.

Hints & Kinks

---

from January 1980 *QST*, p 53:
CURING HIGH-POWER TVI

I completely eliminated a very serious TVI overload problem that affected my RCA XL-100 TV set whenever I operated my 2-kW PEP rig in the 20-meter band. There has been no problem, however, with low-power operation. My method may help other amateurs faced with a similar situation.

With a dip oscillator meter set in the absorption mode and tuned to the band causing the greatest TVI, turn the transmitter on and "sniff" along the TV feed line with the dipper. (Start near the TV set.) No doubt you will find a very noticeable indication on the meter as a result of the standing wave developed on the TV feed line.

Add approximately 1/4 wavelength (based on the band you have chosen) of twin lead to the TV line and reconnect it to the TV set. Now sniff along the line until you find a minimum reading on the meter. This indicates a low-voltage point of the standing wave. Cut the line at this point and reconnect the TV set. You should find that the overload is minimized or eliminated. Be sure the antenna is connected to the TV set whenever you sniff with the dip meter.—*Sam Peck, W6CQR*

from November 1984 *QST*, p 55:

LOCATING CATV INTERFERENCE SOURCES

I have found what I believe to be an effective technique for locating leaks in cable TV systems. The method involves listening for the Doppler shift of the leaking signal as you drive by in your car.

The Doppler shift, in hertz, is simply the number of signal wavelengths that you cross per second. At 40 miles per hour, the Doppler shift on 2 meters is about 9 Hz. This is a small frequency change, but remember that you will hear twice this change as you pass the source. The frequency will be 9 Hz higher than transmitted as you approach, and 9 Hz lower when you pass it. I have found this shift to be plainly audible with my mobile SSB rig.

To track down a leak, tune your mobile SSB receiver to the channel-E video carrier (near 145.25 MHz). Switch on the noise blanker (if you have one), and adjust the RIT control to provide a low-pitched beat note (just high enough in frequency to be audible). This makes the frequency change easier to detect. When you drive past the leaky tap, the signal strength will peak on the S meter and there will be a sudden change in the beat note. If you pass close by the leak (such as on a pole by the street), the shift will be more rapid and the signal will be stronger than if the leak is farther away (as within a subscriber's house). If there are multiple leaks nearby, there may be an interference pattern that will be a little more confusing. This may take some experience to sort out.

With a little practice, you'll soon find that you can quickly isolate a CATV leak. At that point, you should report your results to the cable company so they can correct the problem.—*Phil Karn, KA9Q*

---

from September 1987 *QST*, p 42:

VCR TVI

I was recently informed that my 15-meter signal was causing RFI to a neighbor's VCR located several hundred feet away. While a friend operated my station, I verified the interference. We maintained contact by telephone and noted that the VCR motor would change speed, following the SSB modulation. Turning on my linear amplifier increased the severity of the interference.

To pinpoint the source of RF entry, I disconnected the coaxial cables from the TV receiver and operated the VCR by itself. The interference persisted. In an attempt to rule out signal entry via the ac line cord, I wrapped several turns of the cord through a 2-1/2 inch ferrite toroid; that didn't help. Apparently the RF energy was entering through the plastic VCR cabinet.

The customer service department of the VCR manufacturer (Mitsubishi, a model 105 CH) admitted that the motor control circuit (which is synchronized to the video signal for proper picture display) is very susceptible to HF interference and would require good shielding to eliminate the interference. There appears to be no simple solution to the problem. My neighbor's only recourse is to take the VCR to a Mitsubishi dealer who may be able to coat the inside of the plastic case with conductive paint, a project that I am not willing to undertake.—*Dale P. Clement, AFIT*

Chapter 7

Telephones

By Pete Krieger, WA8KZH
PO Box 82
Randolph, OH 44265

Before we heat up the soldering iron or start trying to fix telephone equipment, let's get some perspective on the conditions surrounding telephone interference and the applicable FCC regulations. Two major events have moved the old problem of telephone EMI into a new arena: Telephone industry regulations have been relaxed, and much modern telephone equipment contains sophisticated electronic circuitry.

History

Years ago, the telephone industry was heavily regulated. Before "deregulation," all telephone equipment in the US was owned, installed and maintained by the telephone company. Since the telephone system was proprietary, there was no question who would diagnose and correct interference problems. The telephone company was clearly responsible for all phases of the telephone system and its proper operation, from the main office to the customer telephone. Procedures to fix EMI problems were developed in telephone-company laboratories and documented for use by field-service personnel. *The Bell Systems Practices Plant Series Manual,* for example, provided detailed information on treatment of telephones in use at the time. Some of those procedures are still valid.

Then, it was easier to solve telephone EMI problems. The telephone equipment consisted primarily of rotary-dial, electromechanical telephones which could be desensitized to RF with simple bypassing methods. EMI problems in such older telephones can still be cured by installing 0.001-µF ceramic-disc capacitors across the carbon microphone elements.

The telephone company produced filters that could be installed at the customer's home, although they were more effective at broadcast-band frequencies than the ham bands. In stubborn cases, the telephone company provided replacement phones with RF-resistant transmission networks or used shielded twisted pairs for the line-drop and house wiring.

THE PRESENT SITUATION

Now, in most cases, the phone company no longer owns the telephones or the wiring in your home—you do. Because you own it, you are responsible for maintenance. You are now responsible for telephone equipment that, a few years ago, you weren't allowed to touch.

Telephone Company Responsibilities

The redefined role of the telephone company (now broken up into many service providers) makes telephone EMI a problem some companies choose not to confront. In most cases, the telephone company is responsible for the proper operation of the telephone system only up to the point where the telephone wire is connected to the lightning arrestor at your home. (The wire, whether overhead or underground, and the connection point at the house, are usually referred to as the "drop." In telephone-company jargon, this wiring point is often called the "demark.") Leased telephones and wiring maintenance contracts are exceptions (but not automatic assurance that you'll receive EMI assistance).

Even with the help of professionals at the telephone company, there's no guarantee of a cure for interference. The days of extensive laboratory testing and documentation of EMI remedies for company-owned home telephone systems are over. You may have noticed that very little new telephone-interference information has become available to amateurs since deregulation. It's a good bet that your local telephone-service department hasn't read anything new either! When repair personnel use outdated procedures on contemporary interference problems, EMI problems are usually not cured.

Customer-service policies vary widely, so you need to ask your local telephone company what they can and will do to help you with an EMI problem. If assistance is available, be sure to ask what it will cost. Telephone companies usually charge for service to customer-owned wiring or telephones. Interference investigations can fall into the same billing category as repair service.

The Role Of The FCC

The Federal Communications Commission does not require that telephones have any interference protection. In their publication *Interference Handbook* (1990 ed.), however, the FCC states that telephones (and other audio devices) that receive interference from radio transmitters are improperly functioning as radio receivers. See Fig 1.

The FCC *encourages* telephone manufacturers to consider radio-frequency interference (RFI, a term used for EMI caused by radio use) susceptibility in the design of their equipment. The Commission also *encourages* manufacturers to help customers resolve interference problems that occur after the telephones have been purchased.

Modern Telephone Equipment

Today's new generation of electronic

PART II

INTERFERENCE TO OTHER EQUIPMENT

CHAPTER 6

TELEPHONES, ELECTRONIC ORGANS, AM/FM RADIOS, STEREO AND HI-FI EQUIPMENT

Telephones, stereos, computers, electronic organs and home intercom devices can receive interference from nearby radio transmitters. When this happens, the device improperly functions as a radio receiver. Proper shielding or filtering can eliminate such interference. The device receiving interference should be modified in your home while it is being affected by interference. This will enable the service technician to determine where the interfering signal is entering your device.

The device's response will vary according to the interference source. If, for example, your equipment is picking up the signal of a nearby two-way radio transmitter, you likely will hear the radio operator's voice. Electrical interference can cause sizzling, popping or humming sounds.

Fig 1—Part of page 18 from the FCC *Interference Handbook* (1990 edition) explains the facts and places responsibility for telephone EMI.

telephone equipment is remarkably sophisticated in performance capabilities. With a host of special features, new phones have made obsolete the less glamorous (but much less interference-prone) instruments made in the past. As is all too common in consumer electronics, these devices usually include little or no internal protection against RF.

Electronic telephones offer many convenience features like DTMF pads, displays, memories, clocks, calendars, electronic ringers, audio amplifiers and so on. Such added features mean additional circuitry. Modern phones contain ample components to build several basic radio receivers! Ironically, a phone that costs more may be more difficult to provide with EMI immunity. Higher cost often indicates a greater number of special features using numerous diodes, transistors and integrated circuits.

Most telephones for the US market are now built overseas. ARRL Lab Engineer Ed Hare, KA1CV, recently purchased an inexpensive desk telephone that had only a manufacturer name and an FCC identification number. If there were an EMI problem with this telephone, there

would be no easy way to contact the manufacturer for assistance.

The FCC can supply manufacturer information from their files about registered ID numbers.[1] If you supply the FCC with the ID number, they can give you information about the manufacturer or US distributor.

Modern telephone equipment can be quite sensitive to RF signals or fields. When a telephone that is not affected by RF energy is replaced with a different telephone that is susceptible to RF, telephone EMI can result where none existed. Or, previously insignificant interference levels may become significant.

A Typical Telephone Installation

Fig 2 shows a typical residential telephone wiring system. Since deregulation of the telephone companies, telephone wiring is often installed by contractors.

[1]Write to Federal Communications Commission, Manager Part 68 Rules, Room 6106, 2025 M St NW, Washington, DC 20554, 202-634-1833.

While there are many competent qualified contractors, they do not have the support of dedicated labs and engineers as before deregulation.

The Lightning Arrestor

Telephone service enters a house at a grounded, fused lightning arrestor located outside at the house end of the telephone company drop. Years of exposure to weather, ground water, or basement moisture can cause corrosion or discoloration (onset of corrosion) of wires, junction boxes or components inside the lightning protector housing. If it is accessible, a good visual inspection usually reveals potential problems.

The Telephone Ground

Correctly installed telephone systems use their own ground rod (which should be tied to the power company safety ground) or are tied directly to the power safety ground rod. In the past, it was common practice to tie the telephone ground to a cold water pipe, either at the protector or inside the service entry. In fact, the telephone ground is usually a safety ground for the lightning arrestor only.

There have been some installations where the telephone ground is tied to a cold water pipe at one end of the home, and the water pipe is then tied to the power safety ground at the other end of the house. This very bad practice creates a large ground loop. Such ground loops are more susceptible to interference than a proper installation.

If you see evidence of an improper ground installation or a problem at the protector, contact the telephone company repair department. Local phone companies generally correct these installation problems without charge. If you're unsure who pays for a particular phone-system repair, ask first.

The Service Entrance

The service entrance is also referred to as an interface or connector block. From this point, telephone wiring is distributed outward to the phone jacks, usually in one of two wiring configurations: straight cable runs (parallel wiring) or loop series.

Wiring Styles—Straight Cable Runs

Parallel wiring (Fig 2A) uses a separate cable run to feed each jack in the house. The wiring style can be identified by looking at the service-entry connec-

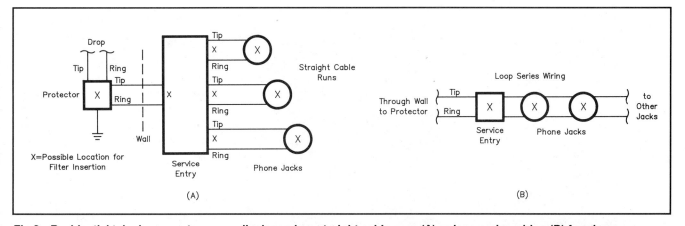

Fig 2—Residential telephone systems usually depend on straight cable runs (A) or loop series wiring (B) for phone interconnection. The owner's system connects to the telephone company via the pole-to-house *drop* (which may run underground in some installations); a grounded, fused *protector* that minimizes lightning damage to house wiring and telephones, and the *service entry—the terminal block* at which the phone wires actually enter the house. This drawing shows possible insertion sites (X) for the RF filters discussed in the text. *Tip* and *ring* are phone-company nomenclature for the talk-circuit wires, which date back to the days when telephone operators used phone-plug patch cords for call routing.

tions. When the number of used wire pairs leaving the service entrance equals the number of phone jacks in the house, the house is parallel wired.

Wiring Styles—Loop Series

Loop series wiring (Fig 2B) is the usual method. A single wire pair connects the service entrance to the nearest phone jack and continues to each of the system jacks. Except for the most distant jack (where the wires terminate), the cable enters and leaves each jack location.

Wiring Color Code

The color code is universal in modern telephone installations. The red and green wires are used for single-line telephones. For a second line, the yellow and black wires are used.

Unused Pairs Should be Grounded

Good telephone wiring practice calls for all unused conductors inside active cables to be grounded. This is done at the service entrance. Simply attach all unterminated wires to the system ground.

Ribbon Cable v Twisted Pairs

Flat-cable wiring (see Fig 3) is becoming common for telephone service. Such cable is not as good as a twisted pair for EMI control. It is prone to interference when subjected to radio signals. Use twisted-pair telephone wiring if possible.

CURING TELEPHONE EMI

WORK SAFELY! The risk of electrical shock from working with telephone wiring is minimal, but use good judgment when coming in contact with any active circuit. In the US, 24- to 48-V dc appears across the telephone pair at all times, and incoming ring signals are 20-Hz ac at 90 V or more.

Telephone wiring is often hidden from view as it travels inside walls, ceilings and so on. Although chances are slight, hidden wires could come in contact with live ac wiring. As a precaution, use a voltmeter to verify that the talk-circuit voltage is not abnormally high before working with a circuit. Protect yourself from any incoming rings by taking all phones off-hook.

Don't work on telephone wiring when lightning storms are in your area or if you have a pacemaker.

What Can Be Done?

As mentioned earlier, the telephone

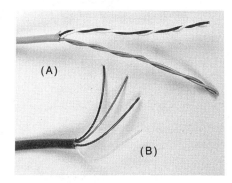

Fig 3—Flat telephone cable (B) is rapidly replacing the traditional twisted pair (A) in residential installations.

owner is now responsible for installation, maintenance and repair on the customer side of the lightning arrestor. The owner can connect external (plug-in) RFI filters to telephones or accessory equipment like answering machines and modems.

Do not attempt to internally modify a telephone instrument, or filter telephone wiring, that is owned by the telephone company. Only telephone company personnel may service telephone-company owned equipment.

Phone Modifications

What about home-brew EMI treatments inside a phone? There are several good reasons for owners to avoid such modifications:

1) Many new telephones are mass produced as "throwaways" that are not easily disassembled. Others permit partial access, but the "guts" are encapsulated or extremely difficult to reach.
2) By law, only registered telephone refurbishers may repair or modify phones in ways that affect telephone operating characteristics or requirements for FCC registration.
3) Anti-EMI techniques used on older phones may not work. Internal measures such as bypass capacitors may adversely affect modern telephone operation.
4) External (outside-the-telephone, Fig 4) RF filtering works nearly as well as internal modifications. This is an alternative to replacing or modifying most telephones. Ferrite-core chokes can

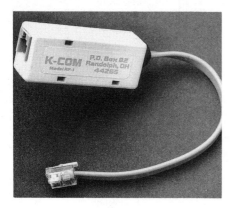

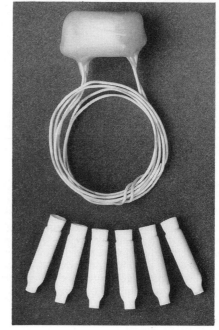

Fig 4—Typical telephone EMI filters. The filter at A has modular connectors for installation at a telephone. B shows a filter used behind wall jacks, at the service entrance and other places where modular connectors are inappropriate.

(A)   (B)

eliminate RF current on the wiring before it reaches the phone, or at least decrease the current to an insignificant level.

It is best to simply avoid internal modifications to telephones. Internal modifications should only be performed by registered telephone service personnel.

Good Phones

It may be worthwhile to replace EMI-prone telephones with less-sensitive older models. There are also new telephones that work well. The ARRL Technical Department maintains a Telephone-EMI Resource List of EMI-resistant telephones, filters and so on. A copy of this list is available from ARRL HQ Box RFI for an SASE.

Telephone Interference

Telephone EMI is the interception of radio signals by telephones. Although it is not intentional, telephones can act as simple radio receivers when components rectify (AM detect) radio signals and produce unwanted audio.

Desired telephone signals (voice, control and power) all share the same two wires (referred to as the "pair"). These are differential-mode signals which use one of the pair wires as the source and the other as the return conductor.

All radio services can cause telephone

EMI. Amateur Radio is only one facet of the problem. Persons living near AM-broadcast stations are familiar with interference from that source. Telephone EMI can result from any radio transmission of sufficient power on a frequency that telephones can detect.

There are exceptions, but the greatest likelihood of telephone EMI is from signals within the 0.5- to 30-MHz range. This results from several factors:

1) Telephone components that detect RF usually have a frequency cutoff somewhere in the upper HF or lower VHF range.
2) The resonances of the telephone wiring are usually more pronounced in the HF range. (Even small homes can contain hundreds of feet of telephone wiring.)
3) The field strength of VHF and UHF stations is generally weaker than that of HF stations. VHF/UHF stations usually run less power, and the attenuation offered by buildings and trees decreases field strength even further.

There are several components that contribute to telephone EMI: Telephones and accessories that improperly function as radio receivers (some worse than others), telephone wiring that functions as a radio antenna and radio transmitting sources. In some cases, improper RF reception (audio rectification) can occur outside

telephone instruments, as in faulty wiring.

Simplify The System

The fundamental principles of good troubleshooting are discussed in the Troubleshooting chapter. Take the time to review that chapter before you continue. Many generalized troubleshooting principles are not repeated here.

As outlined in the Troubleshooting chapter, *simplify the problem.* Start troubleshooting by disconnecting (unplugging) all but one telephone.

If there are different models of phones in the house, make comparisons to determine which are the least sensitive to RF energy on the telephone line. Modular connections make this easy. Test each of the phones at the same phone jack. By the process of elimination, you may identify a single susceptible telephone that is more trouble than it is worth—a result of either inferior design or component malfunction.

If EMI is present with each phone or accessory, suspect a wiring fault or external device. Once the simple system is EMI free, add devices back one at a time, eliminating problems as they appear.

At each step, listen for improvements on all telephones in the building. If, for instance, EMI is eliminated by reducing the RF field strength, no further work or expense is necessary. In some cases, one bad telephone (or other device such as an answering machine) may feed detected audio to all other telephones in the house.

There are a great number of variables from one site to another. You may need to spend time troubleshooting the fundamental cause of the interference problem. The good news is that solutions are possible and seldom technically complex. You need not be an electrical engineer or professional telephone troubleshooter to understand telephone EMI and apply the appropriate cures. With good information (found in this book) and a healthy dose of patience (you must provide that yourself!), you can solve interference problems.

Telephone Defects

Telephones can continue to operate despite some component failures. Some of these failure modes can affect the entire installation.

Faulty Telephone Equipment or Installation

Telephones can continue to operate despite defects in the telephone wiring—

CORDLESS TELEPHONES

Cordless telephones use radio frequencies. They are actually small two-way radio systems. As with any two-way radio, they are susceptible to interference from nearby radio transmitters, including other cordless phones. The FCC does not protect cordless telephones from interference. (They are Part 15 devices.) They are required to carry a label indicating that: (1) cordless telephones must not cause any interference to other services and (2) cordless telephones must accept any interference that is caused to them. The owner of a cordless telephone may be unhappy about any interference problems that may occur, but the law is clear about the whole matter. The owner should contact the telephone manufacturer for assistance. Some manufacturers may help.

Cordless telephones come in many styles and have several different features. Older cordless telephones, operating under a FCC waiver, use frequencies between the upper end of the AM broadcast band and the 160-meter amateur band. They are very susceptible to EMI and also cause harmonic interference to amateurs. Newer cordless phones use frequencies near 49 MHz. These phones work much better than the older ones, especially from an EMI standpoint.

Cordless telephones often include other features. For example, some are two-line phones, some include automatic answering machines, intercom systems, hands-free speakerphones, three-party conference features, memory features and so on. Such additional features are subject to EMI.

Cordless telephones can interfere with other cordless telephones when operated on the same frequency.

The illustration shows a block diagram of a typical cordless telephone system.

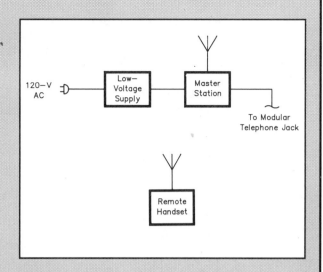

Cordless EMI Cures

The most common source of interference to cordless phones is through the telephone wiring. The unwanted RF is almost always a common-mode signal. To solve this problem, install a telephone RFI filter in the telephone line that connects to the modular jack on the master station. This normally solves the interference problem. If it does not help, contact the manufacturer, leave the filter installed and proceed to the next step.

The Power Supply

Master stations operate from the ac line, and common-mode RF signals can enter via that route. The power supplies contain rectifiers, which are possible sources of audio-rectification interference.

The power-supply lead is often about 7 ft long, and it may be resonant on 10 meters. If there is any excess length, wrap it in a small loop, which is less likely to act as a good antenna. If the trouble persists, install a common-mode choke where the power cord connects to the master station. Wind 10 to 20 turns of the power-lead wire on a (no. 43 material) ferrite toroid or rod. This should solve the problem. Some people have eliminated EMI by installing an ac-line filter at the ac outlet and a telephone RFI filter at the device. The chapter on Power Lines shows an appropriate ac-line filter.

Direct Radiation Pickup

Both the master station and the cordless unit may be subject to direct radiation pickup (especially if the unit is close to a transmitting antenna other than its mate). If the unit contains an audio amplifier, that can also pick up RF. Direct radiation pickup is difficult to solve, but sometimes the master station can be relocated to another area of the home (as far away from the other transmitter as possible). If this does not solve the pickup problem, contact the manufacturer for a remedy.

The Cordless Handset

The cordless-telephone handset of 49-MHz units is not often a problem. Older units (on 1.8 MHz) may have more problems. Problematic older systems should be replaced with newer equipment.

Other

Cordless telephones can have other problems that might be confused with actual interference. When the batteries are low, range is reduced, audio sounds distorted, and the phone is more susceptible to interference. A broken antenna reduces the range and produces "scratchy" audio. This also occurs when the handset is operated too far from the base station.—*John Norback, W6KFV, Santa Barbara Section Technical Coordinator*

AUTOMATIC TELEPHONE ANSWERING MACHINES

Answering machines come in many styles and often include many features. Some are stand-alone devices that are used with an external telephone; others have a telephone instrument built into them. The illustration is a block diagram of a typical installation.

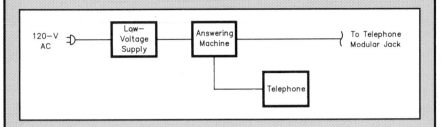

EMI Cures

The Power Supply

Answering machines operate from the ac line, and common-mode RF signals can enter via that route. The power supplies contain rectifiers, which are possible sources of audio-rectification interference.

The power-supply lead is often about 7 ft long, and it may be resonant on 10 meters. If there is any excess length, wrap it in a small loop, which is less likely to act as a good antenna. If the trouble persists, install a common-mode choke where the power cord connects to the machine. Wind 10 to 20 turns of the power-lead wire on a (no. 43 material) ferrite toroid or rod. This should solve the problem. Some people have eliminated EMI by installing an ac-line filter at the ac outlet and a telephone RFI filter at the device. The chapter on electrical devices shows an appropriate ac-line filter.

Direct Radiation Pickup

Radiation pickup may affect the answering machine or an attached telephone. A shielded enclosure should help, but an enclosure is inconvenient when the machine includes an integral phone. Shielded enclosures are often unacceptable for convenience and appearance reasons.

Other

RF signals can be picked up by the line that connects the telephone to the answering machine (watch for resonance). Install a telephone EMI filter in the line. It may be necessary to install one near the telephone and another near the answering machine.—*John Norback, W6KFV, Santa Barbara Section Technical Coordinator*

---

or installation errors made recently or decades ago! (Telephones with internal component failures can also continue to operate!) If the twisted pair reaches the telephone unbroken, the telephone will usually function normally, except for some noise. A telephone user may never notice deficiencies that increase interference susceptibility.

As a result, interference can be more severe at one location than another when both are exposed to the same RF field. For example, EMI might be worse at a neighbor's house than at a ham's—even though the antenna is closer to the ham! The neighbor could have a number of wiring and installation problems, but still have telephone EMI only when the ham transmits. Unfortunately, this sometimes leads to the erroneous conclusion that the amateur is at fault.

Telephone EMI and Nearby Transmitters

Check to see if the interference is really related to transmissions. Sometimes telephone EMI problems are related to power lines or crosstalk within the telephone-company circuitry. If the EMI does not follow transmission patterns, and you can demonstrate that to the complainant, relax. At this point, the matter is the telephone company's problem.

When interference begins and ends with RF transmissions, RF is somehow involved in the problem. This does not mean that the transmitter is *causing* the problem. Since telephone equipment is not intended to receive radio signals, the affected equipment is not functioning properly. Continue troubleshooting to discover the exact telephone or device that causes the problem and apply a remedy there.

In some strong-RF situations, changes at the transmitter can favor the telephone system. (These changes are discussed later under "Strong RF Environments.") A station licensee may *elect* to make changes to improve the situation, but that decision does not establish cause or responsibility.

Bad Hook Switch

A defective cradle microswitch (in older phones, a mechanically complex "switchook" driven by a cam) can cause EMI without disabling the telephone. When the handset is on-hook (hung up), switches are supposed to disconnect the voice circuits from *both* sides of the wire pair. This completely removes the telephone transmission network and handset from the system. By design, a telephone that is hung up leaves only the ringer across the line.

If one of the cradle switches fails and the telephone remains connected to one side of the pair, watch out! Most telephones use a 12-ft coiled cord for the handset. If one of the switches fails to open, this leaves 24 ft of wire connected to one side of the line. This unbalances the system and leaves RF-sensitive components in the telephone connected to the line! The telephone may detect RF when it is hung up and put the rectified audio on the line. The audio then appears throughout the entire installation.

Further, the telephone could operate properly when the handset is lifted off the cradle (off-hook). Therefore, it is of no value to pick up the phone and listen for interference while troubleshooting.

"Talking" Telephones

A neighbor may complain that voices are heard on one or more telephones that are hung up. A phone should not produce audio when on-hook because the handset should be completely disconnected from the lines. Such phones are either subject to direct pickup or still connected to the lines.

Simplify the system to one phone, and then reconnect one instrument at a time. Listen for interference as each instrument is added, and pay particular attention to minor variations in signal levels. Telephones with ac-line connections are prime suspects, followed by those phones with the most electronics. Some electronic telephones keep talking even when disconnected from the telephone line!

Dirty Contacts

Dirty hookswitches or dirty microphone contacts cause noise that is often interpreted as interference. The noise sounds like static or a hissing/frying sound. It is most bothersome in humid areas. The cure is to clean the contacts.

The Twisted Pair— Built-in Interference Rejection

Aside from technically advanced telephones and the switching equipment, the rest of the system might be described as "low-tech." Telephone engineers, nonetheless, established some brilliant design criteria for basic telephone systems, particularly with respect to interference rejection.

Consider the telephone cable itself— commonly referred to as the "pair" or "twisted pair." Twisting the wires results in nearly complete cancellation of induced differential-mode signals. (A common-mode signal may be induced on the pair, but such signals are easily removed by common-mode chokes.) When a twisted pair is combined with the common-mode rejection inherent in old-style telephones (because all of their circuits are balanced), external signals are effectively rejected. This rejection of external electromagnetic fields is significant because telephone wiring is often installed near electrical wiring, motors and so on.

Twisted-pair phone systems with old telephones can pass through noisy environments and still remain quiet and interference-free. This benefit is lost if the twist in a phone wire is physically or electrically eliminated by damage, unbalance, or installation of non-twisted wire. The benefit is also lost if the phone responds significantly to common-mode signals, as do many modern phones.

In telephone jargon, "straight-line induction" then takes place and unwanted signals may be heard on the phone. Straight-line induction can also be the result of not-for-telephone conductors such as thermostat wire, speaker wire or ac "zip" cord. Unqualified personnel may have used such wire to add extensions or to replace telephone wire! Use only matched pairs inside multi-conductor telephone cables.

Unfortunately, much residential telephone wiring does not use twisted pairs. Untwisted wires are sometimes used even by telephone companies.

What To Listen For

Anyone with a "trained ear" (that can pull a weak DX signal out of the "mud" on an Amateur Radio receiver) can hear symptoms of straight-line induction in telephone circuits. Induced power signals, for example, are often heard when the amateur station is not even switched on. Such signals in the telephone system will sound like a hum or sometimes a "pop" accompanying current inrush to an appliance.

If the ac circuit feeding a transmitter induces current on the pair, load variations can modulate the induced signal. This can make it appear that the interference is a result of the transmitted signal.

Determine whether the interference is a result of the transmitted signal by connecting the transmitter to a well-shielded dummy load and transmitting. If the interference remains, it's not associated with the presence of RF, and must be caused by something else.

System Unbalance

Telephone systems use balanced transmission lines to efficiently transfer audio and control signals over long distances. When a condition in a telephone installation upsets critical system balance, the system is more susceptible to EMI problems.

As described earlier, a defective hookswitch can unbalance the system by leaving many feet of wire connected to one side of the pair. It's a coincidence, but Murphy's Law is always working in Amateur Radio: Popular telephone cord lengths come very close to resonant antennas for ham bands between 10 and 20 meters! It may help to change the cord length.

Other causes of unbalance include permanent or intermittent connection of a ground or unused cable wire to one side of the pair. Such connections can be caused by wiring errors, wire-insulation breakdown, staples driven through a telephone cable or poor connections at joints or terminations.

Audio Rectification

This is the most common form of telephone EMI. It occurs when some component attached to the phone line rectifies (detects) RF and places the detected audio on the phone line. Filters are usually quite effective at eliminating interference caused by RF on the phone lines, once the detector is found.

RF signals are likely to reach the phone through the telephone wiring because the telephone wires form a large radio antenna (Fig 5). Induced RF signals usually travel along telephone wires in the common-mode (with all the pairs in the cable acting as if they were one wire), with the return path through earth ground. The radio signal is traveling on the telephone wiring but it is not part of the balanced circuit.

The detection process that results in

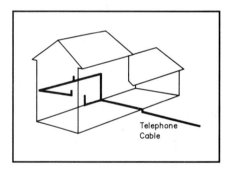

Fig 5—The wiring of a typical home installation forms a large random-length antenna.

telephone EMI can occur anywhere in the telephone system. A poor or corroded connection can actually convert an RF signal to audio with no help from a telephone instrument. If detection occurs outside the telephone (on the telephone-company lines or in a line amplifier, for example) "foreign" differential-mode audio signals are heard along with the desired audio. It is impossible to filter such foreign signals because they are at the same frequency as the desired signals.

Locate the Detector

It can be difficult to discern the difference between externally rectified signals and those generated in a telephone in the

TELEPHONE MODEM/FAX INSTALLATIONS

Telephone modem and fax installations are becoming commonplace in the home. (Modern fax machines exchange digital data, so their EMI problems can be treated just as those of modems.) Modems are most often associated with computers. EMI to computers is covered in another chapter of this book. Here we will deal with modem EMI only as a telephone accessory. Modem installations fall into two categories: internal and external.

Internal modems are located inside the computer cabinet. They draw power from the computer supply and their data is placed directly on the computer bus. So long as the modem was properly designed, it should suffer no EMI from its computer. An internal modem is connected to the outside world only through the phone line, so that is the only place where EMI can enter.

External modems are located outside the computer cabinet. They are usually powered from the ac line. The phone line carries audio to the modem, and a serial-communications cable carries data from the modem to the computer. Thus, most external modems can receive interference over three paths.

"Pocket" modems are a special class of external modems. They are enclosed in a case that is nearly as small as the data connector (small enough to carry in a pocket). Although external to the computer, pocket modems are battery powered, and they connect directly to the computer serial port with no cable. From an EMI viewpoint, they are equivalent to an internal modem.

EMI Cures

The most common source of interference to modems is through the telephone wiring. The unwanted RF is almost always a common-mode signal. To solve this problem, install a telephone RFI filter where the telephone line connects to the modular jack on the modem. (Modems sometimes use two telephone lines to send and receive data simultaneously. In such cases, filter both lines.) This normally solves the interference problem. If it does not help, contact the manufacturer.

The Power Supply

Many external modems operate from the ac line, and common-mode RF signals can enter via that route. The power supplies contain rectifiers, which are possible sources of audio-rectification interference.

The power-supply lead is often about 7 ft long, and it may be resonant on 10 meters. If there is any excess length, wrap it in a small loop, which is less likely to act as a good antenna. If the trouble persists, install a common-mode choke where the power cord connects to the device. Wind 10 to 20 turns of the power-lead wire on a (no. 43 material) ferrite toroid or rod. This should solve the problem. Some people have eliminated EMI by installing an ac-line filter at the ac outlet and a telephone RFI filter at the device. The chapter on electrical devices shows an appropriate ac-line filter. If interference persists, consider switching to an internal modem.

The Serial-Communications Cable

EMI can enter external modems through the data-communication cable. Since the cable carries data that could be affected by filters, a shield is the only option. Most quality cables are shielded. If the cable is not already shielded, replace it with a shielded cable. — *John Norback, W6KFV, Santa Barbara Section Technical Coordinator and Bob Schetgen, KU7G, ARRL Staff*

---

building. In both cases, the audio varies with the radio transmitter modulation. The major difference is that externally rectified audio is present in the telephone circuit even when there are no telephones or accessories connected to the line. Keep in mind that any telephone or accessory can couple detected audio onto the telephone lines, resulting in interference to all other devices connected to that line. External audio-rectification interference can still be heard on a completely RF-proof telephone!

If audio rectification is present, the detector must be located to effect a cure. The simplification procedure should indicate defective telephones and accessories.

If no instrument or accessory is indicated by simplification procedure, visually check the physical integrity of as much of the system as you can. Start at the lightning protector, move to the service entry and continue to the connector block. Then check all accessible phone-cable runs and the conditions inside each phone jack.

Look carefully for physically or electrically poor connections, corrosion, moisture, wire-insulation breakdown, staples driven through cables, stretched or otherwise damaged cables. You may also find wire other than telephone wire. Unqualified personnel may have installed speaker or thermostat wire, or other not-for-telephone conductors. Improper wire will pick up more RF energy than a properly balanced twisted pair. Replace such incorrect wiring if possible.

Devices that are not FCC approved should not be connected to the telephone system wiring. They may cause interference by unbalancing the line or acting as an RF detector. If practical, eliminate the defect that causes the problem; if not, install a telephone EMI filter to prevent RF from reaching the detector.

If the telephone installation uses parallel wiring, it might help to disconnect cables to unneeded telephone jacks. Dis-

connect unneeded pairs from the service entry and connect them to the telephone ground. This practice reduces the size of the telephone-wire antenna and usually the amount of RF picked up (especially if one or more of the unused wires is resonant on the transmit frequency). This may result in an immediate solution to the problem.

Direct Radition Pickup In Telephone Systems

It is possible for components in the telephone system to demodulate the RF energy produced by a radio transmitter. Some telephones and accessories are so sensitive that they produce audible interference even when on-hook or disconnected from the telephone lines. If equipment demodulates RF (audio is heard from the earpiece or speaker) when hung up, it may be:

1) defective. The voice circuits are not

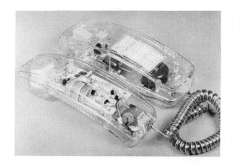

Fig 6—This phone is transparent to light, so we can see through it. If we could see RF, most phones would look like this!

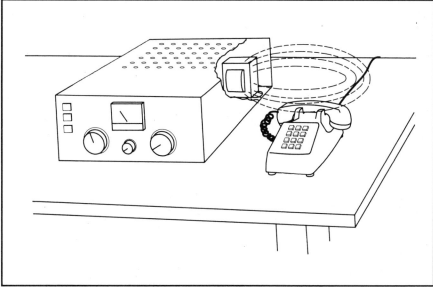

Fig 7—Magnetic induction can couple signals into telephone equipment. Separate the phone and the magnetic-field source to eliminate the problem.

disconnected from the line when on-hook. Repair or replace the instrument.

2) directly picking up RF (see Fig 6). Little can be done to remedy this. A shielded enclosure and telephone EMI filters should help, but an enclosure may be impractical. Replace the instrument with one that does not directly detect RF fields.

3) picking up RF from the ac line (if the instrument is ac powered). An ac-line filter should help.

Magnetic Induction

Audio-frequency electromagnetic fields can couple into not-for-telephone wire (and equipment) by magnetic induction (Fig 7). The most common magnetic-field sources are the house ac wiring or transformers inside electrical equipment. When induction occurs in telephone wiring, the best cure is to replace improper or damaged wire with a good quality telephone cable. If fields are coupled into equipment, physically separate the equipment and field source.

An Example—Telephones Near Amateur Equipment

Fully outfitted amateur stations often include a telephone. One operator experienced telephone interference for years—always assuming his transmitted signal was getting into the phones. Upon investigation, the problem turned out to be magnetic induction (affecting the whole telephone system) through one telephone on the operating desk!

The phone was physically close to a high-power amplifier. Hefty magnetic fields are generated by transformers in high-voltage or high-current power supplies, and some models of telephones are

easy targets for induction! (Older telephones that use induction coils for electromechanical bell ringers are especially prone to pick up nearby fields.) The telephone was picking up the 60-Hz electromagnetic field from the high-voltage transformer. The phone was modulated as the operator spoke, thus disguising the true nature of the problem. The simple solution to this sneaky problem is just to move the phone away from the transformer.

The Amateur Station And "Hot" RF Environments

Things can be done at an amateur station (other than going off the air or reducing operating power) to reduce RF on telephone lines. In no case is telephone EMI the fault or direct responsibility of the amateur. Sometimes, however, the easiest cure is the best cure. If moving an antenna a few feet can solve a neighborhood (or family) dispute, that may be the best solution.

We amateurs create our own devil, so to speak, with antennas near the house, on the house and sometimes in the house. This results in strong RF fields at the telephone installation. It is much better to send the signal skyward (to the other end of the QSO) than to a neighboring telephone!

Even a slight increase in the distance between antenna and house could make a big difference. It is not practical to move a tower to solve an EMI problem, but moving one leg of an inverted-V is an easy way out. This is not an ideal solution

because it does not cure the fundamental problem, which is a defective telephone or installation. It is mentioned here so a ham can consider all alternatives.

Telephone wiring picks up RF energy from nearby transmitters. In some cases, telephone wiring acts as part of the transmitting antenna! Telephone wiring that is less than about 0.3 λ from an antenna and nearly parallel to it may act as a parasitic element in the antenna system. The telephone wiring would then contain a lot of RF energy.

Cable runs in the upper levels of a home are particularly susceptible to RF pickup. Place telephone wiring near ground level (in the basement or crawl space) to reduce RF on the lines.

Transmission-line radiation can contribute to an EMI problem. Do not place transmission lines close to telephone wiring. Feed-line radiation is more of a problem with dipole antennas, when the feed line does not leave the feed point at 90° to the antenna. If the feed line must pass close to telephone wiring: (1) Use high-quality coaxial feed line. (2) Use a balanced antenna with a balun at the feed point. (3) Ground the feed-line shield where it enters the house. (4) Effectively ground all station equipment. All of these practices may help reduce feed-line radiation.

THE SOLUTION—RF CHOKES

Chokes made with ferrite materials are the most effective devices for reducing common-mode RF current on tele-

Home security systems are often connected to the telephone network by means of an automatic dialer, which can be a two-way source of EMI. The automatic dialer and alarm system master station are normally located in a hidden, secure area of the home. Alarm cabinets are usually locked. Security-system details are beyond the scope of this book, so we shall consider them as single-port systems. The illustration shows a typical alarm system connected to an automatic dialer.

Alarm/Telephone EMI Cures

The figure shows great potential for two-way RFI. These systems can *cause* TVI and EMI to broadcast radios and telephones. The main sources of TVI/EMI are the rectifiers and (in the case of the automatic dialer) a polarity-guard circuit. It is best to solve the telephone EMI problem first because of effects from the polarity-guard circuitry in the automatic dialer.

The telephone circuit to the automatic dialer is normally hard wired to terminals inside the unit (it does not use modular jacks), so modular telephone EMI filters cannot be used. Fortunately, appropriate filters are available for use with wall telephones and installation behind telephone jacks. Since the EMI-causing electronic equipment is located inside the automatic dialer, install filters as close to the dialer as possible.

Next, filter the two-wire circuit between the alarm master station and the automatic dialer. Use a ferrite choke (43 material), and install it as close to the dialer as possible. These two steps solve most alarm/telephone EMI problems.

If EMI remains, apply common-mode chokes or ac-line filters at the line cord. Where separate low-voltage supplies are used, common-mode chokes may be needed on the low-voltage leads as well (see the sidebars about cordless phones or answering machines for details of low-voltage supply filtering). Since alarms are constructed to operate during ac power failures, there may be complex

and tenuous battery charging and switching circuits. The chief EMI threat in such circuits comes from audio rectification in switching or rectifying diodes. Read the

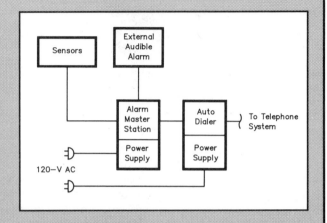

discussion of audio rectification in this chapter and the chapter about external causes of EMI for information about locating and treating audio rectification.

Alarm-TVI/BCI Cures

To solve alarm-TVI/BCI problems, install a ferrite choke (43 material) at each alarm sensor (as close to the sensor as possible). This also prevents common-mode RF signals from falsely triggering the various alarm sensors. Be sure to treat all alarm sensors in the home.

If the lead from the external audible alarm to the alarm master station suffers from common-mode RFI, install a ferrite choke (43 material) at the master-station end of that line. This prevents RFI from reaching the alarm master station by that route.—*John Norback, W6KFV, Santa Barbara Section Technical Coordinator*

phone wiring. Chokes accomplish this by presenting a high impedance to RF in series with the telephone wiring. In order for a choke to act as a filter, there must be a dissipative RF path available with lower impedance than the path through the choke and telephone equipment. (Often the dissipative path is the phone line.) If the EMI is not severe, simply wrap the telephone cord around a ferrite rod or toroid core (about 10-15 turns in a single layer). This forms a low-performance RF choke.

Bifilar Chokes

Bifilar chokes are preferable to separate chokes for each side of the pair because bifilar chokes maintain phone-circuit balance. Core permeability varia-

tions are of no consequence in bifilar chokes because both conductors are wrapped around the same core. Hence, circuit balance is ensured.

A bifilar choke has no adverse effect on telephone operation. This is an important consideration in moderate and severe cases of EMI, where several chokes may be required (spaced several feet apart) to eliminate resonances in the telephone wiring. Fig 8 shows a broadband bifilar choke with inductive characteristics appropriate for phone-line filtering in the 3- to 30-MHz range.

Installing Filters

The most convenient and effective filter location is usually the modular

female connector at the telephone or phone jack. Some telephones have most of the circuitry (including the keypad) built into the handset. Locate a filter close to the handset with phones of this type.

If interference is partially reduced by installing a filter at the telephone, additional filters placed in series may help. If interference is only slightly reduced by installing a filter at the telephone, this is probably because the telephone (and filter) are at an RF high-impedance point in the telephone-wiring "antenna." Additional filters will be required to reduce the interference, and they should be located several feet (ideally 1/4 λ) away. Additional filters placed at a RF low-impedance point are more effective. Find effective filter locations

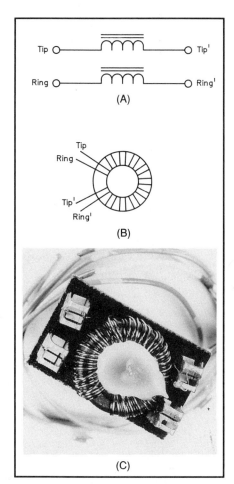

Fig 8—Simple RF chokes like this can help stop RF flow on telephone wires. Schematically (A), the choke consists of two 1.7-mH windings (25 turns of no. 30 enameled wire wound over 80% of the circumference of a 1/2-inch OD, ferrite core). Use mix-75 cores from 500 kHz-10 MHz, mix-43 cores from 3-30 MHz.

B shows how the two windings are wound together *(bifilar)* on a single ferrite core. To make construction easier, twist the wires together (five to ten twists per inch) and wind them as a single wire. Once the choke is wound, use an ohmmeter to check for continuity through each winding and label the ends.

C shows a commercial bifilar choke mounted on a board for potting. A dab of silicone adhesive (such as GE RTV) holds the completed choke to its carrier board. K-COM also sells phone-line filters (model RF-1) equipped with modular phone connectors; contact K-COM for details.

by trial and error.

Don't Give Up

When the first filter is not effective, some people conclude that the problem can't be solved and give up—Don't!

Filters that are properly designed and built do offer RF attenuation, and enough attenuation will resolve an induced-RF problem. There must be some other reason that the filters do not reduce interference.

There are two likely explanations. The filter might be (by chance) at an RF high-impedance point in the circuit. The filter impedance is in series with the system impedance at the point of installation. In order for a filter to attenuate unwanted signals, the filter must have an impedance much greater than its termination impedance. This is not the case at a high-impedance point in the circuit.

There is another possible explanation. The RF-current level at the filter could be so great that the filter core is saturated. When saturation occurs, any current increase diminishes the inductance. For example, a filter that works quite well at low power levels may not work when the RF level at the filter exceeds a few watts. If you need to filter a telephone line in a high-RF-power environment, ask the filter manufacturer for information about the power-handling capability of the filter.

It is possible to build filters for high-current applications, but the required core size can be prohibitive. Another approach is to reduce the current-handling demand placed on the filters by reducing current level in the telephone wiring.

As you recall, telephone wiring acts as a large antenna. Our goal is to find a way to make this "antenna" absorb as little RF energy as possible. There is a way to do this.

Breaking Up The Phone Line Electrically

The task is to make the phone line less responsive to RF. Let's take an approach often used with conductive guy wires. Tower-guy resonance at amateur frequencies is eliminated by breaking the wires into electrically short, nonresonant lengths. A telephone system must be continuous at AF, so we'll discourage RF-current flow in the system with high-impedance RF filters.

A Shotgun Approach

In a parallel-wired telephone system, a "shotgun" approach places filters at the beginning and end of each cable run (the fast way). A similar approach places a filter at each jack in a loop-series system. (Pay attention to the wiring at the jack to make sure your filter is installed in the series loop, not just in the wires terminating at the jack.) This approach solves nearly all induced-RF problems quickly, but it is costly: chances are that not all of the filters are necessary.

A Systematic Approach

A systematic approach identifies and treats specific RF-entry points; it is relatively slow, but more precise. This approach is better because it draws attention to sections of wiring (and/or connected telephones) that contribute to, or cause, interference.

If there is a telephone EMI filter at the telephone or telephone jack, leave it for now. Install the first additional filter at the telephone service entrance.

Check the interference level on all bands. If the interference is gone, congratulations! If not, don't give up! Even if a service-entry filter doesn't reduce or cure the EMI, that filter narrows the search by eliminating RF (but not AF) signals that might enter via the service drop. Continue by grounding all other cables (except the one already filtered) at the service entrance. The filters already installed should solve all but the most severe telephone EMI cases. If interference remains, physically break any long cable runs at one or more locations and insert additional filters. (When cutting a long cable to insert a filter, remember to maintain the dc continuity of all cable wires through the filter.)

By this time, there are several high-impedance filters between the RF signal and the telephone. Even severe interference should be eliminated. If not, suspect direct radiation pickup or audio rectification.

When interference is removed from the single cable, it's time to reconnect, one at a time, the cables removed earlier at the service entrance. Continue filter installation as described above, listening for interference on the reference phone and other phones as they become active. Work your way through to the final cable and the job's done!

Summary

Past troubles in the treatment of modern telephone EMI resulted from a lack of understanding. Solutions designed for carbon-mike rotary-dial telephones (such as bypass capacitors) do not usually apply to modern telephones, which are more sensitive to RF signals.

Modern phones exhibit EMI for three

reasons: (1) direct radiation pickup, (2) audio rectification and (3) magnetic induction. Direct radiation pickup is either a design defect or malfunction of the telephone instrument, which should be taken to the manufacturer. Audio rectification results from a defect on the line; it can be eliminated by correcting the defect or proper application of common-mode RF filters. Magnetic induction is a problem of proximity that can be eliminated by moving the magnetic source away from the phone line.

TELEPHONE EQUIPMENT MANUFACTURERS

Here is a list of some telephone-equipment manufacturers. Several of these companies provide modular AM filters, free of charge. Most of the filters are *not* effective against HF, however, because they are designed for AM-broadcast frequencies. Many manufacturers make internal repairs to their phones to eliminate RFI. Most do this free of charge (except shipping charges) while the phone is under warranty. If the warranty has expired, there is a charge for repairs. Contact the manufacturer by phone before shipping any instrument. Find out exactly what information is needed to make the repairs. *Many manufacturers caution that some interference problems CANNOT be fixed, or that*

the cost of the repairs may exceed the value of the phone. In such cases, it is best to replace the faulty phone with another that is less susceptible to interference.

For manufacturers that are not listed here, contact the main branch of your public library. The reference librarian can look up addresses and phone numbers of most corporations in Standard and Poor's *Directory of Corporations*. Or, try the Electronic Industries Association (EIA).[2] The EIA also has a Consumer Complaint Assistance Program and acts as liaison between manufacturers and customers in service disputes.

Telephones

AT & T
   Contact the nearest Phone Center Store, or call the National Sales & Service Center at 1-800-222-3111.

Radio Shack (Tandy)
   Contact the nearest dealer for assistance.

General Electric (GE)
   Contact the GE Answer Center at 1-800-626-2000 (open 24 hours, 365 days

[2]Electronic Industries Association (EIA), 2001 Pennsylvania Ave NW, Washington, DC 20006, 202-457-4977.

a year).

Phonemate
   Call 1-800-247-7889 or 213-320-9810.

Sanyo
   Call 1-800-524-0047 extension 502 (Customer Service).

ITT
   Call 1-800-526-4262 or 601-287-5281 and ask for "technical assistance."

General Telephone and Electric (GTE)
   Contact the nearest GTE Phone Mart, or call 611 Repair Service.

Telephone EMI Filters and EMI-Proof Telephones

K-COM
Box 82
Randolph, OH 44265
216-325-2110
(telephone EMI filters)

TCE Laboratories
5818 Sun Ridge
San Antonio, TX 78247
512-656-3635
(telephone EMI filters, and EMI-proof telephones)

Hints & Kinks

from June 1988 *QST*, p 50:
TELEPHONE RFI—THANKS!

I experienced telephone RFI, which I minimized by using homemade toroidal filters. Severe interference from a local broadcast station transmitter, however, defied all curative attempts. After reading Matthew Bell's letter[1], I investigated my telephone setup. I found at least 20 wires (of which only two were used) running through each of seven telephone jacks, all of which were dead-ended at the kitchen wall-phone jack. Connecting

[1]M. Bell, "Telephone RFI," Technical Correspondence, *QST,* Nov 1987, p 44.

together all the unused wires and grounding them at the kitchen wall jack eliminated the BCI problem. Thanks so much, Matthew Bell! — *Ralph C. Williams, W4REO*

from April 1988 *QST*, p 44:
ANOTHER TELEPHONE RFI CURE

A hint by Miko Maruya, WA6BSJ, in April 1984 *QST*, page 43, suggested the addition of line filters to eliminate telephone interference. Each of WA6BSJ's filters contained four RF chokes and three capacitors. After experiencing telephone RFI with my Yaesu FT-101EE transceiver,

I found a simpler—and possibly more universal—cure. First, I installed an ac-line filter in the transceiver power cord within *2 inches* of the rig. Next, I connected the filter case to a cold-water-pipe ground by means of heavy wire. This simple solution entirely eliminated my interference problem.—*Dave Zinder, W7PMD*

Editor's Note: The use of plastic pipe fittings in modern plumbing makes the "cold-water-pipe ground" an increasingly unsure option. Before depending on cold-water plumbing for electrical grounding of any kind, be sure that the system is conductive between your intended ground point and the main water inlet for the house.

Chapter 8

Stereos and Other Audio Equipment

By James G. Lee, W6VAT
1060 Big Oak Court
San Jose, CA 95129

The size and economy of integrated circuits (ICs) and other semiconductors have brought about a revolution in consumer electronics. (Let's refer to all semiconductors as "transistors.") However, several characteristics of solid-state devices make modern consumer equipment more susceptible to EMI than some older equipment.

Transistors are generally less linear than vacuum tubes. They are also more sensitive to operating conditions than vacuum tubes. Transistor bias conditions are critical for linear operation. These conditions combine to make transistors particularly susceptible to RF-signal detection.

The trend toward small size and low power consumption has made it difficult (or impossible) to apply some EMI-reduction techniques used on vacuum-tube circuits. An IC is a hermetically sealed component. There is no access to the internal circuitry. The basic EMI solution for an IC: isolate it from RF.

As ICs get smaller, so do the consumer products in which they are used. Reductions in the size and cost of consumer electronics has greatly increased the number of such devices near transmitters. The small size of some equipment makes it difficult to install EMI cures. For example, personal AM/FM cassette players usually have built-in antennas and are used in such a way that it is impossible to apply EMI cures.

AUDIO/STEREO SYSTEM DESIGN

While today's audio and stereo equipment is mostly solid state, vacuum tubes are still in use. This chapter covers both technologies. EMI-reduction techniques apply to both semiconductor and vacuum-tube audio devices, but semiconductor circuits have some unique characteristics that preclude the use of one traditional EMI cure.

Transistor Power Amplifiers

Fig 1 shows a typical transformerless audio amplifier with a complementary-symmetry output stage driving a speaker. (Complementary symmetry uses of a combination of PNP-NPN transistors, Q4 and Q5, to match the speaker.) For clarity, much of the bias-stabilization networks have been omitted. The feedback path (from the output through R3 and R2) is shown, along with R1 (the collector load for Q1). The schematic is an "ac schematic," in that it

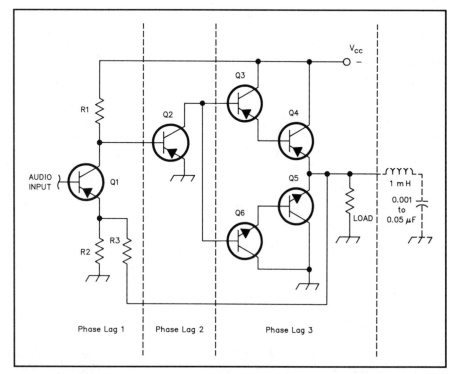

Fig 1—A simplified (ac) schematic of a typical audio amplifier. Phase lags occur between the dashed lines that determine the frequency response of the amplifier. The choke and capacitor (shown dashed) may be added to suppress EMI entering the amplifier via the speaker cable.

shows how the circuit appears to signals, or ac conditions.

Without going into great detail, the figure also shows three phase lags that contribute to the overall frequency response of the amplifier. The amount of lag in each section depends upon both the alpha and beta cutoff frequencies of the individual transistors. The feedback path (R2 and R3) from the output to Q1 is adjusted to provide the desired frequency response characteristics.

Traditional EMI cures would add a bypass capacitor across the output. *Several factors make such cures dangerous to transistor amplifiers.*

Do Not Apply Bypass Capacitors to the Output of Transistor Amplifiers

Transistors are extremely sensitive to bias variations. Their gain typically decreases as frequency rises and increases with temperature. Feedback circuits are normally used to compensate for these traits. Because of the gain-frequency relationship, the feedback circuits have a high-pass characteristic.

Transistor amplifiers can provide low-impedance outputs. This lets them drive speakers without matching transformers that would attenuate frequencies above the audio range.

A high-pass feedback network (with no output transformer to limit high-end response) permits potentially dangerous oscillations. The feedback network is adjusted to prevent such oscillations with a resistive load. When a capacitor is added across the amplifier output, however, such

---

INTERCEPTION OF RADIO SIGNALS

This discussion applies to interception of radio signals where the condition is observed on all channels or all across the dial. These units, and similar units, are referred to as "audio devices":

    phonograph
    CD player
    hi-fi or stereo amplifier
    tape recorder
    public-address system
    home-music system
    telephone
    the audio section of a television or radio receiver

Audio devices are designed to amplify audio signals such as music or speech and are not intended to function as radio receivers. The FCC does not give any protection to audio devices that respond to signals from a nearby radio transmitter. The problem is not caused by the improper operation or technical deficiencies of the radio transmitter. Strong radio energy gains entry to the audio circuitry, "overloads" the amplifier, is "rectified" and amplified and appears at the loudspeaker as undesired sound. The only real cure is achieved by treatment of the audio device. Owners should contact a qualified technician, the seller, or the manufacturer of the audio device (or the telephone company, for a leased telephone) for assistance.

Why The Owner?

Owners may reasonably ask, "Why should I do something to my audio device? It works fine except when the radio station is transmitting. Why is it my problem and not the responsibility of the radio operator?" The answers lie in policies concerning the economics, design and sale of these devices in a highly competitive market. The audio device has two objectives: (1) to reproduce a desired audio signal, and (2) to reject unwanted signals that may degrade the overall performance of the device (at a reasonable cost).

In the 1990 edition of their *Interference Handbook*, the FCC encourages consumers to "be sure that their TV, stereo or other electronic equipment can sufficiently reject undesired radio frequency signals." The consumer should also be sure that any newly purchased electronic equipment meets the voluntary Standards or Interim Standards for immunity to interference as published by the EIA.

Relatively few audio devices are located near strong RF fields and respond to unwanted radio signals. Those that exhibit interference require added filters or shields or both.

Manufacturers believe it is unfair and unnecessary to burden all consumers with the added costs of special circuits and designs, inasmuch as the number of devices affected is relatively small. Many manufacturers, dealers and technicians have devised procedures to improve the RF rejection capability of audio devices.

The conditions at a particular location are not necessarily stable. A consumer residing in an area with no strong unwanted signals may begin experiencing EMI in three ways:

1) Local unwanted signals increase because a new transmitter is installed or an existing transmitter increases power. (Remember that "transmitters" may be consumer devices such as computers, touch-controlled lamps and light dimmers.)
2) The consumer moves into an area where strong unwanted signals are prevalent.
3) The affected audio device malfunctions.

In cases one and two, the situation has changed from the majority case (where special treatment is not necessary) to the minority case, where special treatment is needed. In case three, the equipment should be repaired or replaced.

The foregoing information is not widely distributed or fully recognized by all manufacturers, dealers, and technicians. *The statement of a salesman or dealer that he sells a "good quality" device, or that there is "nothing wrong with the unit," is insufficient; it avoids the issue.*

Those who possess an audio device incapable of rejecting unwanted signals must fully understand the situation: *An audio device may perform the task for which it was designed with excellence and still require special treatment to improve its capability to reject strong unwanted signals in some locations.*

An audio device that responds to strong unwanted signals requires added filters, shields or both. The consumer is urged to bring this information to the attention of the electronics technician, dealer or manufacturer.—
John Norback, W6KFV, SB Section Technical Coordinator

oscillations may begin. Since the oscillations are above AF, the only noticeable effect may be overheating or destruction of the amplifier. If you cure the EMI but the amplifier fails as a result, you will have no credibility with the complainant. Fortunately, there is a way out of this dilemma.

The outputs of transistor amplifiers can be safely treated in two ways: Ferrite toroid common-mode chokes do not place a capacitor across the output. A bypass capacitor may be placed across the output if it is preceded by an RF choke. The choke and capacitor form a low-pass filter and present an inductive, not capacitive load to the amplifier.

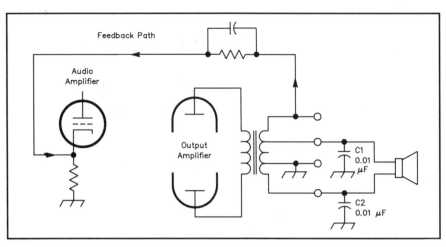

Fig 2—A simplified schematic of a vacuum-tube audio amplifier. C1 and C2 bypass EMI entering from the speaker cables.

Vacuum-Tube Power Amplifiers

It is possible to design audio amplifiers with bandwidths exceeding 100 kHz without much difficulty. Vacuum-tube amplifiers are high-impedance devices though, and they need a transformer to match the tube impedance (usually several thousand ohms) to the speaker (4-16 Ω). Most use large iron-core transformers, which are not broadband devices. The transformers have an upper frequency limit of about 20 kHz.

Fig 2 shows a simplified diagram of a typical vacuum-tube audio amplifier. Note the multiple output taps on the transformer and also the feedback path to a driver or preamplifier stage. The multiple taps are there to match a range of speaker impedances, which can vary from 3.2 Ω to 32 Ω on individual speakers. One of the output terminals is shown grounded, but this may not be so in every amplifier.

It is not uncommon for more than one speaker to be connected to one channel of an amplifier in a stereo system. Even monophonic systems have been wired to drive two speakers. Two 16-Ω speakers in parallel are an 8-Ω load, and as long as the amplifier has the power capability to drive them, it is a very practical arrangement.

Other output taps provide various levels of feedback to preamplifiers or drivers (to control the frequency response of the amplifier). As a result, any RF picked up by the speaker cables is fed directly back to low-level stages of the amplifier. Those stages amplify any unwanted signal along with the desired signals, until one or more stages are overloaded. The placement of the feedback tap on the secondary winding is the only restriction on the unwanted signal fed back to the early stages.

EMI entering vacuum-tube amplifiers via the speaker cables can be controlled with bypass capacitors (C1 and C2) connected from speaker output to ground. Assuming that the ground terminal offers a good ground, the two capacitors should prevent any unwanted RF from reaching early stages in the amplifier. In general, the capacitance should be as little as possible, but large enough to cure the EMI. Start with small capacitors, say 0.001 μF, and increase the value if they do not work. The output impedance of power amplifiers is usually low enough so that the capacitors can be relatively large without affecting the high-frequency response of the amplifier.

Nonetheless, all audio/stereo EMI problems share some characteristics. Fundamental overload is just that, and all EMI arrives via conduction, radiation or magnetic induction. The only thing that changes is the particular audio/stereo system layout and environment. Each situation is unique, but the EMI solution always includes the same troubleshooting techniques and standard cures.

Patience and understanding are important tools for solving audio/stereo problems. Use patience to keep on good terms with an impatient complainant. Strife causes more problems than it solves, so "keep your cool." Try to see the scene from the other viewpoint, and do your best to build confidence in the mind of the complainant.

COOPERATION

Several parties should cooperate to solve an EMI problem efficiently. Once you convince the complainant that you are both on the same side, and you both want the problem solved, you're halfway home. Cooperate with dealers, technicians or manufacturers who become involved (in the context of your responsibility). The Electronic Industries Association (EIA, a manufacturers' group) may be able to help if a manufacturer is unresponsive or uncooperative.[1] If the problem seems to be a tough one, ask other hams about it. They may have good information waiting to be tapped.

RESPONSIBILITY

Each Amateur Radio operator is responsible for the proper operation of his or her station. If spurious or harmonic emissions generated at your station cause interference to another radio service (such as broadcast AM or FM radio), the FCC requires you to reduce those emissions as much as possible. Beyond the FCC obligation, good character compels us to consider our neighbors and reduce spurious and harmonic emissions that interfere with electronic equipment (even when it is not used for communications). If your station transmitter is clean by FCC standards, however, the problem lies solely with the affected product.

The FCC Interference Handbook states that when an audio device receives interference from nearby radio transmitters, that device is improperly functioning as a radio receiver. The equipment owner

[1]Electronic Industries Association (EIA), 2001 Pennsylvania Ave NW, Washington, DC 20006, 202-457-4977.

should have the equipment repaired by a qualified technician. The dealer who sold the unit may help if he understands the problem. (After all, he wants to keep his customers happy.) Some manufacturers can assist with filters and advice about problems unique to their products. There may be a manufacturer's representative nearby who can help with the problem. The owner should check local phone directories for the manufacturer's phone number. When the complainant can't get satisfaction from the dealer, manufacturer or repair person, you may choose to help. Most of the time help is appreciated, and an excellent relationship with the neighbor results. Occasionally, there are pitfalls, which can be very deep and nasty. Remember that an Amateur Radio operator is not responsible for the cost of modifications or filters required to restore operations of a neighbor's equipment. The choice is up to you.

IDENTIFYING THE PROBLEM

The first step in identifying a problem is to characterize it as completely as possible. Ask the complainant, "What does the interference sound like?" Get the best description possible from nontechnical people. Certain kinds of interference can be identified by characteristic sounds. Amplitude modulation, for example, is readily detectable and audible on an AM radio or audio/stereo system. Single-sideband is also an AM system, but it usually is not intelligible. Instead, it produces a "muffled" sound on an AM radio or audio/stereo system. See Table 1 for a list of characteristic interference sounds.

What time of day is the interference noticed? How long does it continue? Does it vary from day to day, or is it consistent? Don't be afraid to ask questions; it demonstrates a genuine interest in solving the problem. After all, the ham station may not be involved! Remember that there are many other RF devices "on the air" today that can cause problems (see Table 2). Hams and CB operators are suspected first because their antennas attract attention.

When you have as much information as you can get from the complainant, check the station log for the complaint times. While the FCC has relaxed logging requirements, it's a good idea to keep a detailed log in case of interference complaints. If it appears that the station was active during the interference periods, say so. Let the complainant know that you may be a source of help, and diplomatically ask for cooperation in solving the problem. Explain that you might need to run tests, perhaps with the help of another ham. When you are unable to enlist the help of another ham, have the complainant help you (use the telephone to communicate while testing).

The telephone works well for coordination, but VHF/UHF hand-held transceivers are better (if you are both licensed) because there may not be a telephone near the affected system. Low-power simplex communication usually works fine and does not contribute to

Table 1
Characteristic Sounds Produced by Audio Rectification of Various Signals

Source	Sound
AM	The voice or music is heard as any normal signal applied to the amplifier. The interfering signal may be extremely loud and slightly distorted.
SSB	Voices are garbled and unintelligible.
FM	Usually no sound is heard, but the amplifier volume decreases during transmissions. Clicks may be heard as a transmitter is keyed.
TV	The audio rectification of a TV signal makes a loud buzzing noise. The buzz changes as the TV picture changes.
Data	Data transmissions may affect the receiver in the same way as FM. There may be an audible "tweedling" sound, constant or intermittent or constant-level static.
CW	Usually a rhythmic clicking with keying.
Motor	Constant-level static that varies with motor speed.
Atmospheric	Uneven crashing sounds.

Table 2
Possible Interference Sources

Communications Service Transmitters
CB
Amateur Radio
Police
Cable TV (CATV)
Business or other two-way
Aircraft (near airports only)

Electrical Noise Sources
Doorbell transformers
Toaster ovens
Electric blankets
Fans
Heating pads
Light dimmers
Appliance switch contacts
Aquarium or waterbed heaters
Sun lamps
Furnace controls
Smoke detectors
Smoke precipitators
Computers (and video games)
Ultrasonic pest-control devices
Lights: fluorescent, mercury vapor and
    touch-controlled
Neon signs
Power Company Equipment
    Defective line insulators
    Loose or unbonded hardware
    Discharges from defective lightning
        arrestors
    Defective transformers
Electric fences
Alarm systems
Loose fuses
Sewing machines
Electrical toys (such as trains)
Calculators
Cash registers
Lightning arrestors

interference. Don't tie up a repeater when troubleshooting. Do mention the problem on a repeater though; you'll probably get quite a bit of advice. Somewhere in all the advice there may be one or two solutions to the problem.

If transmission times do not agree with the reported interference, say that as well. Here also, offer your help. At this point you have no obligation, but the neighbor may not be convinced. Have the neighbor give you a call when the interference appears, then go to the house. Your presence while the problem is occurring gives you first-hand knowledge and demonstrates to the complainant that you are not the cause. Nonetheless, the neighbor may not believe it until you locate the actual cause.

Remember that EMI can be cured with patience, understanding, and know-how. Often the biggest obstacle to a complete cure is the complainant. Remember that the cause may be something in the complainant's home. Most hams remember EMI from older TV sets, which radiate 15-kHz horizontal-sweep harmonics well into the HF region.

With the proliferation of computers, calculators, cellular phones, and other such devices, the potential for EMI has multiplied. While the FCC has strengthened the rules for Part 15 Incidental Radiation Devices, many older models are still in use that do not conform to the latest rules. (This is why keeping a good log for your own station, and finding out the times the EMI occurs is important.) For example, a youngster in the complainant household, who normally does his homework on a computer while his parents are listening to the stereo, could well be the source of the problem.

EMI MECHANISMS

There are three ways that EMI is experienced in audio equipment: fundamental overload, audio rectification and direct radiation pickup. The energy that causes these phenomena may arrive by three essential paths: radiation, conduction and magnetic induction. The best cure for a given situation relates to both the experience and the path.

Fundamental Overload

Fundamental overload is a condition, not a frequency range or a mode of operation. All operating electrical devices (such as transistors, ICs, and vacuum tubes) respond to all electrical signals at their inputs. When an input signal is too large for a device to handle, the device is overloaded. When the signal overloading a device is the desired signal (fundamental) of a lawfully operating transmitter, the condition is called "fundamental overload."

When fundamental overload occurs, one or more signals at a device input are strong enough to change the operating characteristics of the circuit. The strong signal causes "gain compression" (the affected amplifier loses the ability to amplify), which may also be called "desensitization" or simply "desense." "Saturation" is another term related to overloaded devices. A saturated device has more drive than it can faithfully reproduce. In other words, the device is overwhelmed by its input and does not perform properly. Thus, fundamental overload is a basic disruption of proper circuit operation because the affected device is incapable of rejecting strong unwanted signals.

Proximity is the usual cause of high RF levels at audio and stereo systems. The normal RF input levels of these systems range from millivolts up to a few

---

INTERCOM FEEDBACK AND GROUND LOOPS

Some years ago, I was the project engineer on part of a large data-processing system. It required an intercom system and intercom access to a group of audio recorders. During the preliminary tests, there were severe hum and multiple feedback problems when more than two intercom stations were connected and when intercom stations accessed the tape-recorder system.

All audio cables were twisted-pair shielded lines, which should have been clear of both hum and feedback. Since the system was a prototype, no special care had been taken to keep leads short or provide optimum grounding. While troubleshooting the feedback and hum problem, I noticed that the number of recorders installed directly affected the hum level; more recorders increased hum.

Ring Around Feedback

When I disconnected the recorders completely, the hum problem was drastically reduced, but the feedback was still there when more than two intercom stations were connected. The feedback problem was determined to be a familiar problem known as "ring around." (This problem first showed up in cross-country microwave relay circuits, where a complete loop was generated by several site operators using the service line simultaneously.) The effect is similar to the feedback in PA systems when the gain is too high. In my case, the connection of the intercom stations was provided "through" each station, rather than by a common node. Once this was determined, phase-delay networks (with slightly different delays) were installed at each station. This interrupted the ring and solved the problem.

Ground Loops

Next the hum problem was examined (it occurred when the stations were connected to any of the tape recorders). A quick look at the recorder schematic revealed the cause. There were eight separate grounds at the output connector of each recorder. The solution to the problem was obvious. Each recorder was internally modified to provide only one ground at each connector. The hum disappeared, and the intercom system operated normally.

The moral to this story is: Too many grounds can spoil an otherwise good system. With eight grounds exiting the recorders, the potential for ground loops was enormous; any hum or noise on those grounds could enter the system. A single ground from each unit (connected to a common system ground) completely solved the problem.—*James G. Lee, W6VAT, San Jose, California*

tenths of a volt. The RF input path is not, however, the only way that an undesired signal can enter the system. Any stage in the signal path can be affected by fundamental overload, so a process of elimination is used to isolate the affected stages and apply corrective measures.

AUDIO RECTIFICATION

Audio rectification occurs when an RF signal is strong enough to saturate the amplifier stages in the audio equipment. It may seem puzzling that a transmitter at 1.8 MHz can affect audio equipment with an upper frequency limit of about 100 kHz. Audio rectification works equally well, however, at 30 MHz and all radio frequencies. Transistors are essentially combinations of diodes arranged according to a specific internal metallurgical scheme. You cannot "make" a transistor from individual diodes, but a transistor contains "forward biased" and "reversed biased" diode junctions which act like individual diodes. Since diodes are good "envelope" detectors (or rectifiers), they detect any unwanted ac signal of sufficient strength. This occurs in spite of any desired signals that may be present.

Thus, audio rectification occurs even though the stronger-signal frequency may be well outside the audio range. This is a form of fundamental overload (a very strong signal at the transistor or vacuum tube) which disrupts the normal operation of the device. The unwanted signal can arrive at the transistor or tube via several paths.

EMI PATHS

Signals are propagated by three modes: radiation, conduction, and induction. Radiation and conduction are the primary interference modes in audio equipment.

Radiation

Radiated interference arrives at its victim in the same fashion as any other radio transmission. There is a radiating antenna, a transmission path, and a receiving antenna. The antenna of the victim device is often the interconnecting cables between various elements of the audio/stereo system. Fig 3 shows a typical layout of an audio/stereo system which consists of at least two input devices (a tape player, and a CD player), a main stereo amplifier and at least two speakers.

Cables from input devices are often

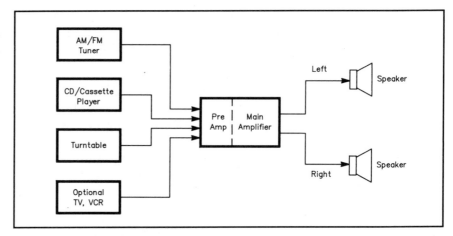

Fig 3—A typical modern stereo system.

shielded, but sometimes cause difficulty anyway. Speaker cables are generally unshielded and (worse) much longer than necessary. The excess length is usually just stuffed in any convenient nook. Long cables make good receiving antennas for any RF fields that may be present.

"Direct radiation pickup" is a special case of radiated interference. The term describes cases where a radiated signal is received by conductors within the affected equipment. EMI that arrives via radiation can be blocked by shielding the affected device (if that is possible; antennas, for example cannot be shielded). When shielding is not possible, direct radiation pickup can be treated as a conducted signal at the circuit-component level (such modifications should be left to the manufacturer or a qualified technician).

Conduction

Conducted interference arrives by wire (or other conductor). Conducted interference may be controlled by filtering the conductor. Conducted interference may appear as either common-mode (all conductors except ground act in common; that is, as one) or differential-mode (the signal arrives on a pair of conductors, with a 180° phase difference between the pair) signals. Common-mode signals are most prevalent, but have both common-mode and differential-mode filters on hand for testing. Some cases require both kinds of filter.

Very seldom is conduction the sole path of an interfering signal. The incoming signal is usually radiated to a conductor near the affected device, which then passes the signal to the affected device. In those cases, interference may be

reduced by shielding, or somehow detuning, the conductor. (Detuning a conductor might include changing the length, or inserting RF chokes to break its apparent length at RF.)

Magnetic Induction

Magnetic-induction interference requires a very close proximity of the interfering device (a coil or transformer) to its "victim." Hams are not normally confronted with this situation, even in their own homes. (Except in the case of telephones. Look at that chapter for more about "magnetic induction.") Magnetic induction can be reduced by physically separating or magnetically shielding (as by an enclosure of soft iron) the interacting devices.

AUDIO EMI CURES AND THEIR APPLICATION

Historically, amateur treatment of EMI problems has included detailed information about circuit modifications to affected equipment. These days however, such measures are seldom appropriate. Nearly all EMI problems can be solved by means of filters and external shields. Also some states have answered safety and consumer concerns by requiring anyone who works on electronic equipment to hold a state-issued license. Some in-circuit modifications are shown here as guidance for licensed technicians who are unfamiliar with EMI cures.

Ferrite Chokes Are Your Friends

Modern metallurgy has provided powdered-iron and ferromagnetic materials which may be shaped into toroids, beads, bars, rods, and E-I cores much in the same way as steel has been used for chokes and transformers.

Powdered iron is normally used as a core material at RF frequencies because of its low permeability. Ferromagnetic materials, called ferrites, are made from different alloys that have greater permeability and allow stronger magnetizing forces to be generated than does powdered iron. Ferrites were developed in the 1930s, and one of the earliest uses of ferrites was in the core of television receiver "flyback" transformers (which operate at about 15 kHz).

Simply wrapping speaker cables (or an ac cord) around a ferrite rod often eliminates EMI as well as do bypass capacitors. Fig 4 shows how to make ferrite chokes. As a general rule, more turns yield more rejection of unwanted energy. For severe EMI cases, more than one core can be used (the effect is additive). Use ferrite with a permeability (μ) of at least 850. Higher permeability is okay, so long as it does not lessen at the frequency to be rejected.

Chokes work by presenting a high impedance to RF in series with the conductor. In order for a choke to act as a filter, there must be a dissipative RF path available with lower impedance than the path through the choke. (Often the dissipative path is the conductor and source.) If the choke does not cure the EMI, add a second choke at the other end of the cable. In long cables, add another choke 8 to 10 ft from the first.

Place the choke as close to the affected chassis as possible. A choke at a speaker is not usually as effective as one at the chassis terminals. Excess speaker cable is generally sufficient to wind a choke. Power cords may not be long enough unless the power outlet is nearby.

Keep a variety of short choke-cables on hand with different connectors on the ends to match the usual audio/stereo component connectors and terminals. If you have had TVI or other EMI problems in your own house, you may already have pretested chokes on hand. Note that if you need ferrite chokes to tame EMI in one unit, you will likely need more for others. (You won't be able to remove any such chokes from previously cleared units because their EMI will return.) Therefore you may need two or three ferrite chokes per source as you continue clearing the system. You look more creditable if you

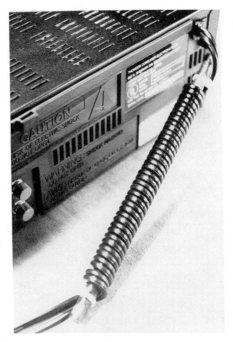

Fig 4—Wrap the ac-line cord around a ferrite rod to break ground loops that include the ac line.

come prepared with effective "gadgets" to quickly solve the complainant's problems.

Ferrite Beads

These dowel-like ferrite cores have one or more holes through them. When placed on wires, they act as RF chokes. The beads are made from several different ferromagnetic materials, in order to vary their permeability. Beads work very well where their larger brothers won't fit. Fig 5 shows four different bead configurations. As the figure shows, one need only thread one or more beads on a lead, or wind one or more turns through them (Type 2 and 3 beads) to impede unwanted RF signals. Table 3 shows design data for many popular ferrite beads.

To install a ferrite bead in a transistor amplifier built on a PC board, unsolder the base lead, lift the lead clear of the board, place a ferrite bead over it, reinsert the lead and solder it. This places an inductively coupled impedance in the signal path without adding resistance. Remember you are trying to eliminate RF, and the desired signal is AF.

Bead type 4 is a split bead that can be used over cable assemblies (1/4 to 1/2 inch diam) without disassembly of the cable. The split bead is mounted around the cable to enclose it. Flat, split beads are available for use on ribbon cable.

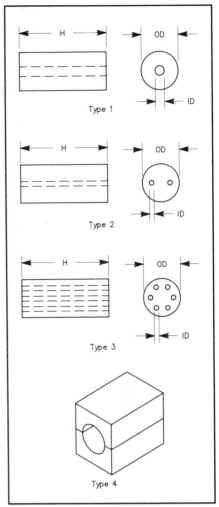

Fig 5—Typical ferrite bead configurations. Types 1 through 3 are one-piece beads. Type 4 is a split bead for assembly around cables or wire bundles.

The impedance coupled into a circuit by a bead is a function of the bead material, the signal frequency, and the bead size. As frequency increases, permeability decreases, resulting in a band-reject response. Fig 6 shows the variation of impedance with frequency for several materials, based on a "101" bead size. Mix 43, 73 and 75 ferrite beads are most commonly used for EMI suppression.

Impedance is directly proportional to the length of the bead, and it increases as more beads are used. The magnetic field is totally contained within the beads, and it does not matter whether they touch one another or not. Ferrite beads need not be grounded; in fact, take care with these high-permeability materials because they are semi-conductive. They should not touch uninsulated wires or ground.

Beads come in various sizes and permeabilities, so manufacturers' literature should be consulted to determine their size, permeability, and optimum frequency range. Manufacturers usually publish this data, along with an "impedance factor" (related to the permeability), which relates impedance to bead size and the number of turns.

Shields and Grounds

In rare cases, the previously discussed cures will not completely remove EMI. Such cases may require more stringent techniques to eliminate the last traces of EMI. Grounds and shields should be investigated. Grounding techniques are important to eliminate ground-loop currents and help eliminate conducted interference. Shields serve to protect circuits from direct radiation pickup.

Shields

Amateur equipment has shielded cases to prevent chassis radiation. Transmit filters permit only the fundamental to reach the antenna. Most consumer devices do not have shielded cabinets, nor do they have any internal shielding.

It is possible to shield an affected device, its enclosure, or cabinet. The manufacturer or a qualified technician may be able to install rigid (often foil bonded to cardboard) or spray shielding inside the case.

The owner or ham should not attempt to install shields inside an affected device. There are safety, regulatory and other issues involved. For example, spray shielding on transformerless consumer equipment could create a shock hazard.

The owner may make (or have made) a shielded enclosure to hold the affected device. A suitable enclosure is reasonably easy to find or construct. For example, a stereo system might be placed inside a metal entertainment-center cabinet. The main drawbacks are appearance and convenient access to the equipment inside the enclosure.

A suitable enclosure may be made of

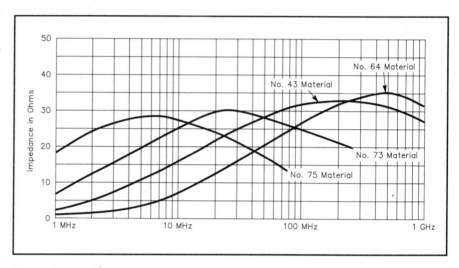

Fig 6—A plot of attenuation v frequency for "101" size ferrite beads. The impedance of different beads can be approximated by multiplying the impedance shown here by the appropriate "impedance factor" from Table 3.

Table 3
Ferrite Bead Data

Part No.[†]	Bead type	Dimensions (in)			A_L v Mix No. (nH$_t$[††])					Impedance factor
		OD	ID	Ht	43	64	73	75	77	
FB-(xx)-101	1	0.138	0.051	0.128	510	150	1500	3000	—	1.0
FB-(xx)-201	1	0.076	0.043	0.150	360	110	1100	—	—	0.7
FB-(xx)-301	1	0.138	0.051	0.236	1020	300	3000	—	—	2.0
FB-(xx)-801	1	0.296	0.094	0.297	1300	390	3900	—	—	2.5
FB-(64)-901	2	0.250	0.050	0.417	—	1130	—	—	—	[†††]
FB-(xx)-1801	1	0.200	0.062	0.437	2000	590	5900	—	—	3.9
FB-(xx)-2401	1	0.380	0.197	0.190	520	—	1530	—	—	1.1
FB-(xx)-5111	3	0.236	0.032	0.394	3540	1010	—	—	—	[††††]
FB-(xx)-5621	1	0.562	0.250	1.125	3800	—	—	—	9600	7.4
FB-(xx)-6301	1	0.375	0.194	0.410	1100	—	—	—	2600	2.1
FB-(43)-1020	1	1.000	0.500	1.112	3200	—	—	—	—	6.2
FB-(77)-1024	1	1.000	0.500	0.825	—	—	—	—	5600	3.7
2X-(43)-151	4	1.020	0.500	1.125 split bead, no. 43 only						
2X-(43)-251	4	0.590	0.250	1.125 split bead, no. 43 only						

[†]Complete part no. by substituting material no. for "xx".
[††]Based on low-frequency measurements.
[†††]Based on a single "U-turn" winding.
[††††]Based on a 2¹/₂-turn side-to-side winding.

Information courtesy of Fair-Rite Corp and Amidon Associates.

metal, or a nonconductive enclosure may be lined with metal or a conductive spray.[2] Copper and brass offer ease of connection, aluminum is cheaper (use lockwasher solder lugs for electrical connections to aluminum). A small metal case should be perforated (such as decorative screening) to allow adequate ventilation. Use piano hinge for good shielding at openings. Opening covers should fit closely and overlap the opening edge. If the enclosure cures the EMI without a ground connection, leave it that way. A good earth ground helps in some situations.

Grounds

A ground is supposedly a "zero-potential" surface or point, but most float at some small voltage. In reality, our best hope is that there is little circulating current (particularly RF). Most households have two ground reference points: the ac power ground and the plumbing system.

Water pipes are buried, but their ground resistivity can be quite high, and there are no precautions to ensure electrical continuity. As a result, the plumbing is usually tied to the ac ground return at some point in the house. This point is often at an outside faucet, where a wire is attached from the ac ground to the water pipe with a clamp.

Improper grounding can introduce unwanted audio hum and may violate the National Electrical Code or local regulations. An added ground for an audio system could create a big ground loop, which could act as an antenna and worsen the EMI.

Ground Loops

In nature, it is very unlikely that any two grounds are at exactly the same potential. Therefore, when a conductor is connected to two different ground points, a "circulating" current flows through the (supposedly grounded) wire. By connecting the wire to a second ground, a ground loop is created. The currents circulating in ground loops can cause or worsen EMI problems.

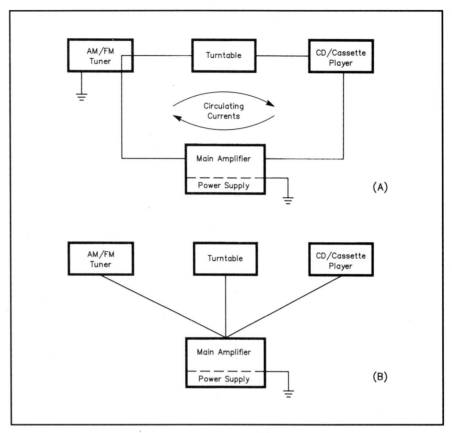

Fig 7—Ground loops in stereo systems. A shows a system with multiple grounds. Circulating currents may flow through the loop formed by the ground connections between the tuner and phonograph, phonograph and CD player. There is also a loop between the tuner's separate earth ground and the main amplifier earth ground. Ground loops are eliminated at B, where each component has only one path to a single system ground via the main amplifier.

A "single point" grounding system is usually necessary to avoid circulating currents between components of a system. Fig 7 shows how to avoid ground loops between system components. In the figure, a turntable that has a shielded interconnect cable should not have a separate ground lead to the amplifier. Other components such as AM/FM tuners and CD or cassette players should be treated likewise.

To avoid ground loops, there should be one (and only one) path from each point in the system to earth ground. (Visualize the ground system as a tree, with one trunk and many branches.)

When each component has its own three-wire ac cord, it may be difficult to avoid ground loops. It may be satisfactory to connect them all to the same ac outlet. If not, try a ferrite choke on each ac cord. Then add ferrite chokes on any shielded interconnect cables. In some cases it may help to use unshielded interconnect cables. There are no hard and fast rules, because each audio/stereo system is

unique. Try all combinations of potential cures; some are bound to work. Don't give up! When the final solution is achieved, it will hold lessons that can help in future cases.

What does this mean for an audio/stereo system, which may be located on a second story, or in a far corner of the house? Establish a reference ground at the system itself. Then connect the reference ground to the main ac ground with as short a lead as possible. This may not be easily done, but it may be the only solution in severe EMI cases.

AC-Outlet Grounding Practices

All this may sound difficult, but often it is not. Many homes have three-wire grounded outlets near water sources, while the rest of the home has only two-wire outlets. In such cases the home is usually wired with three-wire cable, but the ground wire is not connected at the two-wire outlets (see Fig 8). Then, outlets can be changed to a three-wire style, and the ground wire connected to its

<hr />

[2]A spray EMI-RFI coating is available from GC-Thorsen (see the Supplier's List at the end of this book). It is catalog no. 10-4807. Beware: Such sprays do not adhere well to all plastics. Some cases may require special primers so that the conductive coating does not flake off and cause short circuits.

proper terminal. Of course, the ground wire must be continuous to the ac ground point. For safety reasons, the work should be done by a professional electrician.

Where To Place Internal Components

Be warned! It is unwise to add components to internal circuitry. If a qualified technician has the understanding and willingness to solve the problem, internal modifications may be a viable solution. (Even then, consult the manufacturer first.) In all cases, strive to protect the complainant's equipment and interests. You and he (or she) are on the same side.

Make sure that cures clear the interference, but have no chance of harming the affected unit. Remember that transistor power amplifiers may not tolerate the same cure that vacuum-tube amplifiers accept. See "Do Not Apply Bypass Capacitors to the Output of Transistor Amplifiers" earlier in this chapter.

There is another reason to avoid internal changes. They can void any remaining warranty or service contract. Point this out to the owner; it shows your concern for his/her protection. Volunteer your services as a consultant to the dealer or his technician if they are unfamiliar with EMI problems.

If a technician is to add components, placement may be critical. Fig 9 shows components added to the phonograph, tape-deck, or auxiliary connectors of an amplifier to remove unwanted RF. Since this occurs at a connector, there is usually room to attach these components without difficulty. As before, these components should be placed as far "upstream" (closer to the source) in the signal path as possible to prevent amplification of unwanted signals by affected circuits.

Vacuum-Tube Circuits

Figs 10 and 11 show ferrite beads applied to vacuum-tube and transistor amplifiers. Fig 11 shows two different types of RF chokes (RFC) used as filters. The small square box containing a hyphen (-) is the symbol for a ferrite "bead" choke. (Ferrite beads are discussed in more detail later.) Vacuum-tube and transistor circuits require somewhat different approaches to install the components.

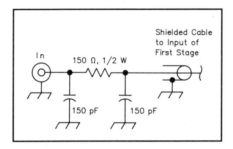

Fig 9—A filter for use at the input of audio equipment. The components should be installed inside of the chassis at the connector (by a qualified technician, see text). Use the shortest leads possible. Use RF capacitors with a minimum voltage rating of 1500 V (in case "ground" is inadvertently connected to 120-V ac).

Most vacuum-tube amplifiers are constructed with "point-to-point" wiring. That is, wires are hand installed between various points of the circuit. It is relatively easy to add components to such circuits. In Fig 10, for example, the lead to the tube grid is simply unsoldered from the socket terminal, the components added, and the end of resistor R1 is resoldered to the socket pin. A small terminal strip may be added to support the extra components, but often this is not needed. Transistor amplifiers are not so easy to modify.

Solid-State Circuits

Essentially all transistor and IC circuitry is mounted on PC boards. The only leads that can be removed are those holding the components to the circuit board. On densely populated boards, it may be necessary to break a foil trace on the PC board to insert a component. This is not a desirable solution.

There is no problem installing C1 in Fig 11 at the base of Q1, but RFC1 cannot be easily inserted unless the foil trace is cut. It is doubtful that a dealer's service technician would do this, and the manufacturer would certainly dislike this solution. Fortunately, the little "square" RFCs with the hyphen inside can save the day.

TROUBLESHOOTING AUDIO SYSTEMS

As mentioned earlier, options with

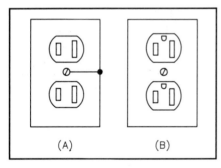

Fig 8—Old and new ac-outlet wiring styles. The old style, at A, has the colored (hot) wire connected to the small blade and the white (neutral) wire connected to the large blade. The green (ground) wire, if present, should (but may not) be connected to the metal outlet box via the outlet center screw. The new style, at B, has the colored and white wires connected to the small and large blades as on the old style, but the green (ground) wire is connected to a D-shaped hole in each outlet. The outlet box of the new style may be plastic.

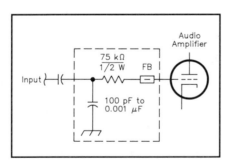

Fig 10—Partial schematic of a vacuum-tube audio amplifier with EMI-corrective components installed (inside dashed line). The added capacitor should be suitable for RF, with a dc working voltage appropriate for the circuit. FB is typically an FB-43-101 ferrite bead. This method is sometimes called "resistor-capacitor bypassing."

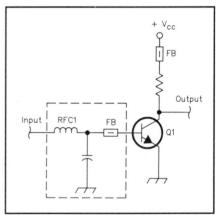

Fig 11—Partial schematic of a transistor audio amplifier with EMI-corrective components installed (inside dashed line). RFC1 is a 1-mH RF choke. The capacitor should be suitable for RF, with a dc working voltage appropriate for the circuit. FB is typically an FB-43-101 ferrite bead. A second bead was added on the positive-supply line.

small systems are few. A table-top radio contains its own antenna. Battery-powered equipment has no line cord to filter. It is not possible for the typical owner to add external filters to such self-contained elements. Fortunately, possible EMI paths are eliminated with the options. Any external connections, such as the power cord on a table-top radio can be treated in the same manner as the corresponding part of a large, complex system. Therefore, let's discuss a large system with the understanding that the techniques can be applied to smaller systems as well.

Simplify The System

When you begin troubleshooting keep an open mind and don't be afraid to experiment. Make your approach to the problem a logical one, rather than a hit or miss proposition. Begin troubleshooting by simplifying the system. With the system power off, disconnect all separate components from the main amplifier. Disconnect the speakers as well, and listen with headphones on a short cord. If a radio tuner is part of the main unit, disconnect the radio antenna. If the interference persists, it is entering the main amplifier through either the power cord or by direct radiation pickup.

Interference that enters by the ac cord is conducted interference. Strong RF fields are inducing currents in the ac wiring, and the wiring then conducts RF to the amplifier. Since nothing useful can really be done with the suspect house wiring, action must be taken at the power supply cord. The interference is probably a common-mode signal, so apply a common-mode filter as close to the amplifier as possible.

Direct radiation pickup may be treated by placing the amplifier inside a home-built shielded enclosure. Perforated aluminum and metal screen are suitable materials. Be sure that the enclosure provides enough ventilation to cool power amplifiers. Alternatively, the manufacturer or a qualified technician can add shielding or a conductive spray inside the amplifier cabinet or make circuit modifications to reduce RF susceptibility.

Once the minimum system is free of interference, begin adding the components that were removed. Begin with the speakers. If interference returns, then the speakers and their cables must play a part. One solution is to reduce the length of the leads, but often that is not enough. Shielding the leads may help, but only if the shield is connected to the equipment chassis. A common-mode filter placed where the speaker cables connect to the amplifier may solve the problem. In severe cases, install a low-pass filter in each speaker lead. (*Do not use bypass capacitors at the output of solid-state*

(GROUND-LOOP) TROUBLE IN PARADISE

As a rule, RF circuitry requires extensive, multi-point grounding to keep circulating ground currents from affecting circuits. On printed circuit boards, a good ground plane is mandatory for this purpose. If the system consists of several large racks of equipment, good low-inductance ground strapping is necessary not only between racks, but also to the best possible "earth" ground. More grounds, however, are not always better.

Such was the case a number of years ago when a fellow engineer and I were field testing a spread-spectrum system in Hawaii. The spread-spectrum system was a prototype that occupied three cabinet racks. Since it was a prototype, the interchassis cabling had not been optimized (although the cables were shielded). This might not seem a problem, but the transmitter boosted our low-power, frequency-hopping signal to 8 kW (peak) output.

No special shielding had been added to the prototype so we expected to have problems. In addition, we were operating inside an FAA transmitter site on Oahu, with row upon row of multikilowatt transmitters. Yes, we definitely expected problems. So when we initially "fired up" the equipment, their appearance was no surprise.

The Hawaiian islands are a combination of volcanic lava and coral. Neither is a good conductor, but the extensive station ground system appeared to be quite good. Our problem was to make a low-inductance connection with minimum ground loops. We decided to apply a little "overkill" by running half-inch-wide copper-braid straps between all racks, the transmitter, and the station ground system.

The transmitter was a 10-channel Collins Autotune HF rig modified for pulse operation. Sure enough, all ten HF channels showed EMI problems that distorted and smeared the pulse train. While my associate switched channels and watched the oscilloscope for EMI changes, I went about removing one end of the various ground straps. Where no effect was noticed by my cohort, I removed the other end of the braid and discarded it.

Finally, I removed one end of a 15-ft strap, and suddenly all channels were clear of EMI. After a recheck of all channels, it appeared we were set to begin testing. When I removed the other end of the braid, my associate immediately asked me what I had done. I told him, and he said "Put it back." I did, and he said "Leave it there." The EMI had reappeared when I removed the second end of the strap.

This is similar to using 1/4-λ wires as a "counterpoise," except for two things: We had to stay near the strap, and it couldn't lie loose on the floor. First, I coiled it up on the floor next to the rack, and that worked fine! Since the braid was not insulated, each turn of the coil shorted out against the next one; that should change the counterpoise or ground radial effect, but it did not. Next I loosely arranged it on the floor and put a blank rack panel over it with the same result: no EMI on any channel. In the end, we simply stuffed the braid under the rack it "grounded" and ran our "round-the-clock" tests for a full week with no EMI problems at all.

What's the moral to all of this? Don't be afraid to experiment. Ground-loop currents are difficult to predict, and it may take an unusual approach to eliminate them. Good layout and circuit design can only go so far; it cannot anticipate every EMI problem. Have the courage to use a little ingenuity; go with what works; it can make all the difference in the world.—*James G. Lee, W6VAT, San Jose, California*

"Oh no, here it goes again!"

amplifiers! Look at the Transistor Amplifiers section for more information.)

Reconnect Step By Step

Continue troubleshooting: reconnect a component, test for interference, consider likely paths and apply likely cures until the interference is gone. (Let's call each reconnected piece a "source" and its cable to the amplifier the interconnect cable.) Repeat the process for each component (work back to the signal inputs by reversing the path that the signal travels to the speakers) until the system is again complete and all interference is gone. If there are multiple sources such as CD players, turntables, and FM tuners, hook up only one source at a time. If possible, have the complainant connect and disconnect the units. Then the complainant cannot accuse you of damaging the equipment. If you must do the job, first establish that you will not do anything that affects the "inside" of the equipment, or its operation.

When The Volume Control Has No Effect

Normally, each reconnected source is ahead of the volume control. When interference returns, check to see if the volume control has any effect on the interference. If it has no effect, then the interference acts after the volume control. Interference after the volume control is usually audio rectification in the amplifier

output stage. While such interference should have been cured in the early stages of troubleshooting, the last source added may provide an interference return path or ground loop that affects the previously cured amplifier problem.

For example, a turntable that: (1) is grounded through its power cord, (2) uses a shielded connecting cable and (3) also provides a ground connection to the system amplifier, adds ground loops (alternate ground paths) to the system. The problem might be cured by disconnecting the turntable ground from the amplifier, by placing an RF choke in that lead or by electrically breaking the shield connection of the turntable signal cable.

Common-mode toroid chokes are often effective against RF in cable shields. Wrap any excess interconnect cable around a ferrite core (such as FT-200-43) and see what effect it has on the interference. If the source has an ac cord, apply the same technique to the cord, particularly if a choke on the interconnect cable reduced any problems in the main amplifier. In severe cases, wind both the interconnect cable and the ac cord around ferrite cores. Don't be afraid to experiment. The sidebar titled "Trouble In Paradise" is a good example of experimentation in EMI work.

When The Volume Control Has Effect

If the interference is affected by the

volume control, then it is almost certainly entering on the signal lines and not the ac cord or ground loops. At this point the problem can get sticky. The interference may be acting in the amplifier between the signal input connector and the volume control, or it may be rectified in the source and passed to the amplifier, as AF, along with the desired signals.

Check the interconnect cable between the two units to see if it is shielded. If it is not, and you have a shielded cable you can substitute for it, do so. If the interference disappears with a shielded cable, then the solution is obvious. If you don't have a shielded cable to substitute, then apply the same ferrite core treatment to this cable as you did to the speaker cables. Even if the existing cable is shielded, try a common-mode choke; it should protect the amplifier against RF on the cable shield (within the limitations of chokes, covered earlier). If a shielded cable and chokes do not help suspect the source.

Audio/Stereo System Components

When the amplifier is clear of all EMI, add the next component upstream. Repeat the same troubleshooting techniques as you begin to progress back to the program source. Each source may need its own EMI susceptibility cured as part of the overall fix. Turntables, tape decks, CD/cassette players, microphones, and AM/FM tuners may all be connected as inputs to audio systems and all may need treatment.

If you are able to get the system cleared of EMI with a turntable as the input, but the EMI reappears when an AM/FM tuner is the input source, then the troubleshooting must continue to solve the new problem. At least when you have one complete EMI-free path through the system, it demonstrates to the owner that you can solve the problem, even though other components may still need work. The complainant can then see that—with his cooperation and understanding—there is a satisfactory solution for all concerned.

Individual components such as turntables, tape decks, and CD/cassette players can have the same basic problems as the main amplifier. Just as speaker cable can be resonant at amateur frequencies in the HF range, cables from the pick-up stylus in the turntable tone arm can be resonant at VHF/UHF. A phono preamplifier has very high gain, and it is just as susceptible to rectification as is the input to the main amplifier.

EMI commonly acts in sources as direct radiation pickup or audio rectification (when the source contains active circuits in the signal path). If there is a level control in the source, it can indicate the location of the EMI as described above: When the control has no effect, interference is rectified after the control; with control effect, affected circuits are ahead of the control. In sources, EMI acting before the control is usually direct radiation pickup (in tape recorder heads and preamplifier circuits, for example). In-circuit modifications could cure the problem, but leave such action to the manufacturer or a qualified technician. Unless you own the equipment and are qualified to do circuit work, use a shield enclosure to reduce RF reaching the source.

As you continue adding units to the system, review other chapters in this handbook for potential solutions to the problem. For instance, if interference reappears when an AM/FM tuner is connected back into an otherwise clean system, other techniques may be required to solve this problem. This is because the problem may move from the AF range to the RF and IF ranges of the tuner.

Shielding

Some system components are poorly shielded. When they suffer from direct radiation pickup, the owner should contact the manufacturer to request shielding modifications. If the manufacturer and service technicians are unwilling to correct the problem, report the problem to the ARRL and the EIA. The owner's only remaining options are a shielded enclosure or replacement of the affected equipment.

Phonograph Tone Arms

If the cables to the turntable tone arm are not shielded, it may be necessary to replace them with shielded wiring. If you do so, be sure that you use wire designed for use in tone arms. Wire that is too stiff interferes with the proper travel of the tone arm, resulting in skips and ruined records.

A tone arm must swing back and forth easily, so it cannot be well grounded through the mechanism. The shield of the stylus wiring should be grounded to the base at the nearest point that allows free tone-arm movement. Use lock washers under ground lugs, unless it is possible to solder the shield. Lock washers help avoid problems with oxidation or lubrication buildup, which occurs when drive motors run for long periods of time. The average user does very little maintenance to keep the inside of the system clean. Any EMI corrective measures must consider this and withstand similar neglect.

Tape Decks and CD Players

Tape decks and CD players do not have the long input cable associated with turntables. Normally, head assemblies in these components are shielded to minimize ac hum and other noise pickup. So any EMI from these components usually comes through the signal cables, or possibly through their ac power supply. The same cures apply to these units as to the main amplifier. The same troubleshooting techniques should give you the evidence you need to solve the problem.

Microphones

Many hams have experienced feedback from their transmitters to their microphones. If a microphone is used with a victim audio/stereo system, the same cures apply here as well. Microphones and their cables should be shielded, and properly grounded to the main system.

Public Address Systems, Intercoms and Organs

These systems are lumped together because they share the EMI problems of audio/stereo systems, but the fixes are slightly different. Public address (PA) systems are found in churches, meeting rooms, and assembly halls everywhere. The lengths of their cables and interconnecting wires set them apart from home hi-fi systems.

Church systems can be large, with multiple speakers and high-power amplifiers. Organs may feed church PA systems through long cables. Most vacuum-tube amplifiers are in older churches. Start by directly bypassing the speaker cables at the vacuum-tube amplifier. Common-mode chokes may be needed at the speakers as well (there is usually no ground available for bypassing at the speakers). If the equipment is solid state, use common-mode chokes or a capacitor-choke low-pass filter as described for transistor amplifiers.

Shielded Cables

Shielded conductors may be mandatory for the long cables found throughout churches and assembly halls. However, access to these cables may be a problem. In severe cases it may be necessary to replace one or more of them. Replacing them in existing structures may be difficult, and any such cures should be done only as a last resort. (This is another place where professional help is required, if it comes to such replacement.) The telephone EMI chapter has some good discussion of techniques that can be used to eliminate resonances in long wiring runs. Refer to that chapter for some ideas.

It is beneficial that resonance and cable length are unimportant in AF work. We need not be concerned about adding wire to place chokes. An extra 10 ft of speaker cable wound around a ferrite core has no effect on the sound when the speakers may be mounted many times that distance from the main amplifier. Home audio/stereo system cures should apply here. You may just need a few more of them. If one common-mode choke does not remove interference at a speaker or amplifier, add about 10-20 ft of cable and a second choke. (This ensures that both chokes are not mounted at an RF high-impedance point.)

Intercoms

Intercoms can suffer the same problems as large PA systems. Unfortunately, there are three different kinds of intercoms to treat:

"Wired" systems carry audio signals over dedicated wires. They may operate from ac or battery power.

"Wireless" intercoms place audio signals on the building ac wiring. (They are only called "wireless" because the owner need not install wires.) These intercoms are essentially low-frequency FM transmitters and receivers that operate in the 100- to 500-kHz range. They take their power from the ac wiring, and simply couple the FM signals into and out of that same wiring.

The third type of intercom uses normal telephones and their associated wiring to carry signals. Although this system is most often found in business operations, it is sometimes found in the home as well. Any EMI affecting this type of intercom should respond to the same cures as ordinary telephone systems. For more information look at the telephone-EMI chapter.

Wired Intercoms

Wired intercom systems, with their

EMI COMES HOME TO ROOST

Because many consumers are not technically oriented, they sometimes cast blame where it doesn't belong. Such was the case when I was approached by a work associate whose ham husband was accused of causing EMI in an apartment house across the street from his station. The complainant's television and stereo were experiencing severe interference, and he was sure the ham station across the street was to blame.

The ham's wife approached me because I lived nearby and could provide the assistance needed to solve the problem. Investigation showed that other apartment-house dwellers also noted severe interference. The strange thing was that no one on the other side of the street noticed any interference. The accused ham did not interfere with any of his next-door neighbors' televisions or stereos, and he had been living there for a number of years. There was a church three doors down that had never experienced EMI on their organ or PA system.

Further investigation showed that the accused ham couldn't be the cause because he was standing beside the complainant during the interference. Other apartment-house residents mentioned that the interference had only started recently. I asked the complainant when he moved in, and it was about the time that the interference began. I asked him what electronic devices he had connected to the power line. When he mentioned his hand-held calculator (he was charging the battery), I asked him to unplug it.

The complainant was causing the problem! He kept his calculator plugged in all the time to keep the battery charged. After we explained why only his apartment house was having the problem, he red-facedly agreed to unplug his calculator. He switched his charging times to overnight instead of round-the-clock, and one more EMI puzzler was solved.—*James G. Lee, W6VAT, San Jose, California*

---

own dedicated wiring, should respond to normal audio/stereo cures. Simply determine whether the EMI is being radiated or conducted into the system, and apply the appropriate cures. Long cables are definitely a possible EMI path in these systems.

Solid-state systems may have an output transformer to drive the interconnecting wires. Many systems consist of a central main amplifier, and outstations that simply function as additional microphone/speakers. These systems often use low-impedance lines that couple back into a matching transformer to complete the communications circuit. Obtain a schematic diagram to find out exactly how the intercom functions.

Wired intercoms may have long cables between the central amplifier and the system outstations. A combination of bypass capacitors (vacuum-tube systems only!) and ferrite chokes may be needed to cure any interference. If the wire runs to outstations are not shielded, it may be necessary to shield them. Again professional help may be needed if structural work is necessary. The simple expedient of adding 10 or more feet wound on a ferrite core may be all that is needed.

Wireless Intercoms

Wireless intercoms may require a combination of techniques to correct EMI problems. Use the same filter and ground techniques used with other audio systems. Remember, however, that you are working with circuitry that is directly connected to an ac power source. Be careful while troubleshooting, and make sure that any cure applied does not increase the likelihood of shock.

Harmonics from amateur transmitters should not be a problem in wireless intercoms. Fundamental overload is the most likely EMI cause for wireless intercoms, but audio rectification can also occur. The EMI may desensitize the intercom receivers and prevent desired signals from coming through. The standard techniques for troubleshooting and solving the problems for both radio receivers and audio apply here.

Organs and Other Electronic Instruments

Electronic musical instruments, such as organs, vary in size, shape, and complexity. Fortunately, many organ manufacturers have recognized EMI problems for a number of years, and have developed "TV chokes" (as they are known) to help solve EMI problems. Tests can help isolate the cause of any EMI. Each function on an organ such as the "swell" pedal, the "band-box" volume or "draw bars" often have their own amplifiers. Each function can be adjusted individually, and the effect on the EMI noted.

When the EMI changes as a control is manipulated, EMI is entering before the control. If a control has no effect, then either there is no EMI to the function, or the EMI is introduced after the control. Once you have determined where the interference is entering the organ circuitry, filters usually solve the problem. Once the EMI has been identified, request "TV chokes" from the manufacturer. If the manufacturer is not acquainted with EMI, or how to cure it, other manufacturers may help by supplying information on how to cure the problem in their product.

ADDITIONAL INFORMATION

This book includes numerous sidebars that describe solutions to unusual EMI situations. Read them to build up a general EMI information background. Remember! EMI can be solved with patience, logic, and understanding.

About the Author

James G. Lee, W6VAT, holds a Bachelor of Arts degree in Mathematics and Physics (Electronics) from the University of California at Los Angeles. A member of the Coastside Amateur Radio Club (Pacifica, California), he holds an Extra Class license and has been licensed since 1944. During his 36 years in industry, he spent approximately 12 years as a Test Engineer. He has also held Commercial Radiotelephone and Radiotelegraph licenses since 1943. Now retired, he pursues writing, DX, and home building his station and test equipment.

Chapter 9

Power Lines and Electrical Devices

By Roger A. Laroche, N6FOU
Santa Barbara Section Emergency Coordinator
1155 Hetrick Ave
Arroyo Grande, CA 93420-5917

and   Robert Schetgen, KU7G
ARRL Assistant Technical Editor

The world population of electrical and electronic appliances is continually increasing. Any appliance that switches current or oscillates is a potential source of EMI. The first part of this chapter discusses many, but not all such sources (additional sources will come to market in the time it takes to print this book). The techniques described for specific appliances may be carried over to others that operate similarly, but are not described.

In addition, any appliance that senses, amplifies or reproduces audio may suffer from EMI. Most such devices (telephones, televisions, computers and so on) are covered in other chapters of this book. Nonetheless, a few notable exceptions are discussed at the end of this chapter.

This chapter provides a working knowledge of how EMI is generated, as well as effective measures for reducing it to an acceptable level. Some of the more common "offenders" and appropriate cures are explained. This section, however, is not all-inclusive. With a good understanding of how interference is created and an idea of where to look, you should be able to locate the source.

Since many EMI complaints allegedly caused by amateur transmitters are unfounded, this chapter is worthwhile reading for most amateurs. Unfortunately, some nonhams continue to blame any EMI on the nearest visible amateur. If you receive a complaint and your transmitter is not at fault, it might be wise to quickly check the neighbor's EMI problem. This may:

1) convince the complainant that your equipment is not at fault, since you cannot be at the house and operating the transmitter at the same time,
2) show that amateurs are concerned and willing to help track down EMI,
3) eliminate possible EMI to your receiving equipment. The cause may actually be in the complainant's house.

SAFETY FIRST!

EMI work can be done safely. First, never work on any circuit unless the power has been removed and any capacitors in the circuit have been discharged! (Use an insulated screwdriver.) If you are not sure of what you are doing, never remove the protective cover plate from a distribution box or outlet. The circuits behind the covers can kill you! The Silent Key is not an award! Many state and municipal regulations require that all work on electrical wiring is performed only by licensed electricians.

Do not open any appliance case unless you are qualified to work on that appliance. Even if you are qualified, it is unwise to modify equipment that belongs to another party in an EMI situation. The FCC places the burden of EMI correction on the owner of the source appliance (see "The Rule"). It is best if repairs are performed by the appliance manufacturer or a qualified technician who is not a principal in the EMI dispute.

Power-line poles are the property of the power company, and they *must* not be disturbed! The wires carry dangerous voltages, and a fall from a pole can kill.

You have no business up there. Also, never tap or hit a utility pole with anything. If there is a cracked insulator, striking the pole could cause the insulator to shatter and interrupt service: Loose hardware could fall on passing traffic or pedestrians. Let power-company representatives do all power-pole investigations. It is their responsibility to see that power is conducted safely; we must allow them full control of the situation.

WHAT IS INTERFERENCE?

The FCC provides a very specific definition of "harmful interference" (see

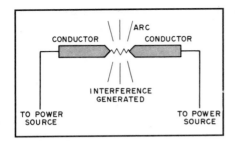

Fig 1—Representation of an electrical arc.

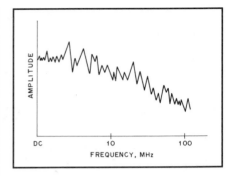

Fig 2— (A) Diagram of an induction motor. (B) Diagram of a brush-commutator motor.

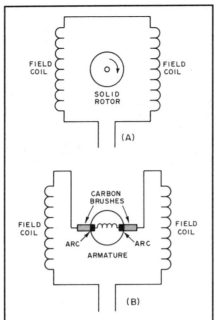

Fig 3—The amplitude of an electrical arc varies as a function of frequency.

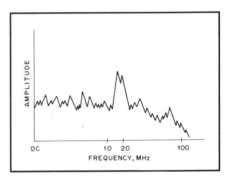

Fig 4—Here the normal frequency/amplitude curve has been modified by a resonant line connected to the interference source.

"The Rule") for the protection of various communications services, including the Amateur Radio Service. That definition applies when an oscillating TV "booster" preamplifier interferes with any lawful communications service.

A complaint about a cycling touch-controlled lamp, however, does not meet the FCC definition. A wider definition would be useful. Let's define interference or EMI as any undesired interaction between electrical or electronic circuits.

Interference may be as simple as natural static, a lightning storm or other atmospherics. There is little we can do to prevent those phenomena. On the other hand, unwanted signals may come from light dimmers, electric fences, the ac mains and so on. These common forms of interference are "man-made." Luckily, such interference can often be greatly reduced or eliminated.

Man-made interference has two elementary causes. The first cause is equipment that produces electrical arcs. Depending on the equipment and circumstances, an arc may or may not be a desired end product. For example, undesired arcs occur in power lines, bimetallic thermostats and switching circuits. Desired arcs are found in welding equipment and fluorescent lamps.

The second elementary cause of man-made interference is equipment that contains electronic oscillators. TV receivers, scanning communications receivers, microprocessor-controlled equipment and microcomputers all contain oscillators that can cause severe interference problems.

Electrical-Arc Interference

Fig 1 shows an electrical arc. An electrical arc generates varying amounts of RF energy across the radio spectrum. These RF signals are completely random,

and appear as "hash" in a communications receiver.

When an electric current jumps a gap, an arc is produced as the current travels through the air. For an arc to occur, there must be sufficient voltage to ionize (break down) the air in the gap. Once an ionized path is established, current flows. The flow of electrons is not smooth because the resistance of the ionized path changes constantly. The instantaneous current varies widely with the resistance. Rapid variations occur not only in the arc itself, but also in the power lines.

There is a definite relationship between the length of the arc, the voltage needed to sustain it, and the amount of interference it produces. In short, longer paths and higher voltage produce greater interference.

Arcs are created in a variety of everyday appliances, especially those with brush-commutator motors. Electric razors, vacuum cleaners, sewing machines and air conditioners are just a few. Induction-type motors, found in record players, clocks and refrigerators, do not normally cause interference. Fig 2 illustrates the difference between the two. Induction motors do not use a brush-commutator system. Since no arcs are normally produced, there is no interference.

Arcs are also common in many ther-

mostatically controlled heating systems. Some use a heat-sensing, bimetallic strip to open and close a set of contacts. As the contacts age they become pitted and prone to arcing. Arcs also occur in light switches, circuit breakers or loose ac outlets. The possibility for an arc exists in almost any piece of electrical or electronic equipment.

Fig 3 shows an amplitude-v-frequency graph for a hypothetical arc. Notice that the amplitude of the noise (interference) decreases as frequency increases. This can be verified by listening to the noise with a communications receiver. As the receiver is switched to higher bands the noise should decrease. It may increase at frequencies where the lines connected to the interference source are resonant. This phenomenon is illustrated in Fig 4; the noise has a definite peak around 20 MHz. Keep this in mind when searching for sources of interference. The frequency of maximum noise intensity may indicate the source.

Fig 5A shows an oscilloscope pattern

of a 60-Hz sine wave. Fig 5B shows the same basic waveform with the addition of noise caused by an arc. Since the noise contains a multitude of different frequencies it makes the original sine wave look "fuzzy."

What About Common-Mode Chokes?

While common-mode chokes are used liberally in EMI treatment, they are conspicuously absent here. Why? Most EMI from arcs radiates directly from the arc or travels the ac lines to the victim in the differential mode. Differential-mode (a separate core for each side of the line) ferrite chokes are appropriate, but common-mode chokes are not. Because bypass capacitors are easier to install and more readily available, they are the cure of choice. Nonetheless, when all else fails, try a common-mode choke: it can't hurt.

Oscillator Interference

Many pieces of electronic equipment use one or more fixed or variable oscillators. Digital circuits use square waves, which are rich in harmonic energy; this adds to the problem. The fundamental of each oscillator, any harmonics and any mixing products (if there is more than one oscillator running) may cause EMI problems. TVs are notable among such devices, and the cures discussed later for TVs apply to most oscillators (be sure that filter cutoff frequencies are appropriate to the undesired frequency).

The horizontal "sweep" rate of a TV is very close to 15.75 kHz. Unfortunately, the sweep signal is rich in harmonic energy. It is common to hear sweep harmonics (which occur every 15.75 kHz) in the lower-HF spectrum. These signals are heard as continuous, rough-ac notes every 15.75 kHz across a receiver dial. A TV set located a few blocks away may cause sufficient interference to overcome medium-strength signals on a communications receiver.

Television color-burst oscillators operate on 3.579545 MHz. This frequency and several of its harmonics (7.159090 MHz, 14.318180 MHz and 28.636360 MHz) fall within amateur bands. Since the relative amplitude of harmonics decreases as the order increases, the chance for interference is less at higher frequencies.

Most modern radios (including CBs, scanners, pagers and broadcast receivers) use frequency synthesis for tuning. It is

common for two or more oscillators to run simultaneously. Any time two or more oscillators operate in the same piece of equipment undesired mixing may result. Frequency combinations can cause interference anywhere in the HF or VHF spectrum.

Microprocessors now control everything from coffee makers to TVs. They can generate large amounts of EMI. Microprocessor clocks usually operate in the HF region: 4 MHz and lower for simple appliances, 33 MHz and higher in computers. This reference may be divided to produce additional frequencies. Square waves are used throughout; these signals are thus rich in harmonic content. There is a potential for wide-band interference across the radio spectrum.

TRACKING DOWN INTERFERENCE

There are two ways that EMI can enter a communications or television receiver. If the interference source is in the same building as the receiver, noise can flow along the ac wiring. For example, interference from a faulty oil-burner motor can "ride" the ac line from the basement to a receiver. This is also true for electric razors, hair dryers, electric knives and a host of other small appliances.

Since communications equipment often makes use of an outdoor antenna system, radiated interference may be picked up by the antenna or feed line.

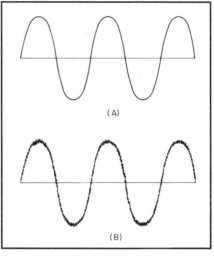

Fig 5—(A) A pure sine wave from the ac mains. (B) Noise interference can be seen as "fuzziness" of the waveform.

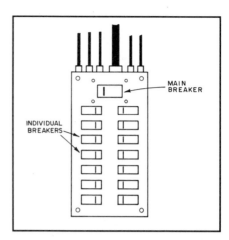

Fig 6—A typical home circuit-breaker box. The main circuit breaker is at the top center with the branch circuits below.

Then the signal is coupled directly to the receiver.

EMI At Home

First, determine whether the interference is generated in the same building. If the interference source is in the building and powered by ac, the EMI should cease when the ac power is switched off (see Fig 6). Obtain a battery-powered receiver for this test because the power will be off! A simple, battery-powered AM receiver is fine (most forms of interference affect the broadcast band). Tune the receiver to a clear channel somewhere near the low end of the band and listen for the interference. Use the main fuse or circuit breaker to switch the house power off and then on, so that the entire building is without electricity. If the interference disappears, the source is probably in the building.

If the source is in the building, a process of elimination can locate the offending device. Remain at the electrical service box, and cut power to one branch circuit at a time. The EMI should start and stop as power to one of the branches is switched. When the source branch has been determined, unplug each electronic or electrical device on the branch while monitoring the receiver.

EMI In The Neighborhood

If the source is not in the same building as the affected device, the search is a bit more difficult. EMI may originate from a neighboring building. Most modern households contain many possible EMI sources, such as washing machines, dryers, TV sets, water heaters, electric blankets and all manner of convenience items.

Should the search lead to a neighbor's house, that can be good news or bad news. If you are friendly with the neighbors, they may let you look around, especially after you explain the interference problem in detail. Don't be surprised to find that they are also experiencing severe interference with a television or AM/FM receiver. They may be more than happy to have the source cleaned up.

On the other hand, the search may lead to a house owned by a neighbor you don't know. There is no harm in explaining the situation to see how far you get. If you meet with a welcome response, fine. If you do not, contact the local power company. Provide as much information

as you can. Tell them that you think the interference is coming from a particular house and the manner in which you were received. It is possible that the power company has had other complaints of power-line noise in that area, especially if the source of the interference is strong. Power-company personnel will no doubt be able to gain access to the house and locate the noise source.

EMI On The Power Lines

As mentioned earlier, EMI can be generated by arcs in the power-distribution system. Since most connections are exposed to weather, the possibilities for corrosion are excellent. Loose or corroded connections, faulty transformers

and rubbing guy wires are all possible sources.

Assistance from the Power Company

An EMI Investigator is an engineer or technician employed by the power company to investigate EMI complaints. The investigator is well-versed in power-distribution systems, RF systems, CATV systems, microwave systems and such. He or she has access to the equipment needed to conduct a thorough EMI investigation. The power company relies on the Investigator to solve power-system EMI problems.

Before Calling for Help

If you suspect power-line EMI, first

POWER-COMPANY RESPONSIBILITY

Who should find and *who* should fix a power-line interference source? Most of us never consider the question. We just assume any EMI that sounds like arcing with a 60-Hz component should be fixed by the power company (let's call it the utility). In fact, much of the noise arriving via power lines comes from appliances attached to those lines in a customer's home. It's important to know *who* is responsible for what in the electric-supply system. When annoying "hash" drowns out a rare DX country, it becomes essential to know whether the EMI is "home made" or "imported."

How can we determine which portion of the electric system is our responsibility and which is owned by the local utility? There is a well defined point at which the change of ownership occurs, but it is defined differently for overhead and underground connections. The definition may also vary somewhat from one utility to the next. First, let's determine the type of drop.

Almost all suburban homes built before the early 1970s and virtually all rural homes are supplied from overhead distribution systems. Most urban and suburban multiple occupancy buildings, and suburban single-occupancy buildings erected after the early '70s, are supplied from underground electric facilities.

If you have an *outdoor* electric meter, you can determine which type of supply you have by looking at the area above this meter. Is there a cable going up the side of the house to a point either on the house side near your roof line or to a mast attached to your house? Does this cable connect to another *electric* (not telephone, CATV and such) cable spanning the distance from a nearby utility pole? If so, you are served from an overhead electric distribution system. If you find no overhead cable, but rather one or more pipes or cables that lead directly from the electric-meter box into the ground beneath it, your service is most likely from an underground system.

If your electric meter is *indoors*, examine the outside of the building for an electric cable coming from a nearby

pole. Overhead electric-supply cables usually attach to a wall or mast at least 16 ft above ground level and connect to another cable that enters the building. If you don't find an overhead feed as described, the electric supply is *probably* underground.

If, after doing all this, you are uncertain, call your local utility. Their records should tell which kind of electric service you have.

Overhead Electrical Service

On *overhead* electric service, virtually all US electric utilities identify the point of "ownership change" as the set of connectors at the upper end of the cable entering the top of the meter panel. These splices connect the ends of *the customer's* service entrance cables to the utility "service drop" (coming in from the pole). The *electric meter* is the only utility-company equipment on the customer side of this point. (A few Utilities also own the cabinet or "meter socket" where the meter is mounted.)

Underground Electrical Service

US electric utilities define the point at which ownership changes on *underground service* in various ways. Most commonly, this location is one of the following:

1) The point where the cables between the main breaker/fuse panel and the electric meter are connected to the electric meter.
2) The point where the underground cable crosses from public jurisdiction onto private property. This location is generally referred to as the "Property Line."

If it becomes necessary to define this point more accurately, contact the local utility for help.

The utility is responsible for locating and correcting interference sources on their side of the "ownership change" point. Responsibility for all such sources on the *customer* side falls to the customer.—*Harry D. Thomas, KA1NH, Windsor, Connecticut*

make sure the ham station is not the EMI source. Ensure that all connections are sound and cables are properly connected or terminated. For example, a loose antenna feed line can cause noise problems in a receiver. An arcing antenna trap could cause area-wide noise that might sound like a power-line problem. Make certain that all individual equipment grounds are connected to a single point in the station, which is then connected to the station ground rod.

Next check the area outside the house using a portable radio tuned to a frequency where the interference is prevalent. Listen for an increase or decrease of the interfering noise. If there is a power pole nearby and the noise increases as you get near it, note the pole number.

Continue the investigation by walking around the block while listening to the portable receiver. Note any changes in the EMI and any other pertinent information such as the location of an operating arc welder. The information you collect will help the EMI Investigator. It is a good idea to tape record the EMI, or videotape any resultant TVI, just in case the noise is not present when the EMI Investigator arrives.

When you are reasonably sure that the EMI is coming from the power-line system or you are unable to make any definite determination, call the power company. The person who answers the phone will direct you to a power-company employee who is knowledgeable in power-line EMI problems.

From this point on, let the EMI Investigator control the investigation. If you think of additional information that might help the Investigator, by all means, speak right up. You might think that a particular incident (such as a neighboring house undergoing renovation) is trivial, but it might be the lead that resolves the problem.

Sources of Power-Line EMI

There are many possible ways that a power line can create interference. Some of these sources have characteristics that allow an experienced EMI Investigator to quickly locate the source of the problem. In all cases of power-line EMI, turn to power-company personnel for help. Your Section Technical Coordinator may have already established contact with the technical people at the power company who can help you.

Corona Discharge

Corona discharge is often present around high-voltage lines. This phenomenon is caused by the ionization of air surrounding a high-voltage gradient. Once ionization occurs, the resistance of the air is lowered and current flow through it can increase. Corona is accompanied by the emission of visible light, usually blue in color. Look for corona near sharp edges or other problem areas in a high-voltage power-line system.

Corona is similar to an arc or lightning, and it produces similar broad-band RF noise. This type of interference is usually more pronounced at MF or HF than at VHF. The noise may be constant, or it might vary if the corona discharge is not constant. Corona discharge is more apt to occur in damp weather areas where moisture affects the characteristics of the insulators, but is found in all climates. The first step to curing corona noise, and most other types of power-line problems, is to make sure that the power-line insulators and related hardware are clean, free from dust and other contaminants.

Sparks

A spark can occur between two electrodes in close proximity when there is sufficient voltage. The sound produced on the radio by the spark is similar to that produced by an arc welder—a buzzing sound that covers a wide range of frequencies. Spark discharge can be caused by corroded metal components on the

pole, the electrical resistance of the pole itself, a leakage path across an insulator, physical damage that decreases the normal separation between conductors or any combination of the above. Clean insulators and hardware help cure spark problems, but it is not always easy to find the faulty component. Special tools and equipment may be required to isolate power-line EMI. *Do not try to perform any corrective action yourself. That task must be performed by utility-company service personnel.*

Power-Pole Transformers

Transformers do not usually cause EMI problems unless they have been damaged. The power-company crews check for problems with insulators, loose connections or the lightning arrestor. If those parts are not defective, the crew may remove and replace the defective transformer.

Pad-Mounted and Underground Transformers

These transformers, thus far, have not been a great source of interference. If they develop a problem, the frequencies involved are usually in the MF and low-HF range. It may be difficult to pinpoint the noise source and be sure that the transformer is indeed the culprit. These transformers are subject to the same kinds of damage as are the pole transformers discussed earlier.

Sensory Clues

You might be able to see the effects of many power-line problems. Corona discharge or sparking are often visible, especially at night. It may be necessary to look at night with a pair of binoculars to spot a problem. (Coordinate this with your neighbors to prevent any misunderstandings!) You may be able to hear arcs if they are particularly bad.

APPROPRIATE EMI TREATMENT

Once the interference source has been found, the next step is to reduce the interference to an acceptable level. Has the source always caused interference or is it doing so because of a malfunction? Suppose that you have been operating the 40-meter band most every day for several years. One day there is an interference problem, and the source turns out to be a light dimmer. It is obvious that something

THE POWER POLE

Two mechanisms cause most of the interference encountered by amateurs: arcing and sparking. Arcs occur when a connection is loose and current passes through the loose connection. Arcs are usually audible as well as visible. Arc EMI may cover a relatively wide area (see the example below).

Sparks are the most common kind of power-line interference encountered by amateurs. It happens when power leaves the system and takes a shunt path to ground: as across the surface of a dirty insulator on a power pole.

Loose metal within the field of an energized conductor can cause spark interference. For example, the metal braces that hold a cross arm to a power pole may be attached with a lag screw (sometimes called a "hard head"), which can become loose and allow sparks from one brace to the other. In such cases the metal braces act as an antenna, and the EMI may radiate several hundred feet.

Pole ground wires (from the top of a pole to ground) may be secured with fence staples. Staples tend to loosen over time and make poor contact with the ground wire. They may then spark and make noise. Basically speaking, any two pieces of metal (within an electrical field) that are not either tightly bonded together or completely separated, can cause interference.

Typical Problems

Loose hardware is the primary cause of interference in sunbelt areas, followed by equipment failure and arcing at loose connections. Other areas of the country have different primary causes. Insulator contamination is a problem near the seashore. Many power companies use "bucket trucks" with high-pressure sprayers to periodically clean insulators. (Salt water contaminates insulator surfaces, which causes "tracking" —arcs across the surface — and eventual destruction of the insulator.) Each part of the country has its own power-line interference problems. Techniques that work in cool, moist climates usually don't work in hot, dry, windy areas.

The sunbelt is where power-line noise causes the most problems. It is windy at times, and periods of rain follow frequent dry spells. Temperatures may change from very cold to very hot daily.

Most power poles are made of wood. Attachments to poles are usually tight when new, but the pole expands and shrinks from rains and dry spells. This loosens nuts and bolts. Staples and lag screws work their way out of the pole, and before long all mechanical connections are loose. (This is a fact of life with wooden poles.)

Poles made of concrete or steel have their own problems. Concrete poles may explode when struck by lightning. (This is apparently caused by moisture in the concrete.) Both steel and concrete poles are very expensive compared to wood.

Arcing Problems Don't Last Long

One day I went out to investigate a complaint. Enroute, I heard a very loud arcing noise while passing under a primary feeder line. The noise was so loud that I started hunting the noise before proceeding to the customer (a good ten blocks away). It was obviously a loose connector on the main line. There was a brilliant spot at a connector about a block away. I drove over to investigate and could hear it arcing from at least 40 ft away.

When a maintenance crew (two men in a double bucket truck) arrived, they went up and attached a temporary jumper around the loose connector. As they attached the jumper, the connector fell apart. A slight delay would have built a very large fire in their faces. It was obvious that repair by climbing the pole would have shaken the line enough to break the connector.

The interference was first observed at about 7 AM. We received the call around 8:30, and the connector failed about 10 AM. This shows how fast an arc can cause total destruction. If you find an arc problem, don't worry. The utility will fix it soon, or it will burn down and the interference will go away. Of course you may not have any electricity either.—*W. W. "Dan" Dansby, W5URI, North TX Section Manager, Fort Worth, Texas*

"I just don't understand it, John. Ever since that family moved into the old Addams house, I've been getting strange QRN every night!"

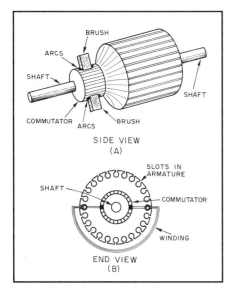

Fig 7—(A) A side view of a typical motor armature. Arcing occurs between the commutator and each brush. (B) An end view of a motor armature shows how the individual armature windings are connected.

"HURRY, before he discovers this path!"

has gone wrong with the dimmer since it never caused a problem before. Alternatively, you may have recently become interested in operating on the 40-meter band. Since you had not operated on that band before, you have no way of knowing whether the light dimmer caused noise problems on that band. Is something wrong with the dimmer? It is hard to say.

The point is: A malfunctioning device should be repaired, while one with inherent EMI needs filters and/or shields. Replacement may be a valid option in either case. Another example: Suppose a furnace motor suddenly starts causing EMI after years of trouble-free service. Rather than apply a filter, it may be better to actually find the cause of the problem. Perhaps the brushes are badly worn and need replacement. Maybe the motor needs a good cleaning, or it's time for a new furnace. A little common sense goes a long way when it comes to solving interference problems.

SOME COMMON INTERFERENCE PROBLEMS AND CURES

Here are cures for an array of specific EMI sources. Cures for several EMI victims appear at the end of the chapter.

External Rectification

A multitude of devices that are not considered RF radiators can be EMI sources because of corroded joints. Cor-

rosion may form a semiconductor junction (rectifier), and produce harmonics of any strong radio signals in the area. Rusty joints in pipes, ducts, fixtures, poor electrical connections, corroded antenna joints and even fences can create and radiate harmonic energy. For example, a fence of metal posts and wire can radiate harmonics when exposed to a strong signal from an HF antenna. Other common suspects are:

Electrical conduit
Heaters
Sheet-metal roof
Electrical wiring
Stove pipes
Furnaces
Water and sewage pipes
Gutters and drain pipes
Lightning arrestors
Guy wires
Radio and TV antennas

Anything that resembles the items on this list has a potential of producing RF harmonics by rectification and cross modulation (IMD). The cure is to locate the "rectifier" and eliminate it by electrically bonding or separating the corroded parts. This topic is covered fully in the External Rectification chapter.

Brush-Commutator Motors

Fig 7 shows a typical brush-commutator motor armature. For clarity, only one

winding is shown. The coil is wound in a slot in the armature, and the ends of the coil are attached to commutator contacts 180° apart. There are usually 10 to 15 separate coils. With two connections per coil, there are twice as many contacts as windings. Contact between the field coils and the armature windings is made through soft carbon "brushes." If the brushes are worn, fit loosely, or if the commutator is uneven or dirty, the brushes bounce as the armature turns and create electrical noise (interference). The constant making and breaking of contacts produces arcs. With properly fitting brushes and a clean commutator, however, arcing is minimized (and so is EMI).

Most small motors of this sort can be disassembled with ease if they are suspected of generating EMI. Check the brushes for correct fit, and replace them if necessary. Clean the commutator; there may be carbon buildup on the copper bars. Emery cloth, steel wool, or an ordinary pencil eraser removes the deposits.

If a brush-commutator motor is an EMI source, and it is in proper working order, install bypass capacitors from each brush to the motor frame. The capacitors should be installed as close to each brush as possible (Fig 8). There may be little room inside the motor housing so use physically small capacitors. A 0.01-µF (or greater) capacitor rated for 125-V (ac) should be sufficient. Disc-ceramic or mylar capaci-

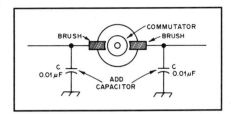

Fig 8—Install bypass capacitors inside a brush-commutator motor as shown.

tors are the best choices. Tantalum capacitors are physically small, but they are not designed for such ac applications.

When it is not possible to install capacitors inside a motor, add them (or a line filter) at the point where the power cord enters the motor frame. Special filters for this application are available from several manufacturers, such as CDE, Erie, Corcom and J. W. Miller. It is also possible to build your own filters as shown in Fig 9. The filter should be enclosed in a metal case with the case connected directly to the motor frame.

If it is not feasible to mount a filter directly at the motor, a plug-in line filter can be used. A commercial unit is shown in Fig 10. The filter plugs into the wall outlet, and the offending appliance simply plugs into the filter. Consider the current requirements of the EMI source, and choose a filter rated for that current or more. A filter installed at the wall outlet is less effective than one at the motor because the cord connecting the appliance to the wall outlet can radiate.

Sewing-Machine Motors

The likelihood of EMI from a sewing-machine motor depends on the age of the motor. Because sewing machines are fairly rugged, there are quite a few old models still in use. Most new machines have been treated for EMI.

Many models use a rheostat to control motor speed. Start by connecting a 0.05-μF, 125-V ac capacitor across the control. If space permits, install two bypass capacitors, one from each brush to the frame, inside the motor. If there is not room for such an installation, mount a line filter as close to the motor body as possible. See Fig 11.

Electrical Lawn Mowers

The motors used on electric lawn mowers are of the high-torque (high-current) variety. Mowers are usually powered by lengthy extension cords (antennas!), and severe EMI is possible. The motors are quite compact, so bypass capacitors cannot be installed directly at the brushes. A line filter mounted directly at the motor is the next best solution. Place the filter as close as possible to the point where the wires enter the housing, and make sure there is a good bond between the lawn mower chassis and the filter case.

Vacuum Cleaners

Most modern vacuum cleaners are fairly well protected against EMI. The inside of the motor gets fairly dirty, however, because not all dust is stopped by the collector bag and filters. The dirt can lead to excessive arcing. Treat the motor as described under "Brush-Commutator

Motors." There should be ample room inside most units for bypass capacitors at the brushes. If there is not sufficient room, use a line filter.

Many vacuum cleaners have an electric rug-beater attachment (power head). This is easily spotted because it requires an electrical connection to the cleaner when the attachment is changed. Most have a small motor in the attachment, which drives a system of bars and brushes. Use the same methods of EMI suppression with this motor as with the main motor.

Electric Shavers

Most currently manufactured electric shavers have built in EMI-suppression capacitors. If the shaver is causing more-than-normal interference, check the capacitors.

Some residual EMI is likely, even when the capacitors are in good shape. It can be reduced to an acceptable level with an external line filter. A plug-in style filter works nicely.

Electric Knives, Mixers, Hair Dryers, And So On

Most small household appliances use compact, brush-commutator motors. Since many of these items are operated only for short periods, only a fanatic would tackle them all. "Brush-Commuta-

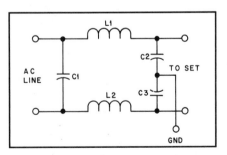

Fig 9—A "brute force" ac-line filter. C1, C2 and C3 can be any value from 0.001 to 0.01 μF, rated for ac-line service (125-V ac). L1 and L2 are each 2-inch-long windings of no. 18 enameled wire on a 1/2-inch diameter form (7 μH). If installed outside the equipment cabinet, enclose the filter to eliminate shock hazard. This filter is similar to commercial ac-line filters.

Fig 10—A commercial plug-in, ac-line filter.

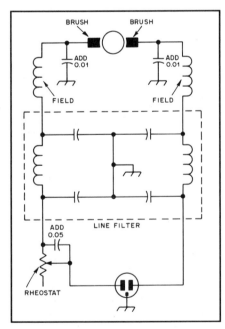

Fig 11—Schematic diagram for a typical sewing machine. Bypass the rheostat and brushes, and add a line filter in series with the power cord.

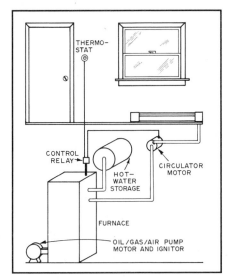

Fig 12—A typical home heating system.

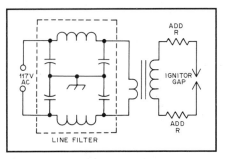

Fig 13—A method for reducing interference from the ignition system used in many furnaces.

(A)

Fig 14—A mercury-switch thermostat.

(B)

Fig 15—(A) Typical electric-heat thermostat-switch assembly. For EMI suppression, add capacitors as shown at B.

tor Motors" treatments apply here as well. If there is no metal motor frame (as is common in small appliances), use a line filter at the motor or a plug-in style line filter.

Office Machines

Some old office machines produce a fair amount of interference. Most new models are well shielded against EMI. Where motors are used, apply "Brush-Commutator Motors" methods. For digital machines (computer cash registers, for example) look at the Computers chapter of this book.

Electric Water Heaters

The typical electric water heater has two heating coils controlled by two separate thermostats. Since the coils draw large amounts of current, the thermostat switch contacts may become pitted. Pitted contacts are prone to excessive arcing; they should be replaced. In some cases, it might be necessary to replace the entire thermostat-switch assembly. The assemblies are not very expensive or difficult to replace. When the contacts are in good shape, a 0.01-μF, 125-V ac capacitor (installed across them) may cure the problem.

Heating, Ventilation And Air-Conditioning Systems

There are several potential EMI sources in "HVAC" systems. (See Fig 12.) "Brush-Commutator Motors" are used in many applications such as pumping air, water,

coolant or fuel. Check each motor for EMI, and treat them as described earlier.

Gas- and oil-fired furnaces may use an ignition system (rather than a continuous pilot light) to light the air-fuel mixture. The igniter draws an arc drawn across a spark gap. The gap is fed with a high voltage (sometimes RF, it strikes an arc more easily than 60-Hz ac) from an ac-powered step-up transformer. If the leads on the primary side of the transformer are not properly bypassed or filtered, EMI can reach house wiring. In high-EMI situations, resistors may be required in each of the high-voltage leads (see Fig 13). Note: The ignition system is normally on for only a few seconds. If the arc is continuous, the ignition control is defective—replace it.

HVAC Thermostats

Thermostats sense temperature and automatically control HVAC systems. Some old thermostats use a set of contacts attached to a bimetallic strip. The bimetallic strip bends slowly with temperature, and contact is made or broken very slowly: chances for pitting and arcing are great. Newer, mercury-switch thermostats (Fig 14) are much quieter from an EMI viewpoint. Programmable, solid-state thermostats are quieter as well. Treat bimetallic thermostats by cleaning the contacts and installing a bypass capacitor (0.01-μF, 125-V ac) across them. Alternatively, install a mercury-switch or solid-state thermostat.

Many heating systems use several relays to control different zones. Because the relays open and close many times they may become dirty and pitted. Replace them as needed, and install a 0.01-μF, 125-V ac capacitor across each contact pair.

Most electric heating systems use a combination thermostat-switch assembly to control the radiators. There is often a separate system (and thermostat) for each room. A typical thermostat-switch assembly is shown in Fig 15A. Since the con-

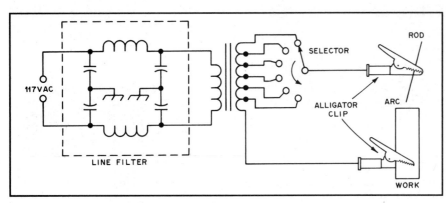

Fig 16—An arc-welding setup with a heavy-duty line filter installed.

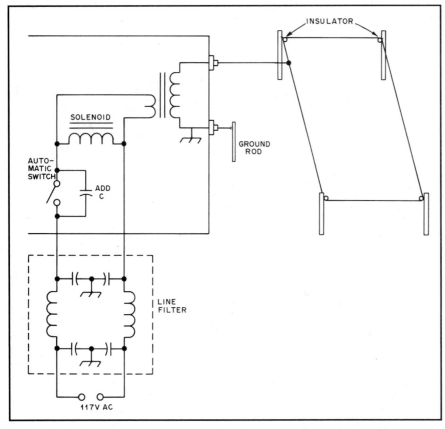

Fig 17—Electric-fence EMI cure: A capacitor is used across the automatic switch and a line filter in the ac line.

tacts switch high currents, there is a good chance they will become pitted. When the contact is opened arcing is possible. Some heating systems utilize single-pole switching; others use double-pole switches. Bypass each set of contacts in the switch assembly as shown in Fig 15B.

Door-Bell Transformers

Some door-bell systems have transformers with a temperature-sensing, shutdown mechanism. If there is a

malfunction in the system, the transformer circuit is opened to prevent further damage. If the contact becomes intermittent, EMI can be generated. Repair or replace the system to eliminate the EMI.

Arc Welders

Arc welders can cause severe interference. Unfortunately, there is no way to suppress the arc and leave the machine functional. Install a heavy-duty line filter

directly at the transformer primary as shown in Fig 16. The filter must be rated for the current drawn from the line. Use 0.01-μF bypass capacitors appropriate to the line voltage. Ac-line bypass caps should be rated for the line voltage. (This allows for spikes at 10 times the peak line voltage.) Wind the coils from at least no. 8 wire.

Appliance Thermostats

These devices control temperatures in electric blankets, automatic coffee pots, toaster ovens, water-bed heaters and just about anything else that requires temperature control. (The thermostats in old aquarium heaters are real troublemakers!) It is fairly easy to troubleshoot thermostat EMI: Switch off or unplug each thermostatically controlled device until the interference stops. Then remove and replace the thermostat with one that does not cause interference. An ac-line filter may help in some cases.

Pipe Heaters

Pipe heaters are used to prevent water pipes from freezing during cold weather. Normally, an electric-heat element is wrapped along the entire length of the pipe to be protected, and powered from a wall outlet. Most units do not use thermostats. If a thermostat is used, treat it for EMI in the manner already described. Replace it if necessary.

Electric Fences

Electric fences are still common in rural areas. A transformer steps up the line voltage, and an automatic switch sends a pulse down the fence once every few seconds. The voltage is pulsed so that it is not a serious shock hazard to humans. A single conductor is used for the fence, and it is connected directly to the switch mechanism. The basic system is shown in Fig 17. Since the wire used for the fence can be very long, it may act as a good antenna. A capacitor across the switch mechanism and a line filter in the ac line should do the job. Also, look for broken or dirty insulators. Clean or replace them as needed.

Fluorescent Lamps

While these lamps operate from 60-Hz current, they are switched on and off 120 times per second. This rapid switching, along with the stream of electrons (arc) going back and forth in the tube, creates rough radio waves that can cause severe

EMI. Most of the EMI is in the broadcast band with secondary peaks in the 7- to 9-MHz region. The noise is coarse and continuous while the lamp is lit.

If the cure cannot be accomplished by reorienting the receiver antenna, RF filters for fluorescent lamps are available by special order from distributors. The filter should be installed on the lamp fixture as close to the terminals as practical.

Neon Signs

The basic circuit of a neon sign is shown in Fig 18. It is very similar to a fluorescent light in that a long arc is drawn through the tubing to excite the gas. A transformer steps up the line voltage to a level suitable for excitation of the gas. Relatively high voltage (approximately 1000 V/ft of tubing) is required for proper excitation.

The cure for neon-sign EMI is a line filter mounted directly at the transformer and resistors (10-kΩ, 1-W) in series with each high-voltage lead. Also, make sure that the transformer case is attached to a good ground. If these measures do not reduce the interference to an acceptable level, try winding thin magnet wire around the tube along its entire length. Six or seven turns per foot should be sufficient. Ground each end of the wire.

Light Dimmers And Speed Controls

Most such controls use SCRs or TRIACs. The controls function by conducting for a short, variable part of each 60 Hz-cycle (with a rise time of about 1 μs). The resultant waveform is rich in harmonics and creates severe EMI well into the HF range. A loud buzzing that covers the lower frequencies is a characteristic feature of SCR EMI. The best cure is to replace the control with a better one that has a built-in EMI filter. (Beware of dimmers in plastic cases!)

A combination of an ac-line filter and a common-mode choke (formed by wrapping about 15 turns of ac wiring through an FT-240-61 or FT-140-61 ferrite core) placed close to the control often helps. It is usually necessary to use two cores, one at the control input and the other at the output, to eliminate the interference. (The cores are rather large; they may not be practical in all cases.)

Light Bulbs

Some 25-W light bulbs emit an RF signal in the 60- to 70-MHz range.

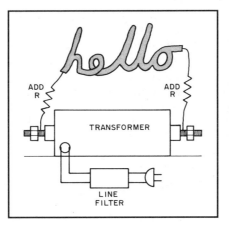

Fig 18—A typical neon-sign setup. Add a 10-kΩ resistor in series with each high-voltage lead and a line filter at the transformer. (The transformer should be well grounded.)

Not many of these offending light bulbs exist, but occasionally one is found. The problem created by these bulbs is usually found on TV channel 2. It appears as two horizontal black lines across the picture.

Electrostatic Discharge (ESD)

Static electricity is the accumulation of electric charges on a surface. These charges of electrical energy can build up on insulators as well as conductors. A discharge, or small spark, occurs when two materials at different potential are brought together. Static electricity is more common in cool, dry weather. Static build-up usually occurs on windy days, during periods of low humidity, during temperature changes and when objects are moved or vibrated.

An electric discharge can be induced in a metallic object that is in the field of a strong power source. In a receiver, this discharge sounds like lightning static, similar to a spark discharge. The nature of ESD makes it very difficult to locate. Two pieces of metal rubbing together, or a piece of sheet metal lying on the ground under a power line on a windy day can cause ESD EMI. The only way to cure this kind of ESD is by electrically separating or bonding the offending objects.

TV "Booster" Amplifiers

In fringe TV-reception areas it is common to employ a booster (preamplifier) mounted at the antenna. These amplifiers may malfunction and oscillate in the upper HF, VHF or UHF frequen-

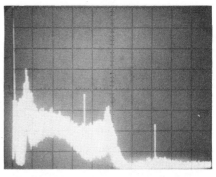

Fig 19—A spectral display of a TV-receiver output as measured at the antenna terminals. Horizontal divisions are 1 MHz; vertical divisions are 10 dB. The top line represents –30 dBm (7000 μV).

Fig 20—A commercially available TV high-pass filter.

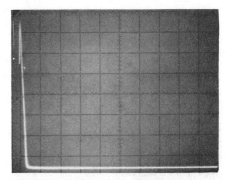

Fig 21—A spectral display of the filtered TV output. The interference is virtually gone.

cies. This can be caused by loose cables, corroded contacts, poor alignment or a defective component. If tightening the connections does not help, replacement is probably necessary.

Television Sets

As mentioned earlier in this chapter, a TV can cause EMI. The horizontal-sweep and colorburst oscillators are usually the cause. Fig 19 shows the spectral output as measured at the antenna terminals of a currently manufactured set. Each horizontal division is equal to 1 MHz with the "pip" at the far left representing zero frequency. Each vertical division is 10 dB with the top line equal to –30 dBm (7000 µV). This energy is radiated from the antenna system. Since the level is great, the chances for interference are good.

To cure this interference, add a high-pass filter at the antenna terminals of the set. A commercial filter designed for a 75-Ω system is pictured in Fig 20. The filter attenuates frequencies below 40 MHz by at least 60 dB. The spectrum-analyzer photograph at Fig 21 shows the difference after the filter was installed. Note the absence of interference.

Interference caused by the oscillators can also be conducted along the ac line. Most sets have built-in line filters. Their effectiveness is questionable, however, because an additional line filter often eliminates the interference. A ferrite common-mode choke may help as well. (Use 10-15 turns of the ac cord on an FT-240-43.)

Microprocessors And Microcomputers

As mentioned earlier, the population of microprocessors and microcomputers is increasing daily. Although a microprocessor may only control a coffee pot, its EMI treatment is similar to that for computers. In appliances, EMI usually exits through a plastic case or the line cord. Use shielding to cure case radiation, a ferrite choke for the second. Application details appear in the Computers chapter.

Touch-Controlled Lamps

Lamps that switch by sensing human contact can be both sources and victims of EMI. They contain a low-Q free-running oscillator that is very rich in harmonic energy. The oscillator is connected directly to the lamp chassis, so that hand contact changes the oscillator frequency.

When the oscillator frequency changes, the lamp switches. In order to sense a touch, the oscillator runs even when the lamp is off. In addition, most such lamps use an dimmer circuit to select several brightness levels.

The lamps can source EMI from the oscillator (about 2 MHz) or the dimmer circuit. (The dimmer should be quiet when the lamp is off.) Once on the ac line, the interference can affect receivers many blocks away.

The oscillators are easily "pulled" by nearby transmitters, and the lamp responds by switching with each transmission. In fact, a 2-meter HT makes a good remote control.

Use the same cures whether the lamp is acting as a source or victim. First, install an ac-line filter. It should block RF entering on the ac line, and that may cure the problem. Next, add a common-mode choke (7 turns on a FT-240-43 core) at the lamp base. Some hams have cured these lamps with a 1- to 4-kΩ resistor and a 10-µH choke in series with the oscillator "sense" line. Have the work done by a qualified technician. Even if you are qualified, it may be unwise to work on equipment you do not own.

Garage Door Openers

Openers use both radio and dc control. Strong RF may enter via the line cord, dc control lines or antenna. Unwanted opening and closing is the usual symptom. Apply a line filter at the line cord and a common-mode ferrite choke at the opener end of the dc control lines. If interference remains, a bandpass filter (in the antenna lead) for the frequency of the opener should cure it. Contact the manufacturer or a repair facility for help.

Ground-Fault Interrupters (GFIs)

GFIs are special circuit breakers that

quickly interrupt the current flow when a power circuit is shorted to earth ground (a "ground fault"). While normal current flows in the differential mode, ground-fault current flows in the common mode. Thus, a GFI is constructed to sense common-mode signals and open a circuit breaker when they are present. They may be very sensitive to common-mode RF on the ac line. EMI appears as false triggering.

Any EMI-suppression components may prevent the GFI from functioning properly. Report any problems to the manufacturer. Any corrective measures should be performed by the manufacturer or the manufacturer's representative.

Smoke Detectors

Smoke detectors suffered some EMI problems in the early 1980s. According to a letter from Charles E. Zimmerman, PE, a Fire Protection Engineer with the National Fire Protection Association (July 1981 *QST*, p 46), Underwriters Laboratories includes an RFI test on all smoke detectors manufactured since 1981. If you experience smoke-detector EMI, *do not attempt any modification.* Replace detectors from the early 1980s with newer models. Report any problems with current models to the manufacturer.

Conclusion

The information contained in this chapter is a summary description of the most common EMI problems caused by power lines and electrical devices. It provides a foundation for the investigation, analysis and cure of specific cases not covered here. Much of the power-line information contained in this section was learned from the *Interference Handbook* by William R. Nelson, WA6FQG. This book is highly recommended reading; it is available from the ARRL HQ Publication Sales Department.

Chapter 10

External Rectification—
The "Rusty-Bolt Effect"

By Mitchell Lee, KB6FPW
172 N 24th St
San Jose, CA 95116

Low-pass filters on the transmitter, high-pass filters on nearby TVs and careful bypassing at key consumer items such as stereos and telephones: These are touted as the "righteous" path to RFI-free hamming. But, even the most thorough application of these techniques can be foiled by a single rectifying joint in the vicinity of the transmitter or receiver.

Nature is a prolific creator of diodes. Weathered joints between pieces of metal (such as TV-mast sections, barbed wire and fence stakes or sections of rain gutter) form crude diodes that are efficient generators of spurious signals. The associated lengths of metal on either side of the joint act as antenna elements. They feed energy to the rectifier. The resulting nonlinear current flow is rich in harmonics, which are reradiated to wreak havoc in nearby receivers. The effects of these harmonics are identical to those of harmonics produced in a transmitter.

Two forms of "external" rectification are troublesome to amateur operators. First (and most familiar to amateurs), rectification of amateur signals causes interference that is associated with amateur-station activity. Signals from other sources may contribute to intermodulation distortion (IMD). This obscures timing relationships between the source activity and interference (because *both* sources must be active to produce the interference) and adds to the insidious nature of this interference.

The second (and often most frustrating) form results from nonamateur transmitters, especially commercial broadcast transmitters. In congested areas (where one or more AM broadcast stations operate in close proximity to buildings and power lines) the resultant interference can be quite strong in the lower HF bands.

This chapter addresses three important points:

- how (and where) rectifying joints are formed.
- how to track them down.
- how to disable them.

In general, any conducting structure (such as electrical, plumbing and antenna systems) can harbor a rectifying joint. Oxides and other corrosion products form crude, but effective, interference generators. Actually, some of these compounds are remarkably efficient. Lead sulfide is the galena crystal of early radio fame. In a crystal radio, small impurities of lead sulfide occurring at the surface of a lead crystal form diodes which are contacted through a "cat's whisker." Copper oxide, a common substance in many antenna and ground systems, was once commercially exploited in power rectifiers.

Any place where two pieces of metal touch is a candidate for corrosion. The process is accelerated if the joint is subject to humidity or weathering.

Rectification may also take place in "real" diodes—diodes contained in equipment around the shack. This includes amateur equipment such as antenna-rotator controllers and VHF transceivers (even when they are unpowered, but connected to an antenna). Also include such nonamateur devices as alarm systems, power supplies, telephone automatic dialers and so on.

The task of locating a rectifying joint is much simpler if you know where to look. Classic rectifying-joint RFI generators are described below. While by no means exhaustive, these examples should give RFI sleuths a good frame of reference.

The Usual Suspects

Guy wires: Metallic guys are used in some antenna and tower installations. The connections at the tower or mast can rust and form rectifying joints. Wire guys are normally broken at regular intervals by insulators, but there is always a piece at the tower end that connects directly to the tower. Even relatively short lengths can cause problems in a strong field; the short length enhances the radiating properties on harmonic frequencies. Egg insulators sometimes break, bringing the otherwise separate wires together and possibly forming a diode. Beware of continuous guys (as used on TV masts); they may be long enough to form a resonant element. Beware of spliced guy wires, guy rings and the tie points; these are likely to harbor rectifying joints.

Occasional broken strands can cause problems described later under "Stranded Copperweld." If guys are properly tensioned, however, broken strands are obvious. They should be replaced immediately.

Jointed antenna elements: The joints between telescoping sections of aluminum antennas must be cleaned thoroughly, coated with conductive grease (available from electrical supply companies under several different brand names), mechanically secured and weatherproofed with a nonacidic caulking product. The ends of the finished element must also be sealed. A diode can easily form if moisture penetrates the joint.

Corroded TV antennas: An old TV antenna, immersed in a strong field, is a likely rectifier. The usual trouble spots are element-to-boom joints and feed-line connections. Replacement is the best option for a deteriorated TV antenna or antenna system.

Towers and masts: The joints between tower and mast sections are subject to the same problems that beset antenna sections. Joints should be mechanically bonded, or short lengths of solid grounding strap should be used to ensure electrical continuity. If the mast or tower is of the crank-up variety, this is impossible.

TV mast sections: The actions for towers apply, but with some interesting twists. TV mast sections are normally erected in multiples of 5 or 10 ft. A 5-ft mast section attached to a chimney renders useless the best efforts at vertically polarizing a 6-meter beam to suppress EMI. A 15-ft mast attracts the nearest 10-meter signal. The 30-ft masts used in fringe areas resonate beautifully on 20 meters. Even though a mast is not an exact resonant length, it can still gather enough current to excite an otherwise marginal diode.

"Stranded Copperweld" antenna wire: This product is available through many retail outlets. Some stranded copperweld is of marginal quality. It becomes an unequaled TVI generator when aged. The strands gradually corrode over time; the steel core is eventually exposed and quickly rusts through. Then, current in a broken strand must bypass the break through adjacent conductors. But, the conductors are covered with copper oxide—once used in commercial power rectifiers! Single-conductor wires are not subject to these effects.

Front-end switching diodes and RF stages: Auxiliary receivers and transmitters are often left permanently connected to their antennas. Strong fields can induce enough voltage in the front end to forward bias switching diodes or an active device in the first RF stage—even when the radio is off. TVI travels right back to the antenna. One of the most insidious sources of trouble is the infamous masthead TV preamp. Strong fields can even send these units into oscillation, wreaking havoc across VHF/UHF TV and amateur bands.

Antenna-rotator control boxes: These often contain circuits that rectify signals picked up by the control cable. Ferrite chokes, ferrite beads and bypass capacitors are the solution.

Remote coax-switch control boxes: Treat them the same as rotator controls. Here RF energy travels in close proximity to steering diodes and control circuitry, compounding the problems.

Ground radials: Nothing beats a copper plate covered with sheet-metal screws as a ground "hub" at the base of a vertical antenna. But, each ground wire must be attached and weatherproofed with care or diodes may form. Ground wires can corrode clear through with time, possibly forming a diode at the break. Even if a radial is buried, harmonic currents are conducted back into the system, where they radiate.

RF "probes" and monitors: These invariably contain a diode or transistor that rectifies the received field. A short sensing wire efficiently radiates harmonics on VHF. Disconnect these instruments except when needed.

Power supply diodes: Especially on the lower HF bands, RF traveling up a power cord can enter a power supply for rectification by the diodes. 0.01-µF capacitors are often included both across the ac line and across the diodes for protection against this phenomenon. On higher HF bands, the supply output leads easily receive RF fields and feed energy back into the regulator circuitry and/or diodes.

RF ammeters: Believe it or not, a 2-A thermal RF ammeter in series with a $3/8$-λ 160-meter inverted L can cause interference in a receiver. A shorting switch solves the problem.

Cold solder joints and crimps in RF connectors: This is a tough one to trace, but a bad joint in an RF connector (or connections at the antenna, for that matter) causes interference of large proportions. Beware of crimped TV-grade coax. Coax can develop corroded breaks in the center conductor (especially where the coax is constantly flexed, see Fig 1) that conduct well enough to transmit power, yet still rectify transmitted and received signals.

Around the House

Burglar alarms and garage-door openers: Both of these contain solid-state devices and plenty of actual diodes. Both

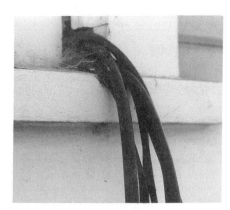

Fig 1—The conductors of cables that are subject to repetitive load and flexing may fracture to create a rectifying junction inside the jacket.

are attached to long lengths of wire. Treat them with common-mode chokes and bypass capacitors near the rectification source.

Metal roofing: Joints in metal roofing can rectify. When a vertical antenna is located directly overhead, the roofing carries return currents. Bond sections together with short lengths of copper strap and suitable soldering materials. Metal siding on mobile homes, motor homes and travel trailers is equally troublesome. Bond aluminum panels with stainless-steel lock washers and straps.

Duct work: Joints between ducts can rectify. Sheet-metal screws are an instant cure in most cases. Alternatively, connect the sections together with short lengths of copper strap.

The story of "talking ductwork," is often told late at night on 75-meter phone: A voice was heard emanating from a register or vent. The strong RF field from an adjacent amateur station (rectified by the ductwork) was responsible. Another version describes an apartment building laundry chute that receives a local broadcast station.

Pipes touching pipes: Service pipes sometimes touch to form diodes. Ground currents are the usual culprit. While it is possible to interrupt the current flow with ferrites or coupled tuned circuits, those solutions are difficult. It's easier to isolate the two pipes with a shim of cork or acrylic sheet, as in Fig 2.

Pipe joints: Where pipes carry RF current, problems similar to those mentioned previously (Towers and masts, Jointed antenna elements, and TV mast

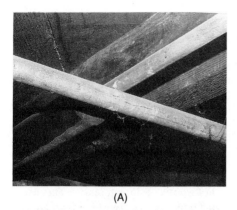

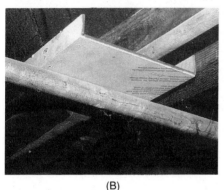

Fig 2—At A, two pipes touch where they cross, to produce a rectifying junction. An acrylic insulator has been inserted at B to prevent them from touching.

(A)

(B)

Fig 3—Cables that rub against conductive poles or pipes may be sources of external rectification. At A, the second cable from the bottom has a protective sleeve. At B, there is no protection between the metal line-amplifier housing and the traffic-signal support.

sections) can occur at joints. Renew or bond suspect joints.

Rain gutters: These are, without question, the most well-known of interference generators. In reality, the likelihood of a rain gutter causing external rectification is small, but ham folklore blames the rain gutter for everything. Rain gutters and down spouts are common components that can function as antennas, counterpoises, or grounds in radio systems. Rain gutters and down spouts are guaranteed to have at least a few suspect joints. Bond them with solder and screws. Better yet, convert to plastic rain gutters and lay in wire for any intended radio use.

GFIs (ground-fault interrupters): These pick up RF energy from the power lines. A properly installed and operating GFI should not cause problems, so replace any units that cause trouble. If proper operation is verified, but the GFI causes EMI, have a qualified electrician install metal conduit and a metal outlet box to shield the GFI and adjacent conductors.

Metal window frames: Where metal frames are set in stucco walls there is a possibility of rectification between the frame and underlying wire. It may be nearly impossible to reach the connection and effect proper bonding.

Plumbing joints: The faucet-to-sink interface, bathtub drain-link joint and various combinations of plumbing and household fixtures produce rectifying joints. Tighten, clean and bond as necessary.

Conduit joints: EMT conduit is joined by sleeves and setscrews. If the joint loosens a diode may be formed. RF is fed to the joint by either the field, ground currents, or RF flowing on the power lines.

Bed springs: This is more popular radio folklore. (Is there anyone who hasn't heard this story?) The ham next door transmits, and the audio is heard coming out of the bed, where RF is rectified between the various springs. Maybe it could happen, especially in apartment buildings where indoor antennas are employed. (This could prove quite interesting if the transmitted signal is full-carrier, double-sideband AM.) For cures, experiment with aluminum-foil shields, bonding springs together or grounding the springs.

Around the Neighborhood

Loose power-line hardware: Loose hardware often produces line noise, but it can also rectify. There is evidence that arcs are modulated by RF energy present on the line. Sometimes broadcast audio can be heard emanating from the arc. Widespread

interference can be the result. The power company must fix this problem.

Utility lines touching other lines, guys or poles: Power poles, overburdened with high-tension lines, 220-V distribution lines, telephone lines and CATV lines offer many sites for dissimilar metal-to-metal contact (Fig 3). One common source of contact is telephone or CATV lines that cross paths with an adjacent metal pole. Utilities appreciate hearing about damaged insulation—before major damage is done. When this occurs, the utility dresses the affected line with a plastic sheath to prevent damage.

Unfortunately, sheaths can wear through, again exposing the wires inside. Protective sheaths also creep from wind action and rubbing; watch for "misplaced" sheaths that don't cover the point of contact. The same problem can occur where lines cross guy wires or each other. When metal touches metal, a rectifier is formed.

Metal fences: Again, pieces of metal that are not securely joined corrode and develop rectifying joints. (See Fig 4.) Bonds at a few strategic points may cure the problem—experiment with a short clip lead.

Elevator shafts: An elevator shaft (probably the vertical beams in the shaft) once caused a severe BCI problem. Not unlike the duct-work phenomenon, the rectification was so severe that audio could be heard in the elevator.

This list should give the RFI sleuth some good ideas of where to start looking for sites of external rectification. It's

Fig 4—These two metal fences are likely sources of external rectification interference.

always easier to find something if you've got a general idea of where to look.

THE SEARCH

Aside from a shotgun approach (tapping every joint in a one-block radius with a hammer), the search for external rectification can be reduced to a semi-science. Use a number of experiments and tools to narrow the field of view. In some areas, similar methods are employed for amateur and non-amateur interference. In most areas they differ significantly. We'll treat them separately, considering TVI first.

Amateur Transmitters

Before testing, *remove* any masthead preamp. Even if the preamp is unpowered, it can easily rectify strong signals when connected to an antenna system. If only one TV is affected, the rectification problem is probably in the antenna system connected to that TV, or the TV has a problem. A random rectifier (downspout, mast, rusty fence and such) usually affects many TV receivers. If the TVI is noted in a CATV system, stop: Either the CATV company has some work to do, or the TV needs some attention. In either case, the problem is probably not the result of external rectification. Look in the Televisions chapter for more help.

Is It External Rectification?

It is difficult to differentiate between TVI caused by external rectification and that owing to other causes. When the TV and transmitter already have filters and other standard precautions, place a step attenuator in line with the affected TV. Gradually increase attenuation while the transmitter runs at full power. If the TVI reduces faster than the legitimate TV signal weakens, the problem is probably fundamental overload (in spite of the high-pass filter). If the TVI level remains constant as attenuation increased, direct injection (common-mode current, via the power cord or feed line) is the path.

In some cases, the TVI level remains constant (relative to the TV signal). If there is no masthead preamp, the interference comes from either the transmitter or a rectifying joint. External rectifiers affect all receivers in the vicinity. (Substitute a second TV as a double check.)

There is another test that is not conclusive: Rotate the TV antenna. If the TVI worsens or peaks in a direction other than that of the transmitting antenna, it may be pointing at the rectifier. The same is true at the transmitting end: TVI should be strongest when the beam is pointed at the rectifier.

Where Is It?

Next test to determine the power threshold at which rectification begins. Reduce power gradually until the problem disappears. TVI with less than 5 W indicates the transmit system. Give the transmit antenna, connectors and ground system a detailed inspection. 10 or 20 W is a more common threshold (higher on 80 and 160 meters, where the average bit of metal around the house isn't of significant length).

It is difficult to exactly locate a rectifier. Make a "first pass" by rotating the TV antenna and looking for a peak in the interference. Be careful, it is easy to mistake a TV signal null for an interference peak.

Rotate the transmit antenna (where possible), and look for a TVI peak in the direction of the rectifier. (Do this test with the minimum power required to cause interference when the transmitting antenna is pointed in the most sensitive direction.)

A portable TV or scanner is required to "home in" on the actual interference source. Check suspicious metallic objects by proximity, one at a time. Don't confuse a drop in legitimate signals with a "hot spot" of interference. The usual clanging, banging, twisting, torquing and pushing may produce recognizable interference changes that lead to the source. Above all else, check the transmit and receive antenna systems thoroughly before beginning a rectifier hunt.

When searching for TVI always remember that rectification in the TV antenna system generates lots of TVI, but it is very difficult to spot. Many a TVI goose has been chased when the source of interference was the TV antenna system.

Nonamateur Transmitters

Even in cases where the amateur is not directly involved, the ham may choose to help track down the problem. The amateur is often the technical "wizard" of the neighborhood. The ham may also be experiencing interference from the nonlinear junction.

Commercial Broadcast Stations

If the transmitter is a commercial broadcast station, the hunt follows a slightly different course. Since the transmitter is always on, and its antenna can't be rotated, the rectifier must be "DFed" from the receiver end. This isn't too difficult—the broadcast signal is coherent, narrow band, and easily identified. After DFing the general direction, set out on foot with a portable receiver and search by proximity. If the interference seems to be farther afield, use a mobile rig or portable receiver in a car to locate the general vicinity of the source.

There are four possible sources of broadcast transmitter interference: (1) receiver overload; (2) IMD in the transmitter (second harmonic of one signal ± the fundamental of the other); (3) rectification at inanimate objects; (4) "active" rectification on power lines.

Fundamental Overload

Especially in contemporary up-converting solid-state radios, broadcast energy can directly overload the receiver. Overload-induced interference is a gross problem that covers whole bands, not just a spot frequency. The interference tends to vary in intensity as antennas, rigs and matching networks are changed.

Differentiate between external rectification and receiver overload by placing a tuned filter in series with the receiver input. External-rectification interference peaks as the network is tuned. Overload-induced interference disappears when the network is installed. Read the Fundamentals chapter for a detailed description of fundamental overload. Cures for receiver overload appear in the Stereos and Televisions chapters.

Intermodulation Distortion (IMD) in Equipment

Intermodulation DFing often leads to a transmitter site. DF bearings from various locations with several methods all point to the transmitting site. Often the second and third harmonics are not strong. To effect a cure, the "mixer" must be located and one or both of the driving signals removed. If mixing occurs in electronic equipment, use filters to remove the driving signal. See the Fundamentals and EMI at the Receiver chapters for more discussion of IMD.

External Rectification

Rectification DFing leads to rain gutters, rusty water pipes and so on. Second and third harmonics are usually evident, as well as third- and high-order IMD between broadcast stations. More often than not, the audio is remarkably clear, but scrambled by the simultaneous presence of two sources.

The antenna or ground system may be at fault. Significant broadcast-frequency currents flow in low-band antennas and towers. In unbalanced systems, broadcast currents are shunted to ground by radials and water-pipe connections (Fig 5). To test: (1) Take an interference reading on a portable receiver some distance from the suspect antenna. (2) Interrupt the possible shunt paths, and (3) take a second reading. When the interference is generated in the antenna/ground system, the second reading is much lower than the first.

In some balanced antennas there is no common-mode path to ground, and rectification may take place across the antenna feed point. Perform the above test, but open the antenna feed point between readings.

Power-Line Rectification

Hum on a broadcast interference signal is a sure sign of power-line rectification. A mobile rig and a quick drive around the neighborhood should pinpoint one or more poles. Report them to the utility company for repair. (Sometimes you can actually hear the broadcast audio while standing near the offending pole.)

Arcs are notorious for producing a wide variety of RF products. Arcs can generate second and third harmonics of the fundamental, as well as second- and third-order IMD of many broadcast signals near and far.

Rectification may occur in passive conductors (grounds, guys and so on), yet a strong hum component may be evident. This is caused by 60-Hz energy in close proximity; it modulates the current flowing in the rectifying junction.

If IMD products are intermittent but seemingly periodic (seconds), suspect a long expanse of wire. Power lines in residential areas swing like pendulums, with a period of 1 to 2 seconds. As the line sways in the breeze, it tightens and loosens or gently rocks the insulators back and forth. This action may be just sufficient to repetitively break or short the offending rectifier, thus switching the BCI on and off.

Any lengthy conductor can be checked with a portable receiver. Couple the internal loop antenna to the conductor under test. The BCI always gets louder (the metal object is an antenna); but in BCI sources it gets disproportionately louder. Test several objects in the area, using an attenuator (or proximity) to control receiver sensitivity. It doesn't take long to get an intuitive feel for levels

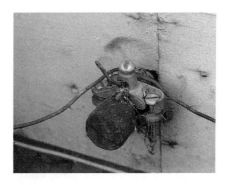

Fig 5—Inspect water-pipe grounds to be sure that the connections are tight and clean of corrosion.

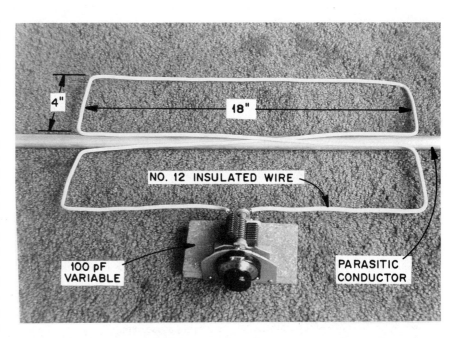

Fig 6—Use a "resonance breaker" such as shown here to obstruct RF currents in a conductor without the need to physically break the conductor. Use a vernier dial for the variable capacitor because tuning is quite sharp. The 100-pF capacitor is in series with the loop. This resonant breaker tunes from 14 through 29.7 MHz. Larger models may be constructed for the lower frequency bands.

that are "normal" and those that aren't.

If the interference is generated in residential wiring or plumbing, check for ground-return currents in the water pipe between the street (meter) and dwelling. In temperate climates (where pipes are buried relatively close to the surface) return currents radiate quite well. By simply "sweeping" the sidewalk near the water entrance with a portable receiver, the relative interference level can be detected. Signals are strongest at the house with rectification problems.

A search by foot and by car is usually required to locate BCI sources. If the BCI is weak, it may be difficult to detect on a portable receiver. Check other related frequencies for a stronger BCI product. The second and third harmonics should be strongest, followed by third-order BCI with other local stations. The search for TVI and BCI is more art than science. A few simple experiments, however, and some practical experience with a portable receiver quickly make an expert out of a beginner.

FIXES

Once the rectifier is found, apply cures. There are four ways to fix a rectifying joint: (1) disassemble the joint, clean, reassemble and weatherproof; (2) short-circuit the joint with a piece of copper strap, a screw or a bypass capacitor; (3) physically open the circuit with an insulator; or (4) impede the RF current with ferrite or coupled, tuned circuits. *If the rectification occurs along a utility line, the utility company must repair the damaged joint.*

Bypass The Rectifier

Where the RF current is intentional (as in grounds and antennas), short or bypass the rectifier. On water pipes, attach ground clamps at either side of the affected joint and connect them with a short piece of heavy wire (no. 10, 12 or 14) or copper strap. If two pipes cross each other, make the connection solid with a grounding clamp. *Do not ground gas pipes to adjacent pipes* because there is a risk of corrosion. Use an insulator to separate gas pipes (see Fig 2).

Where the RF current is unintentional or unnecessary, short the joint or add a component to reduce or null the current. Several methods are common. Install large or multiple ferrite beads, tuned traps coupled to the conductor or some sort of

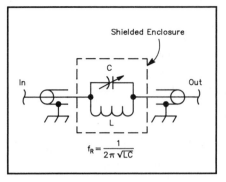

Fig 7—A parallel-tuned trap. Select L and C to yield resonance at the frequency to be rejected. Use a similar trap in each line of balanced feed lines.

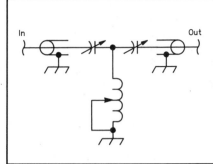

Fig 8—A high-pass Transmatch, such as this "T" network, can help reduce broadcast energy at the transceiver input.

insulator. For example, add egg insulators to guy wires, or use plastic isolators in water pipes.

Clean and reassemble antenna elements and terminals, connectors, radial connections, household fixtures and conduit joints. Replacement may be the best policy with defective antenna wire, guy wires, corroded TV antennas and bad pipe joints (are they leaking?). In cases such as loose tower sections, replace rusted hardware and tighten viable hardware that is loose. Use a shorting strap in difficult situations where it isn't convenient to disassemble the joint. *Leave all power-line maintenance to utility companies.*

Apply a few sheet-metal screws and a coat of RTV to rain gutter joints, TV mast sections, metal roofing and so on. TV masts can be completely sealed at each joint and at both ends with RTV.

If rectification is caused by an electronic appliance (power supplies, antenna relays, rotor servos), bypass diodes with 0.01-µF ceramic-disc capacitors. Similarly bypass wires entering and exiting the product. Use ferrite beads to inhibit the flow of unwanted RF current. Shielded control-cable entrances help impede unwanted current and provide a shunt ground capacitance before the signals reach the control box. Shield the power cord and dc-output lines of power supplies with 1/2- to 1-inch braid. Ground the braid at the supply.

Reduce RF Currents

A totally inaccessible joint calls for special techniques: Stop the rectification by somehow impeding the flow of current.

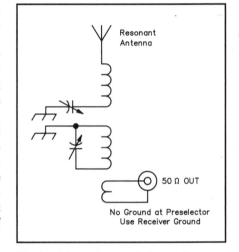

Fig 9—A loosely coupled preselector to reject strong broadcast signals or adjacent-band interference.

Current flow can be altered (in hopes of reducing the interference) by changing the electrical length of the conductor. Add a piece of wire at an accessible point in an effort to make the object less resonant at the fundamental frequency.

A tuned trap is a very successful means of breaking current flow without actually breaking a conductor (Fig 6). It is often employed on pipes, when rectification occurs at a point that is buried. It may be impossible to keep RF current off the pipe, but a coupled tuned circuit can quickly block fundamental current flow. The network requires a bit of wire on broadcast frequencies, but it works quite effectively. The inductor is folded into a "figure-8" pattern. This reduces the efficiency of the inductance as a loop antenna, without affecting the transformer action that couples the system together. The "resonance breaker" shown in Fig 6 was first described by Fred Brown, W6HPH, in the Oct 1979 *QST*.

Rectifiers In Electronics

Unused (but antenna connected) rigs were mentioned as possible rectification sources. In many cases it is inconvenient or undesirable to disconnect a rig, but another means of blocking the flow of current is possible. A parallel trap (Fig 7), tuned to the offending frequency (whether ham or broadcast) is one answer. A high-pass filter or a high-pass T-network Transmatch (two capacitors and one coil) helps reduce broadcast energy (Fig 8). High-pass filters protect VHF and UHF radios from HF energy.

Loose-coupled, double-tuned networks (Fig 9) block energy extremely well, especially on closely spaced frequencies (the broadcast- and 160-meter bands, for example, or any two adjacent ham bands). Unfortunately such networks are tricky to design and adjust. They also have limited power-handling capability. Don't leave them in the line while transmitting.

Hints & Kinks

from April 1986 *QST*, p 41:
TVI MYSTERY SOLVED

Recently, I purchased a new HF linear amplifier. When I operated the amplifier on 14 MHz, my signal caused TVI of the "screen is blank" type on TV Channel 2. To eliminate the TVI, I installed a high-pass TVI filter on the TV tuner. It had no effect. Using a second low-pass filter on the rig produced similar results. I then placed a 75-Ω termination on the TV antenna input. This resulted in a snowy, no-picture screen with little indication of interference. That indicated that whatever the source of the interference, it was coming in through the antenna connection of the TV set.

For several weeks, I looked for some nonlinear device, external to the TV and HF equipment, that might be the source of the problem. Finally, I took a close look at my other radio equipment. I noticed that the S meter on one of my 2-meter rigs would move upscale when the HF amplifier was being used. The squelch did not open, but the S meter would indicate from S2 to S9 + 10 dB depending on the orientation of the HF beam relative to that of the 2-meter beam. Disconnecting the antenna cable from the 2-meter rig immediately eliminated the TVI on all the TV sets in the house, including those not equipped with a TVI filter.

The 2-meter rig causing the problem is an ICOM 25A. Another rig connected to a vertical antenna on the same tower as the Yagi serving the IC-25A showed no S-meter indication and caused no TVI. Installation of standard "TV type" high-pass filter (I use a Drake TV-75HP that is no longer sold) in the coaxial lead to the IC-25A eliminated all traces of TVI and permitted normal operation of the transceiver. Loss through the TVI filter at 144MHz was measured at 0.6 dB. Perhaps this information will save someone else a lot of work.—*Joe Mehaffey, K4IHP*

[The radio that caused no problem probably used a relay for TR switching. The point here is not that the IC-25A has EMI problems, but rather that IMD can occur in semiconductors of inactive radios connected to antennas, and that such IMD can be cured by application of an appropriate filter.—Ed.]

from November 1988 *QST*, p 38:
UNPOWERED COMPUTER GENERATES RFI

Reception from 160 through 10 meters at my location was marred by severe splatter from a "broadcast band" station located 1-1/2 miles away. The interference wasn't continuous, though; it came and went for no apparent reason. After putting up with this for several weeks, I went to work tracking down the interference source. The interference appeared to be emanating from a 30-ft shielded RS-232-C data cable connected to my unpowered computer. Disconnecting the cable from the computer made the interference go away. Problem solved?

No! Attaching a new cable brought the interference back! Further investigation revealed a poorly soldered joint at pin 1 (equipment ground) of the computer RS-232-C DB-25 connector. Evidently, this solder joint was acting as an effective frequency multiplier. So, when hunting for sources of frustrating RFI, consider checking equipment that's not turned on—it just may be the culprit! —*David Barker*

Turned-off electronic equipment can generate such interference even *without* faulty wiring. Investigating interference similar to David's—pops, sizzles and crackles that occurred by day on all medium and high frequencies—I discovered that my unpowered, solid-state general-coverage receiver was generating the junk in step with the modulation peaks of a medium-wave broadcaster 1-1/4 miles away. Hunch: The interference is caused by the unpowered receiver's unbiased input-network switching diodes and/or RF-amplifier MOSFET. (Further hunch: The interference occurs only during the day because the station changes its antenna pattern at night, resulting in a considerably weaker RF field at my location.) Evidence: Turning the receiver on makes the problem disappear. Solution: Disconnect the antenna from the receiver when the receiver is not in use.—*David Newkirk, WJ1Z*

from March 1965 *QST*, p 69:
RECTIFICATION

Technical Editor, *QST*:

In the Handbook chapter on BCI and TVI you point out that harmonics of the transmitter fundamental frequency can be generated at poor electrical contacts between joined pieces of metal, especially if they occur in the antenna itself. You understated the possible harmonic generation problem very much. Yesterday, as I was operating on 20-meter CW with a kilowatt, an exham who cuts records and makes tapes for a living came over and told me I was putting signals over the whole FM band, washing out the white noise between stations, turning Channel 2 and 8 pictures negative and interfering with TV audio. He taped some of the interference; it washed out the FM signal completely and produced a hum like I get in my receiver when I shut off the bfo and listen to my signal without receiver muting. This occurred with his antenna switched off as well, leaving a few feet of twin-lead on the receivers. He is a block away.

My antenna is a quarter-wave ground-plane vertical (on 20 meters) with four radials. The interference existed on 40 meters, but was reduced in amplitude. There the antenna is fed with the coax leads tied together at the transmatch, which is coupled by a coax link to the transmitter push-pull tank link. I went up on the roof and found one radial, no. 14 enameled wire, touching a sewer vent pipe. At the point of contact, the copper was shiny, surrounded by blackened enamel, and the aluminum point on the pipe was charred.

Let that be a warning to anyone who is sloppy about antenna construction, including radial mounting. Now, I wonder what kind of interference was caused by my second harmonic, and the third, and the fourth, and . . . .—*George Tomasevich, K7GKB/K6SBJ*

Chapter 11

EMI at the Receiver

By Joel Paladino, N6AMG
East Bay Section Technical Coordinator
5070 Hilltop Dr
El Sobrante, CA 94803

Most of this book deals with curing interference complaints from others. Here, we will consider the kinds of Electromagnetic Interference (EMI) that are likely to hamper amateur communication. While interference is more often a mutual interaction than a cause-and-effect situation, this chapter looks at the problem from the perspective of the receiver input terminals.

You may be surprised to find that your neighbor's TV can interfere with your equipment, but it's true. As our electronic, microprocessor controlled, data connected world becomes more and more crowded, it is likely that more and more of the signals appearing on our sensitive, high-performance receivers will come from gizmos on our own block than from rare DX stations. We can combat that problem through knowledge, interpersonal skills and flexibility.

The following pages give basic knowledge about how our equipment works with respect to unwanted signals. There is a list of many RF sources that are likely to interact with our equipment. The First Steps chapter covers the interpersonal skills needed to work effectively in the EMI world.

Flexibility is a quality that sets Amateur Radio apart from most other communications disciplines. As amateurs, we have the flexibility to change location and frequency at will. In situations where interacting signals cannot be reduced to acceptable levels, we can change frequency or move our station relatively easily. Amateurs need not run from their problems, but in some situations the most expedient means of equipment isolation is distance.

Let's Get To Work

To work effectively, use a logical, yet intuitive, approach to identify and locate the "transmitter" that interacts with the equipment. Begin by reading "Definitions of EMI Effects" and "Signal Profiles." Those sections should help build your intuitive ability. The EMI flow chart in Fig 1 provides a logical procedure for solving an EMI problem.

Once the signal source is determined, consider the options. "Equipment Protection" lists several of the most effective tools in EMI suppression. In some cases the source is completely beyond your control: static, a nearby broadcast transmitter, military radar. In others, you may have some measure of control: Perhaps the neighbor would disconnect the TV antenna while playing Nintendo. Each of us has complete control over our own appliances and electronic gear. In any case, we can live in an electronically crowded world and still enjoy Amateur Radio.

Definitions of EMI Effects

Co-channel interference—happens when two signals, with nearly the same frequency, mix in a receiver. The mixing product (sometimes called a heterodyne or beat note) may be heard in the receiver output if it is an audio frequency within the receiver passband. Most FM receivers contain high-pass audio filters to eliminate subaudible tones used for repeater access. Such filters remove heterodynes below 300 Hz.

Co-channel interference between amateurs is common on the HF bands, but it is seldom labeled as such. IF shift, variable-bandwidth tuning, narrow-bandwidth filters and notch filters are all means of reducing HF co-channel interference.

VHF/UHF co-channel interference is generally noticed when enhanced propagation allows distant (60 miles or more) co-channel repeaters to interfere with one another. More information about propagation appears in *The ARRL Antenna Book*.

Capture effect—an FM receiver can be "captured" by the strongest of several competing signals if that signal is appreciably (6 dB) stronger than the others. Note that a heterodyne may be heard, even when the weaker signal is 30-40 dB below the capturing carrier.

Adjacent-channel interference is caused by a very strong carrier within either the IF or RF passband of a receiver. It can disrupt amplifier and mixer stages of a receiver and cause receiver degradation or desensitization.

Receiver degradation is a decrease of receiver performance that is caused by any unwanted signal on, or near, the receiver frequency. This problem appears in an FM receiver as an increase in the limiter or S-meter readings. It can be caused by electrical and/or RF noise. Suspect RF noise sources include crowded radio sites (IMD and desensitization), digital equipment and nearby transmitters (modulation side bands or transmitter spurious emissions such as phase noise).

Desensitization results from a very strong signal within the receiver RF pass-

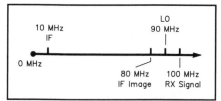

Fig 2—A spectrum illustration of an IF-image problem. Both the desired signal at 100 MHz and the unwanted IF image at 80 MHz mix with the 90-MHz local oscillator to produce signals at the 10-MHz IF. Such problems should be cured by increasing front-end selectivity to exclude the 80-MHz signal.

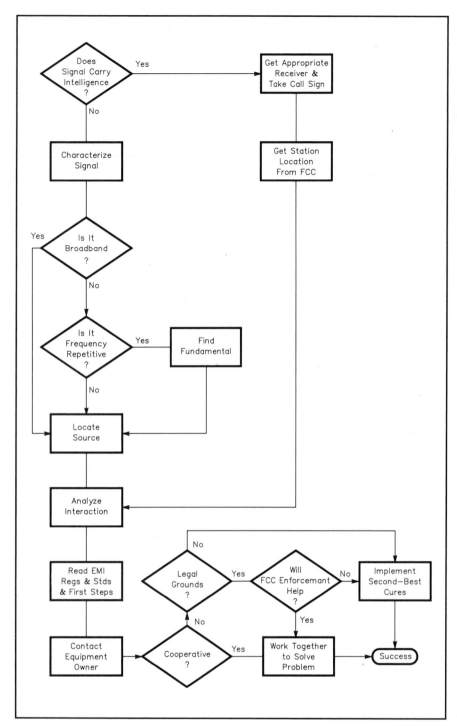

Fig 1—A flow chart for solution of EMI problems. "Characterize Signal" means observe and record any distinguishing attributes, such as pulse rate or times of day when the signal appears. "Analyze Interaction" means consider how the signal affects the receiver and formulate a possible cure.

band. A sufficiently strong signal may force any active stage in the RF or IF sections into gain compression and effectively decrease sensitivity.

In HF receivers, the effect is often called "blocking." Since most HF receivers experience spurious responses well (about 20 dB) before desensitization, the effect is often unnoticed.

When desense happens, the desired signal strength reduces in cadence with the amplitude of the interacting transmitter. Desensitization of an FM receiver may be indicated by a decrease in limiter or S-meter reading.

Sensitivity can be restored by reducing the strength of the incoming signal at the affected stage. This may be accom-

plished with directional antennas or filters if practical. In some cases, the physical or frequency separation between the interacting transmitter and receiver must be increased to provide the needed isolation.

IF-image interference is reception of unwanted signals that results from poor front-end filtering. For example, a 100-MHz receiver uses a 90-MHz LO and a 10-MHz IF (100 − 90 = 10). 80-MHz signals that reach the mixer also appear at the IF (90 − 80 = 10; see Fig 2).

All receivers are subject to IF image reception. The ability of receiver front-end filters to attenuate signals on the IF-image frequency (while passing desired signals) is called "image rejection." Any IF-image problems in multiconversion receivers usually arise prior to the filters in the first IF stage. Those filters tend to protect following IFs from signals on their image frequencies. Receivers with great separation between the RF and IF frequencies have better IF-image rejection.

Intermodulation distortion (IMD or "intermod") is the result of unwanted mixing, which may occur in any non-linear device (such as a diode). Common culprits are overdriven front-end transistors, mixers, class-C amplifiers and corroded connections or metallic junctions. Because IMD is a result of mixing, the products are mathematically related to the mixed signals.

For example, when two signals of frequencies A and B mix, signals at A, A + A, B, B + B, A + B and A − B result. Each signal, A or B, adds to and subtracts from itself (A − A and B − B are each zero, and don't appear as products) and the other signal. Unfortunately, each of the products may mix further with itself,

A, B, or any of the other products. The situation gets very complicated very quickly!

Products are often grouped by their mathematical *order*. The order of a product is equal to the sum of the coefficients of the signals mixed to make the product. The order of the product A + B is two because the coefficients of A and B are each one (and 1 + 1 = 2). When A + B mixes further with B, A + B + B and A + B − B are produced. A + B + B can be expressed as A + 2B (a *third-order* product).

Receiver IMD occurs when two or more unwanted signals mix in the receiver. It becomes a problem when the unwanted result is on the intended receive frequency. The most common example of receiver IMD occurs when two or three strong signals drive the RF-amplifier and first-mixer circuitry into nonlinear operation.

Consider three signals, 146.52 MHz, 146.46 MHz and 146.22 MHz mixing to produce 146.76 MHz (a third-order IMD product, A + B − C). This demonstrates how nearby carriers can cause IMD. Note that most front-end filters do not significantly attenuate commercial signals just outside the ham bands; such commercial signals are common factors in IMD interference.

Whenever a signal is multiplied via IMD, the width of the signal (deviation in FM, sideband width in AM) is multiplied as well. Depending on the modulation methods involved, and the type of detector used, information from the incoming signals may or may not be recognizable in the affected receiver output.

Third-order intercept is a mathematical fiction that predicts the signal level at which the strength of third-order IMD equals that of the desired signal. This figure of merit indicates IMD susceptibility. Higher values indicate better performance.

Spurious responses are unwanted signals *generated within the receiver*. An IF Image is a spurious response that is only related to the mixer injection frequency. There can be other spurious responses on a variety of frequencies that are not mathematically related to the injection frequency. Two important groups of spurious responses are related to the LO frequency. One group is related to LO harmonics, and the other is related to LO spurious emissions (noise).

Depending on various factors in the receiver, some LO harmonics may be more significant than others. Significant harmonics mix with other undesired signals and may translate them into the IF passband.

Nonharmonic spurious emissions in the LO can cause spurious responses as well. Since LO spurious emissions are usually much weaker than the LO fundamental, spurious responses are often relatively weak.

Spurious LO emissions are inherent in all LO circuitry and are usually removed with narrow filters. LO quality generally worsens as the number of multiplier stages increases.

Lastly, LO spurious emissions can appear as broadband noise. This noise is greatest near the fundamental frequency and decreases dramatically above and below that frequency.

Noise figure (NF) is a logarithmic (dB) measurement of wide-band noise factors (F) in a transmitting or receiving system. Here we will discuss receiving systems only. NF is a function of temperature: Colder devices generate less noise. Device structure also affects NF.

NF expresses system noise in dB. System noise can also be expressed as noise temperature (T_e in kelvins; 0 K equals −270° C). A T_e of 0 K is equal to 0 dB NF. The relationship between NF and temperature is shown in Fig 3.

The noise figure of a receiver is predominantly determined by the front-end amplifier stage (let's call it an LNA). As the gain of the LNA increases, the system NF approaches that of the LNA itself. Fig 4 shows some common NFs versus LNA gains. Lower NFs indicate better sensitivity. A typical off-the-shelf VHF/UHF receiver has an NF from 5 to 10 dB. A GaAsFET LNA can improve the system NF to less than 1 dB.

There is a point where increased gain does not appreciably improve the NF. Since increased gain creates a dramatic increase in receiver IMD susceptibility, it makes sense to trade front-end gain for

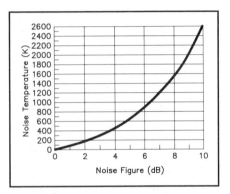

Fig 3—This curve shows the relationship between noise figure (NF) and noise temperature (T_e).

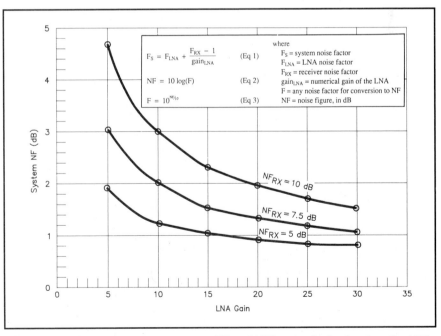

where
$F_S = F_{LNA} + \dfrac{F_{RX} - 1}{gain_{LNA}}$ (Eq 1)

$NF = 10 \log(F)$ (Eq 2)

$F = 10^{NF/10}$ (Eq 3)

F_S = system noise factor
F_{LNA} = LNA noise factor
F_{RX} = receiver noise factor
$gain_{LNA}$ = numerical gain of the LNA
F = any noise factor for conversion to NF
NF = noise figure, in dB

Fig 4—The relationship between system NF and LNA numerical gain when the LNA noise factor is 1.122 (NF = 0.5). Eq 1 yields the total system noise factor. Eqs 2 and 3 are for conversion between noise factor (F) and noise figure (NF, in dB).

lower IMD in some interference cases.

In addition, noise from many of the sources we are about to discuss may not be reduced as low as 1 dB NF. Any sensitivity beyond the ambient noise level is not useful. Therefore, it is valuable to have a means of measuring the maximum useful sensitivity at a given installation. A System Sensitivity measurement is used as a tool to quantify practical sensitivity limits.

Measurement of System Sensitivity

System sensitivity (SS) is the measurement of receiver sensitivity in a real-world installation. This measurement quantifies the system sensitivity with antennas, filters and any peripheral equipment connected. Fig 5A shows the setup for measuring SS.

An RF tap (Fig 5B) is a device that allows one to inject or measure a signal on a feed line without significantly affecting other signals on the line. The RF tap must match the source and load impedances. (The sample terminal generally is not matched.) A 10-dB attenuator at the sample terminal protects the signal generator from the mismatched tap. A directional coupler can be used in place of the RF tap.

Install the RF tap and load at the receiver input. Use a SINAD meter or audio voltmeter at the receiver audio output for an accurate measurement. Connect the signal generator to the tap. Measure and record the receiver sensitivity (remember to subtract 40 dB from the generator level). Now install the antenna in place of the load and repeat the measurement. If the second measurement is 6 dB higher than the first, there is at least 6 dB of receiver sensitivity degradation from noise and the antenna system. This is the SS.

SS measurements can be made on both FM and SSB systems. System sensitivity is an important concept, and the measurement can be used to quantify the effect of receiver interference.

SIGNAL PROFILES

All kinds of signals may interact with our equipment. While noise is any unwanted signal (much as a weed is any unwanted plant) that interferes with our reception of a wanted signal, it is wise to remember that one person's noise is another's livelihood. In truth, the terms

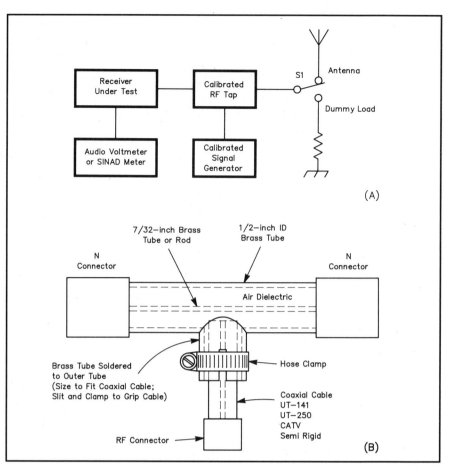

Fig 5—A shows the test setup for System Sensitivity (SS) measurement. Measurements of minimum-discernible signal (MDS) are made and recorded with the system terminated in a dummy load, and then the antenna. The difference between the two MDS measurements indicates the degree of receiver degradation resulting from the antenna system and incoming signals. B shows details of the RF tap.

"noise" and "interference" are subjective.

Here, we place responsibility to facilitate logical thought processes while troubleshooting receiver EMI problems. For each kind of noise we'll list a description, assign responsibility and suggest cures. Don't take the assignment of responsibility as a license for belligerent or uncooperative behavior. Let's consider noise in two classes: natural noise and man-made noise.

Natural Noise

Natural noise is often classified as galactic, thermal or atmospheric.

Galactic Noise

Galactic noise is radio energy that is emitted from the galaxy center. This noise is usually significant only for EME operators, who have receiving systems that are more sensitive that the norm. It is broadband in nature, but it is greatest at 50 MHz

and lessens as frequency increases. Galactic noise is not a problem for the terrestrial and satellite VHF/UHF operator.

Responsibility: No one.

Cures: Effects of galactic noise may be reduced by increasing antenna directivity and/or decreasing receiver bandwidth.

Thermal Noise

Thermal noise is the noise that all objects emanate because of their temperature. It is a consideration only in EME and radiotelescope applications.

Responsibility: No one.

Cures: Thermal noise at the antenna may be reduced by increasing directivity. Circuit thermal noise may be reduced by cooling circuitry to extremely low temperatures.

Atmospheric Noise

Atmospheric noise (static) occurs as

lightning discharge, rain static and corona noise. The massive arc of a lightning bolt emits radio energy all across the spectrum. Most Amateur Radio stations cannot detect lightning static above 144 MHz.

Rain Static

Rain static occurs when charged droplets of water hit an antenna and dissipate their charge to ground through the station equipment. It produces pulses of broad-band noise that usually do not change as the antenna is rotated. Rain static can be heard from LF to VHF.

Corona Noise

Corona noise is prevalent at high-wind, low-humidity sites, when there are nearby lightning storms or the ground has a high static charge (relative to the air). When it happens, there may be visible signs such as St Elmo's fire. It sounds like a squeal, or raspy noise. It can change pitch, usually rising.

Responsibility: No one.

Cures: Antennas protected from rain (as by fiberglass radomes) are not affected by rain static. Noise blankers can eliminate static to some extent. Corona noise calls for lightning-protection measures, such as a grounded $1/2$-λ transmission line stub (which dc grounds the antenna). The grounded stub, however, only works over a narrow range of frequencies.

Man-Made Noise

Man-made noise is a summation of all of the noise that man creates in the environment. Man has introduced many kinds of noise to the world. Here are some of the most common sources.

Digital Noise

Digital noise is rapidly becoming the most common and difficult noise problem. This noise results from switching in logic circuits. The switching produces square waves. Square waves contain energy not only at the fundamental frequency, but also at all odd harmonics thereof. Digital noise can appear as a coherent carrier, wide-band noise or a modulated carrier.

Responsibility: In some cases, the owner is responsible.

The FCC approves some computers for homes and others for business environments only. If a business-class computer is being used in a home environment, the owner is responsible.

Electronic appliances are generally not to interfere with licensed communications. If a device does interfere, it should be returned to the manufacturer as defective.

Cures: The noise generator may be placed inside a shielded enclosure. Bypass leads that exit the enclosure. Be aware that bypass capacitors in data leads may hamper operation of the digital circuit.

The Computers chapter gives more information about interaction in digital circuits.

Power-Line Noise

Power-line noise is often intermittent in nature. Noise from a single pole has a distinct pulse rate that a noise blanker can usually handle very well. (Sometimes power-line noise gets very strong at the beginning of a rain storm.)

In an SSB or AM receiver, power-line noise is noticeable as a 60-Hz buzz. In an FM receiver, power-line noise is not heard until a carrier opens the squelch. Then one hears the buzzing of power-line noise in the FM receiver exactly as with an SSB or AM receiver. (Power-line noise is an AM phenomenon that usually is not detected in a true FM receiver.)

When many poles contribute to the problem over a long distance, the pulse rate is less regular, and a noise blanker may not help. This later noise is called Gaussian noise.

Power-line noise is generally caused by arcing at the hardware on a power pole, which generates 60-Hz pulsating RF from 100 kHz to 1200 MHz. Many variables determine the amplitude and frequency of the noise: weather, temperature, hardware characteristics. Thermostats and light dimmers may display noises similar to power-line noise.

The noise usually increases as frequency decreases, but noise-level v frequency behavior can run counter to the norm. For instance, it is possible to have more noise on 2 m than 6 m. This results from resonances in the power-line hardware.

Power-line noise increases with antenna gain and height. (The effect of antenna directivity on the noise level decreases with frequency.) As antenna height increases, the distance to the radio horizon increases. Thus, the antenna "hears" more power-line noise sources. (It is possible to hear a city's noise from 10 or more miles away, especially on

6 m.) Power-line noise on high-gain non-terrestrial antenna systems tends to decrease as antenna elevation increases.

Responsibility: The power company is responsible to a limited degree.

Cures: It may be difficult to deal with power-line noise. First study the Power-Line Noise section of the Electrical Interference chapter. Rotate the antenna to pinpoint the source if possible. A power line that is very close and very strong may be difficult to pinpoint. Attempt to locate the source and work with your power company to solve the problem.

In cases where that approach fails, a rotatable directional antenna may be the only practical solution. From the noise description, we can see that noise blankers (adjacent strong signals may render noise blankers useless), changes in antenna orientation, design and height can reduce power-line interference at the receiving station

Vehicle Ignition Noise

Vehicle ignition noise is a rhythmic "ticking" that varies with engine speed. Since there is usually only one plug-wire arcing, and that not on every engine revolution, the noise is usually about 5 Hz at idle.

Ignition noise is worst at low frequencies. It lessens at higher frequencies, yet it is not unusual to hear ignition noise from a nearby engine up to 450 MHz, and occasionally it can be heard up to 928 MHz.

Ignition noise may affect an FM receiver in two ways: (1) There may be an S-meter indication of signal strength without breaking the receiver squelch. (2) Noise pulses may compete with a weak signal that does open the squelch.

Responsibility: The vehicle owner is responsible, but ignition noise is so transient in nature that the question is moot.

Cures: First look for bad connections in the installation (especially ground connections). In old installations, check vehicle ground straps at the engine, hood and firewall. Look for arcing at the ignition wires (replace them if needed). Have the vehicle-specified resistor spark plugs been replaced with regular plugs? In a new mobile installation, consider antenna placement (locate the antenna as far from the motor as practical). Route the radio power wires as far from the motor as possible and filter the power leads at the radio.

Consumer Electronics

Home electronic consumer devices

are becoming more of a problem as the number of the devices increases. Many devices from home computers to microwave ovens use digital circuitry that may emit energy on the amateur bands. Low-power unlicensed transmitters in baby monitors, cordless telephones, remote-control vehicles and other devices can radiate on the amateur bands. There may be thousands in any neighborhood.

Consumer devices most often interfere by producing low-level signals or noise on amateur frequencies. This means that you must either filter the consumer device or remove it from service. If you're lucky, your neighbor will remove the device. (Many devices, such as cordless phones, carry labels warning the owner that they must do so if the device causes interference.) If you can convince the FCC that the device is causing harmful interference (this is defined in the Power Lines and Electrical Devices chapter), they can take action. The general consumer-device interference level will likely increase in the future.

Ultrasonic Pest Eliminators

Ultrasonic pest eliminators are an unusual interference source on 2 m. They sweep at audio frequencies and have parasitic oscillations that can go from 140-160 MHz. Interference appears as a repetitive burst of noise with the same rate heard from the pest eliminator. Other frequencies can be affected, but 2 meters is the norm.

Responsibility: Device owner.

Cure: The parasitics can be reduced with bypass capacitors, but the equipment cases are very difficult to open. Remove the device from service, and return it to the seller.

Heating and Lighting Controls

Heating and lighting controls produce noise that is similar to power-line noise, but with pulse rates that vary from 60 Hz. This noise results from power-switching devices that do not switch cleanly.

Responsibility: The FCC rules (Part 15.7 subpart E and Part 15.4 subpart B) say that if the consumer device is causing harmful interference then it must be turned off.

Cures: Low-pass filters on the device leads should help. Remove the device from service and contact the manufacturer.

TV Horizontal-Sweep Oscillators

TV horizontal-sweep oscillators oper-ate at 15.75 kHz. A faulty TV receiver may couple sweep signals to the antenna or ac line. Such interference is evident as steady carriers at 15.75 kHz and all harmonics thereof. Problems are usually very local in nature (several yards to several blocks away).

Responsibility: TV set owner.

Cures: The TV must be repaired or removed from service.

TV Preamp Oscillation

TV Preamp Oscillation can be very difficult to locate because it is intermittent in nature and may vary in frequency. (Environmental conditions usually cause the oscillations to vary in frequency.)

Preamps commonly oscillate in the 70-cm ham band and emit a varying-frequency carrier with 60-Hz modulation. It sounds almost like the buzz from a TV video sideband. It is not uncommon for the owner to experience poor picture quality when the preamp is oscillating.

Responsibility: Manufacturer.

Cures: Poor design is the cause of this problem, and it is best to return the preamp to the manufacturer.

Other Services

TV-Broadcasts

TV-broadcast EMI may appear as either on-channel or IMD products. (Channel 2 is the most common source.) The most common TV RFI problem is on-channel energy that either passes the TV vestigial-SSB filter or some IMD product of the video and audio carriers.

Video sidebands (5 MHz wide) can go across the whole 6-meter band. The sidebands sound like carriers with a 60-Hz buzz on an SSB or AM receiver. These carriers are 15.75 kHz apart.

IMD between the TV transmitter video and audio signals causes a wideband FM carrier about 9 MHz (depending on the TV station offset) away from the video carrier.

Responsibility: TV broadcast stations are nearly always within FCC specifications, so they bear no responsibility.

Cures: Apply appropriate filters at the receiver.

CATV (Cable Television)

CATV (cable television) noise is common to all areas with CATV systems. The noise appears in two forms: TV channel energy and swept signals used to equalize RF frequency response.

TV-channel leakage is both AM and FM in nature. The video is AM, and the audio is FM. There are 15.75-kHz carriers (usually with a 60-Hz AM modulation) adjacent to the video carrier frequency. These carriers extend upward in frequency towards the TV signal audio. These carriers also extend down in frequency but only 1 MHz or so. The FM portion of the TV signal deviates 50 kHz.

Radiation of channel-E energy is the most common CATV problem for hams. The video of channel E is on (or near) 145.250 MHz. This may not show anything but a signal-strength indication on an FM receiver because the video is an AM signal. The Televisions chapter shows the frequencies for the various CATV channels.

Responsibility: CATV operator.

Cures: The FCC has been watching CATV franchises more closely over the past few years and systems have generally improved. Solving the amateur's problem usually solves other problems as well. Therefore, it is to the benefit of the system operator to fix CATV RFI. The most difficult part of resolving a CATV problem is reaching the correct technical personnel.

Do not consider CATV radiation interference unless it interferes with an on-frequency amateur signal. The FCC does not see merely opening the squelch of a quiet receiver as harmful interference.

FM Broadcast

FM broadcast transmitters can cause IMD problems, especially in 6-m receivers. The products are weak signals up to 100 kHz wide. Since broadcasters are usually within FCC specifications, suspect the receiver.

Responsibility: Receiver owner.

Cures: A low-pass filter with an appropriate cutoff frequency solves FM-broadcast interference to 6-m receivers.

Harmonic Interference

Harmonic interference may happen when unwanted transmitter harmonics are not properly filtered. Harmonics are inherent in amplifiers and other circuitry. The FCC specifies allowable harmonic radiation for all services. This problem is more common when the third harmonic of a transmitter falls on the frequency desired in a nearby receiver.

External IMD is a common cause of harmonic interference. External-IMD harmonics are not affected when an appropriate filter is installed at the transmitter. External IMD is discussed fully later.

Responsibility: The transmitter owner is responsible for transmitter-generated harmonics.

Cures: For transmitter harmonics, install an appropriate transmit filter.

Radar

Radar interference is experienced on 6 meters and every band above 1.25 cm. The Amateur Radio Service shares these bands with the US military (except for 6 meters). Radar (which sounds like mosquitoes, sweeping every few seconds) can cause significant interference on the 70- and 33-cm bands in the US and abroad. Any 70-cm repeater within 160 km of a PAVE PAWS radar installation will likely receive interference. 70- and 33-cm repeaters may interact with military shipboard radar. Radar power levels can be 30 dB, or more, stronger than the average UHF user. Tropo-enhanced propagation can extend radar interference out to 500-3000 km.

Radar Activity Summary

6-m: A variety of wind-shear radar systems are being installed worldwide. Their center frequency is below the 6-m band. They run very high power and may affect nearby amateur operations on 6 meters.

1.25-cm radars have been decommissioned and should not be on the air any more.

70-cm has two radar systems at 420-450 MHz. Shipboard systems cause problems in the coastal areas of the US. It is possible to hear shipboard radar from over 1600 km with tropospheric ducting. These systems have a characteristic whine that covers many MHz. The southern California coast has the worst shipboard radar problems because of many military bases and good year-round tropo' propagation.

The other kind of 70-cm radar is called PAVE PAWS. It is an over-the-horizon (OTH) system designed to look for submarine-deployed missiles. It also tracks thousands of objects in low earth orbit, like amateur satellites. This radar uses frequency-hopping techniques, so it sounds like random pulses on a weak or strong signal. All amateurs who live within 150 miles of a PAVE PAWS site have a 50-W transmit-

ter power limit from 420-450 MHz. Ship and OTH radar systems do not usually break a receiver squelch.

33-cm has shipboard radar like that on 70 cm.

23-cm band has radar that sounds similar to that on 70 cm, except that the repetition rate is higher. 23-cm radar comes from commercial aircraft.

There is also radar from military sources on higher frequencies. The Soviet OTH radar is off the air as of this writing. The US is canceling OTH radar systems on the east and west coasts.

Responsibility: No one.

Cures: Directive antenna patterns may help, but reflections from mountainous terrain can negate the effect of narrow-beam antennas. Unfortunately, very little can be done about this problem. Military ships are not supposed to use radar close to shore (but they sometimes do).

Miscellaneous Sources

Phase noise in the LO of an HF transceiver can cause two problems: (1) The transmitter emits a signal with excessive noise on either side of the carrier center. (2) A receiver responds to strong adjacent signals with a noise increase that is a function of the signal amplitude.

The interference sounds like soft white noise (like the noise in an SSB or unsquelched FM receiver with no signal), which is modulated by a voice. The same noise appears when the receiver is tuned to frequencies on both sides of the transmitter. The noise appears throughout the transmitter and at the antenna. Phase noise in your receiver would have the same effect.

Responsibility: Transmitter phase noise is usually within the FCC spectral purity requirements, so there is no assignment of responsibility.

Cures: This problem is inherent in some transceivers; that cannot be helped. In other transceivers, it may indicate some kind of equipment failure. If so, the equipment should be repaired.

Transmitter IMD

Transmitter IMD takes two forms. Internally generated transmitter IMD results when non-linear multiplier stages produce undesired harmonics. Power amplifiers can also generate IMD. Therefore, filters are necessary to remove a variety of spurious signals (which are related to any multiplier stages and the LO

frequency) from the output of a transmitter. The level of internally generated IMD is essentially set when a transmitter is built. Therefore, this form of IMD should only appear as an RFI problem when new equipment is installed or when there is a component failure in an existing system.

Externally induced transmitter IMD is caused by the injection of RF into the transmitter. Such energy can enter via the power supply, direct radiation or the antenna feed line. Good construction practices usually eliminate power supply and direct-radiation induced transmitter IMD.

Responsibility: If FCC spectral-purity requirements are exceeded, the owner is responsible.

Cures: Band-pass filters or directional antennas may help at HF. Cavity filters and circulators have proven good weapons against signals on the feed line at VHF and above. Faulty equipment should be repaired of course.

External IMD

External IMD is generated when one or more signals mix outside a receiver or transmitter. In general, a diode junction may be formed at any joint where there is not a good electrical connection between conductors (the "rusty bolt" effect). Any diode may act as a mixer. Therefore, mixing may occur in any poor or corroded connections, such as connectors, antennas, tower joints, guy lines, gutters or metal siding. The same mixing action may take place in antenna switching diodes of electronic equipment such as two-way radios or VCRs—even when the equipment is not operating.

The symptoms of external IMD are similar to those of IMD in general. Suspect external IMD once all transmitters and receivers in the area have been found faultless.

Responsibility: Equipment owner.

Cures: This IMD only happens when strong RF excites the "diode." In order for this to occur, there must be a significant "antenna" to gather RF energy—usually a conductor approaching 0.1 λ or longer. The effect can be cured by grounding the "antenna" or providing a good electrical path at mechanical connections. If the culprit is a diode inside equipment, install an appropriate filter to isolate the diode from the exciting RF energy.

RF Connector Noise

RF connector noise is often over-

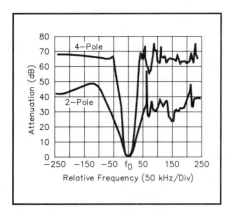

Fig 6—Crystal-filter frequency response. These curves are for 2- and 4-pole filters as manufactured by Piezo Technology, Inc.

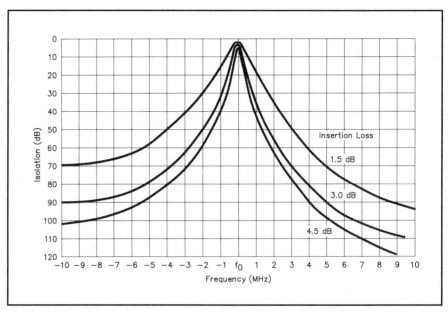

Fig 7—Band-pass cavity frequency response. A three-cavity filter for 450 MHz is shown.

looked. High-power systems should use connectors chosen to minimize connector-generated IMD. IMD is a problem with ferrous connectors. Other connector-generated noise comes from connectors contaminated with water. Power applied to a wet/corroded connector can generate noise over a wide range of frequencies.

Responsibility: Equipment owner.
Cures: Replace the faulty connector.

EQUIPMENT PROTECTION

Shielding

An RF shield is the most obvious kind of RFI protection for transmitters, receivers and peripheral equipment. Ideally, a shield keeps energy from entering and leaving the shielded equipment.

The effectiveness of a metal equipment cabinet (shield) varies with frequency. Although solid metal is the best shield material, screen (or metal with ventilation holes) provides adequate shielding in many EMI cases.

High-quality coax (with either double braid or a solid shield) is important in order to reduce feed-line leakage. Coax leakage increases with frequency. Hence, high-quality coax is most important in the VHF/UHF region. Shield leakage can cause a variety of problems that degrade receiver performance.

Some manufacturers produce shielded plastic cabinets by mixing conductive powders into the plastic before it is molded. Others use conductive coatings to shield plastic equipment cabinets. A spray EMI-RFI coating is available from GC-Thorsen (catalog no. 10-4807). (Their address appears in the Suppliers List, at the back of this book.)

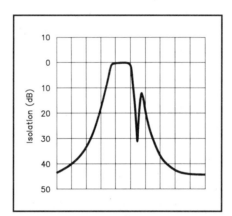

Fig 8—Window cavity-filter frequency response. This filter has a notch at 455 MHz. The plot is centered on 450 MHz with 10 MHz per horizontal division.

Use external shields whenever possible, and do not work on any equipment unless you own it and are qualified to do the work. Avoid installing shields inside existing equipment. Since shields are conductive, they may cause short circuits with the potential for equipment damage or personal injury.

Lumped-Constant Filters

Filters assembled from discrete components such as capacitors and inductors are often called "lumped-constant filters." They include the ac-line filters, high-pass and low-pass filters that are most familiar to hams. Details of ac-line filters appear in the power-line chapter. High-pass and low-pass filters appear in the Televisions and Transmitters chapters, respectively.

Crystal Filters

A crystal filter is a narrow bandpass filter that uses piezoelectric material (crystals) as tuned circuits. Crystal filters are common in the IF stages of HF transceivers. The high-frequency limit of crystal filters is about 200 MHz. Fig 6 shows the bandpass characteristics of crystal filters manufactured by Piezo Technology, Inc.

The filter is designed to eliminate strong adjacent signals, which cause receiver desensitization. The filter is best for signals from about 25 kHz to 300 kHz away from the center frequency. Beyond 300 kHz, cavity filters are more practical than crystal filters. Crystal filters are used to remove interacting signals that are very close to the operating frequency, but which do not cause on-channel wide-band noise. Crystal filters exhibit insertion loss from 4 to 8 dB.

Cavity Filters

The 1/4-λ cavity is the most important protection device for VHF/UHF receivers. It is essentially a 1/4-λ piece of solid-shield coax that is shorted at one end. RF is coupled into the cavity via an inductive or capacitive probe.

Cavity filters are designed to either pass or reject (notch) a narrow band of frequencies. Certain designs pass one narrow band while rejecting an adjacent narrow band. Cavities may be arranged in many configurations. For example, a typical VHF repeater duplexer has three bandpass/reject cavities on each side, while a UHF version usually has only two on each side.

Both cavity filters and cable stubs are sensitive to their placement in the system; reactances of other feed-line components can affect performance. Also, harmonic resonances may be important, a 2-meter 1/4-λ cavity or stub is 3/4 λ at 70 cm.

The *bandpass filter* or *pass filter* is a two-port device that allows a relatively narrow range of frequencies to pass with little attenuation. The bandwidth of this filter is a function of its Q. Q is related to the ratio of the center- and outer-conductor diameters. (The formula is described in *The ARRL Handbook*.) Greater cavity diameters yield a higher Q. Filter Q can also be increased by connecting two cavities (with critical lengths) in series. This actually offers a better pass curve and less loss than is possible with a single cavity. Fig 7 shows the curves for a three-cavity band-pass filter.

A *"window" filter* has many series-connected bandpass filters to achieve a relatively wide bandwidth with very steep skirts. Bandwidth is a function of loop coupling; it can range from 300 kHz to many MHz. Skirt steepness is a function of the number of cavities. The coupling loops and connecting cables are carefully designed to provide the desired characteristics.

When one antenna serves multiple receivers on different frequencies, window filters are used to protect the receivers from adjacent transmitters. One application of a window filter is shown in Fig 8. Window filters on different bands can be connected together with special cables to split off the different bands on a dual-band filter as shown in Fig 9. This latter configuration is called a multiplex filter. Such filters are generally used in receiver-combining systems.

A *band-reject* or *notch filter* is nearly the same as a bandpass filter except that there is only one port on the device and it removes, rather than passes a narrow band of signals. Fig 10 shows a typical notch-filter response curve. All of the effects of Q are the same as in the bandpass configurations. Greater Q gives a deeper and narrower notch. The cable lengths between cavities are critical. The passband must be at least 800 kHz from the notch.

Pass/reject filters use a combination of the characteristics of bandpass and

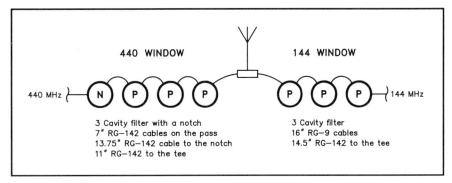

Fig 9—Window filters may be used to provide dual-band coverage with a single feed line or antenna system.

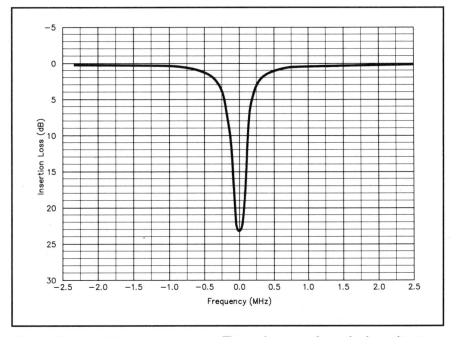

Fig 10—Notch-cavity response curves. Those shown are for a single cavity at 150 MHz.

bandstop filters to make a filter that fills a special need. This filter has a deeper notch per cavity than a notch filter, and its notch can be closer to the passband. The passband characteristics are worse than those of a standard bandpass cavity, especially far (many MHz) from the center frequency. Pass/reject filters are usually two-port devices and have reactive sections added to the coupling probes. An inductive probe raises the notch above the passband, while a capacitive probe lowers the notch below the passband. Fig 11 shows filters with both inductive and capacitive probes.

A *duplexer* is a system of cavities (with critical length cables) that allows a repeater to use one antenna for simultaneous transmission and reception.

Duplexers normally have multiple pass or pass/reject cavities. Fig 12 is a response graph of a Motorola T1504 pass/notch duplexer. Fig 13 shows a T1507 pass/pass duplexer.

It is important to understand the pros and cons of each of these duplexers. A *bandpass duplexer* protects a receiver from its own and other transmitters. It also protects the transmitter against other strong signals on the antenna, which may cause IMD in the transmitter output circuits. The transmitter-to-receiver isolation is about 50 dB.

There are two problems with this duplexer. (1) A GaAsFET preamp can be overloaded by the transmitter's incident signal. (2) Some transmitter noise can interfere with the receiver. On the posi-

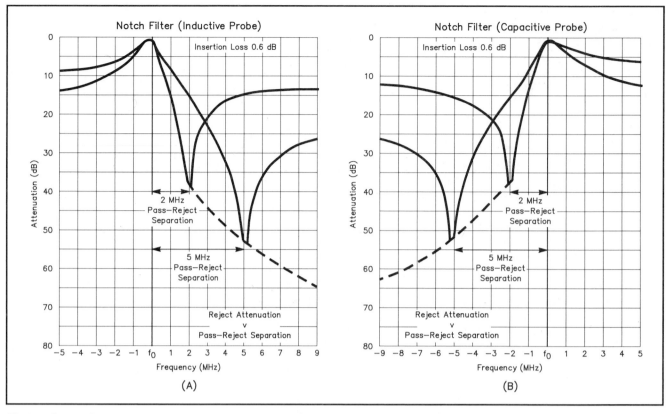

Fig 11—Pass-reject cavity-filter frequency response. 450-MHz cavities shown.

tive side, you protect other users better from your transmitter, and your receiver is protected better from other transmitters.

A *pass/reject duplexer* protects its receiver from its transmitter much better than does the pass duplexer. (Transmitter-to-receiver isolation is around 100 dB.) The negative aspects are: (1) It doesn't protect the receiver well from strong signals 5 to 20 MHz away. (2) It doesn't protect nearby receivers from the transmitter very well. The positive aspects are: (1) Transmitter-to-receiver isolation is good. (2) Insertion losses are low.

The best kind of duplexer for a crowded site uses the positive aspects of both pass and pass/reject cavities in one duplexer. Unfortunately, duplexer manufacturers do not make them.

Such *hybrid duplexers* are made by integrating a pass/reject with a pass duplexer to make a duplexer with 90-dB notch and pass characteristics (20-dB better than a normal pass/reject duplexer). The cables between the pass and the pass/reject cavities must be specially

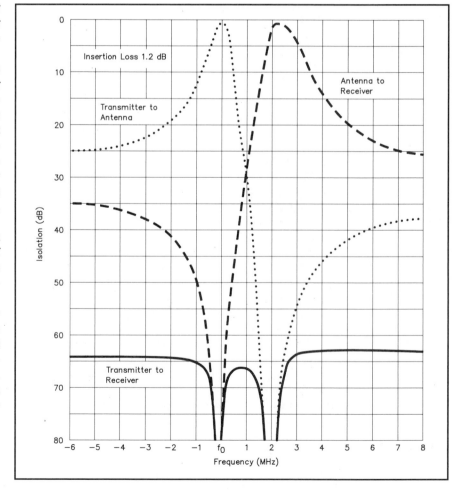

Fig 12—Pass-reject duplexer frequency response. A Motorola T1504A, AF (450 MHz) is shown.

made. Fig 14 shows a hybrid duplexer arrangement. The characteristics are shown in Fig 15. This example uses Motorola T1500 series cavities.

In the repeater situations, a complete comprehension of duplexer characteristics allows one to troubleshoot interference problems in a sophisticated manner.

Coaxial-cable stubs may be 1/4 λ or 1/2 λ long. They are parallel connected to the feed line via a "T" connector. In 1/4-λ form, an open stub is used. In the 1/2-λ form, the stub is shorted. In both cases, the end at the connector presents a short circuit across the feed line at the stub resonant frequency. A 1/2-λ stub can be used to suppress pulses from nearby lightning strikes if the shorted end is grounded. It also protects equipment from static effects, such as corona and rain static. It does not protect against direct or secondary lightning strikes.

Coaxial-cable stub Q is very low compared to that of a cavity. (The notch may be only 10 dB.) Stubs are discussed in the *The ARRL Handbook* and *The ARRL Antenna Book*.

Isolators/Circulators

A circulator has three ports. Power flowing into any one port appears at the next port, with the first port assumed to follow the last. When one port is resistively terminated, an isolator is formed. When properly connected, an isolator prevents reflected power from appearing at the input port. (We will not discuss the internal operation, and from here on, we will make no distinction between the two devices.)

Above 30 MHz, isolators are very compact, with about 20 to 40 dB of isolation. Two or more isolators may be series connected to achieve greater isolation. Isolators are very important tools for the transmitter-IMD troubleshooter. Fig 16 shows a basic single isolator.

Isolators are generally used on transmitters to perform two functions: (1) They eliminate power reflected from the antenna system. (2) They prevent other signals from causing IMD in the transmitter output stage.

Ferrite isolators have drawbacks: (1) Ferrite devices are not linear in nature; they may produce harmonics. A transmit filter is the best way to remove harmonics. (A 1/4-λ cavity does not suppress odd harmonics.) Normally, 10, 6 and 2-m transmitters include low-pass filters. (2)

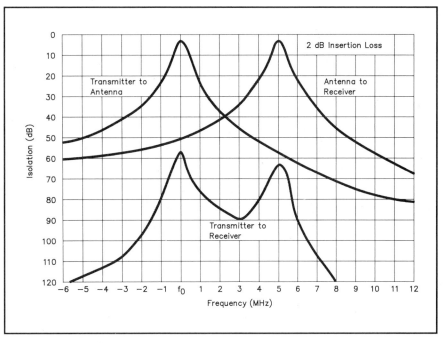

Fig 13—Four-cavity band-pass duplexer frequency response. A Motorola T1507A (450 MHz) is shown.

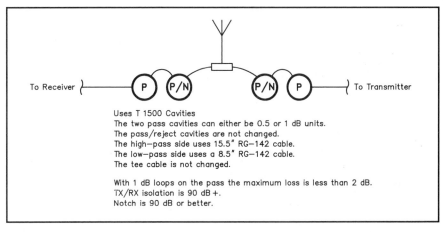

Fig 14—A hybrid pass-notch cavity duplexer arrangement for 440 MHz.

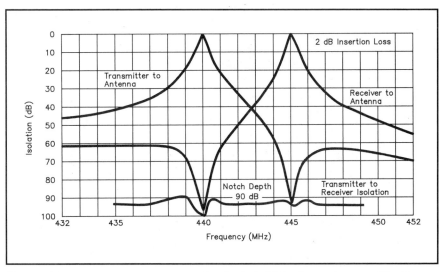

Fig 15—Frequency-response curves for the hybrid duplexer shown in Fig 14.

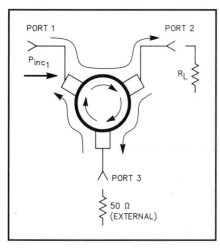

Fig 16—A single-stage ferrite isolator.

Isolators are ferromagnetic devices; they should not be mounted near ferrous (steel) surfaces.

Attenuators

As discussed in the introduction to this chapter, some receivers are subject to interference because they are too sensitive for a given environment. In such cases, a simple resistive attenuator can be a valid EMI cure. Because IMD and desensitization are not linear phenomena, a few dB of attenuation at the receiver input may yield many dB of protection. Information about constructing attenuators appears in *The ARRL Handbook*.

Chapter 12

Computers

By Philip R. Graham, KJ6NN
723 Berkshire Pl
Milpitas, CA 95035

Computers are becoming a standard appliance of the modern ham. They can be purchased for about the same price as a very good typewriter of 10 years ago. It is no wonder that so many hams have them; they are great tools for tracking satellites, logging contacts and digital communications (AMTOR, RTTY, and packet). A computer in a ham shack can present some special EMI challenges.

The problem is that computers use RF energy (in the form of clock signals) to control the CPU, memory, disk drives, keyboards, video, printers and many other devices. Each digital circuit can radiate RF energy that can be heard with amateur receivers (normally called RFI or EMI). Conversely, each of these devices is a path back into the computer for RF from the transmitter. The computer and transmitter are each a threat to the other. This chapter discusses the problems of and solutions to using a computer in a ham shack.

GOVERNMENT HELP

In 1991, there are two systems for measuring computer RFI. The FCC classifies computers for sale in the United States. The German system (VDE) is presently used in most of Europe and other parts of the world. The VDE rules are a subset of the existing CISPR (International Special Committee on Radio Interference) rules. Other countries generally use systems based on either the FCC or VDE standards, with local requirements added. Each of these systems sets methods of measurement and radiation limits for the computer to be classified. The two systems are very similar. Many computers are manufactured

and tested to meet both standards. (New European Standards, prefixed "EN," will be phased in throughout Europe beginning in 1992.)

In both systems there are two levels of compliance. FCC class-A computers are for use in a commercial environment (where interference to other devices can be controlled). If there are interference problems, the equipment user is required to correct the problem at his own expense. FCC class B is more stringent, for computers used in residential environments.

VDE level A requires the owner to notify the government before using the equipment. Level B is a general permit to use the equipment without government notification. Unlike the FCC (which is a government agency), VDE is an independent testing house much like UL (Underwriters Laboratories) in the US. VDE is authorized by the German government to issue the general permit (level B); and VDE issues the test report required for level A.

Figs 1 and 2 show that the level B requirements are at least 10 dB more stringent than the level A requirements for both FCC and VDE. 10 dB is a significant difference, and a level-A device would provide significant interference! The "FCC §15.105" sidebar explains owner responsibilities for both classes of digital equipment.

All computers for home use should conform to FCC or VDE class B. These computers are very quiet and should have minimal EMI problems. Any problems should respond well to the treatments given later in this chapter. If the computer does not meet FCC or VDE level B, don't panic! The solutions work for class-

A computers as well, although it may take more of them to achieve acceptable results.

Classification Criteria

Computers are tested for RF energy in several ways. Both FCC and VDE measure conducted and radiated signals. VDE measures magnetic strength (10 kHz to 30 MHz) as well. For the conducted-interference tests, the computer is placed in a shielded room, and a spectrum analyzer is used to monitor RF leaving the computer through the power cable. This test covers 450 kHz to 30 MHz for FCC tests, 10 kHz to 30 MHz for VDE B and 150 kHz to 30 MHz for VDE A.

For the radiated-interference test, the computer is placed on a turntable. A movable antenna is used to measure radiation at heights from 1 to 4 meters. The antenna is 3 meters from the computer for FCC-B tests, and 10 meters for all others. The computer is rotated and the antenna is raised to measure the maximum RF radiation from the computer. This test is performed from 30 MHz to at least 1 GHz, and it can go much higher depending on highest frequency used in the computer.

The test results are submitted to the FCC or VDE to show compliance. The FCC and VDE often purchase equipment from stores to verify compliance. For this reason, many computer manufacturers have internal test programs to verify that their computers meet all requirements. It is interesting to note that these regulations are not only meant to protect hams from noisy computers. They protect television, radio and other radio services from interference. They also make our lives much easier, because the tests cover amateur frequencies as well as broadcast

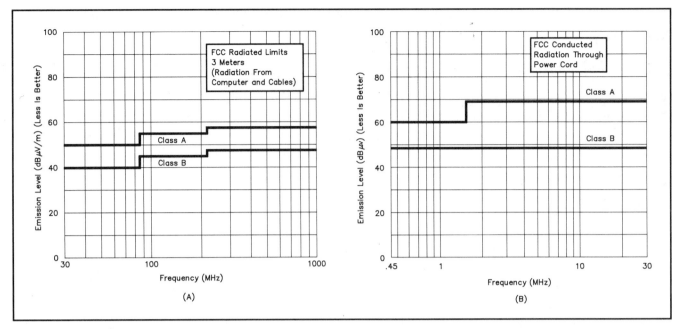

Fig 1—(A) shows the FCC emission levels for radiated emissions from computers. (B) shows the levels for RF conducted through the ac-line cord.

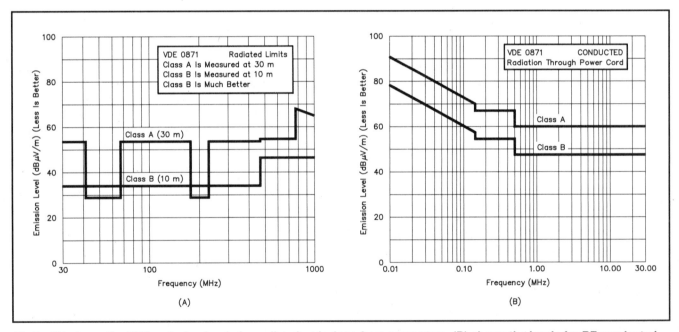

Fig 2—(A) shows the VDE emission levels for radiated emissions from computers. (B) shows the levels for RF conducted through the ac-line cord.

and business frequencies.

EMI CAUSES

Computers use square-wave timing signals to control their various components. Typical clock frequencies (4.77, 8, 10, 12, 16, 20, 25 and 33 MHz) are spread throughout the HF spectrum. Unfortunately, square waves are rich in harmonic RF energy that can disturb a receiver. In the ideal case, no RF energy would leave the computer cabinet to affect radio receivers. In reality, RF often does leave the computer and causes EMI problems. If there were only one clock frequency the solution would be straightforward, but there are usually many different frequencies in a computer. Oscillators may be found on the CPU board, disk drive, disk controller, serial-interface board, video board, keyboard or pointing device (mouse). If you look to see what crystals your computer uses, check every card and plug-in device!

The FCC has a computer bulletin-board system (BBS or CBBS) that provides information about computers legally sold in the US. The system is called the FCC "PAL." It is free of charge except for the phone call, and it gives valuable information about many computers.

Before calling PAL, get the FCC ID number of the computer you are considering. Get the number from the computer dealer, or copy it from the ID tag that is required on all computers. Unfortunately, you need a computer (with a modem) to contact PAL. Ask your friends; you probably know someone with the necessary system.

To access PAL, set the modem to 1200 baud, no-parity, 8 bits, 1 stop bit and dial 310-725-1072 (don't forget to dial "1" if it's a toll call). When the computer connects, move through a few menus to "Access Equipment Authorization Database" and then to "Equipment Authorization Application Status." At this menu, enter the FCC ID number of the computer in question. PAL then responds with the CPU type and the speed. If PAL does not find the ID number, or if the CPU type

and speed do not match what you were told by the dealer, don't buy the computer and advise the FCC! You may have just found an illegal computer.

EMI IN COMPUTER DESIGN

Engineers have been working on this problem since 1981, when the FCC began testing computers for compliance (since 1949 for early VDE). Each area of a computer that emits RF must have its emissions reduced to an acceptable level. *Although these solutions apply to existing systems, don't try them!* Without all pertinent information needed to test all possible combinations, modifications may do more harm than good. (The computer may no longer work after modification.) These techniques *are* helpful for builders

who are designing their own computer board, accessory or other device.

Circuit Design

Engineers have found that the best clock (a square-wave generator) is also a very good RF generator. As harmonics are filtered from the clock signal, it rounds the corners of the square waves, and the computer may not function reliably. Keep this in mind should problems arise. Several EMI-suppression methods can be applied to logic boards with crystals on them:

1. Put a damping resistor in series with the crystal and the components that use the clock. The resistor loads the oscillator and reduces its output. (Larger resistors yield less output.) For clocks lower than 50 MHz, use 33 Ω; faster clocks may need as little as 10 Ω.
2. A damping capacitor may be used with the series resistor in method 1. One or two capacitors "smooth" out the sharp waveform of the clock. The capacitors are usually very small (about 22 pF) and placed at the damping resistor. Connect one from each side of the resistor to ground. *Important: make sure that the clock waveform still conforms to all specifications of components that use it.*
3. Surround the clock PC traces with ground traces. Think of the clock trace as a transmission line and the ground traces as shields.

Shielding

Computer noise can be minimized in two ways. We have already talked about reducing the energy produced on the board, but what if that is insufficient or impossible? Many manufacturers enclose the logic boards in a conductive housing (shield) to contain the RF.

The most common enclosure is a sheet-metal box. (Screws ensure good electrical connections at joints.) Some computers use a shielded plastic housing. Plastic may be shielded with special paint, conductive plating or by mixing conductive particles into the plastic before it hardens. In each case it is important to have large contact areas and solid attachment points to provide electrical continuity at enclosure seams. If there are gaps, RF energy can escape and cause a problem.

Cables are the last important EMI pieces in the computer design. Any cable

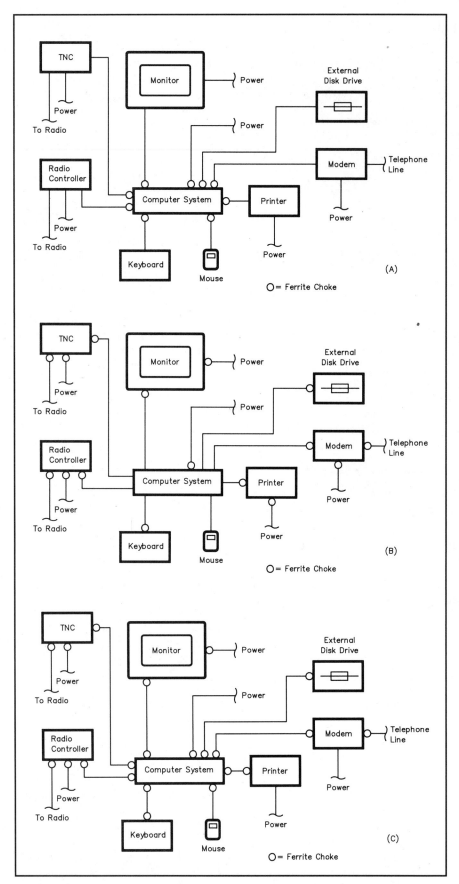

Fig 3—(A) Location of ferrites if computer is noisy, but all other external devices are quiet. (B) Location of ferrites if the computer is quiet, but other external devices are noisy. (C) Location of ferrites if both the computer and the external devices are noisy

that exits an RF-tight enclosure may radiate. For this reason, most computers require shielding on every cable except the power cord (special filters suppress RF there). Shielded cables are usually effective computer-EMI reducers.

EMI SOLUTIONS (INTERFERENCE *FROM* COMPUTERS)

As we have seen, there are things that the designer can do to make a computer radiate less RF energy. It is very difficult to apply those solutions to existing designs, where all of the design information is not available. What can we, as computer owners, do to combat EMI?

First, we need tools and parts to solve the problem. If you have noticed a problem, you already have the first tool: a detector for the RF that you want to eliminate. The work usually goes quicker with a spectrum analyzer, but this tool is not necessary. The goal is not to remove all RF, but to free the affected device (receiver).

If you suspect a computer, but are unsure, switch the computer on and off while checking the affected device. If the interference does not come and go with the computer power, look elsewhere for the cause. Check other computers, televisions, radios, clocks and appliances: anything with remote control, microprocessor control or a digital clock display. Switch each device on and off and see if the interference is affected.

Once the interfering system is found, check any switchable subsystems to see how many are related to the problem. Detach any cables, one at a time. Does the noise level drop? Make notes of what happens, reconnect the cable (tighten any attaching screws) and go on to the next cable. Do this for each cable except the power cord. If this procedure doesn't indicate the noise source, do the reverse. That is, remove all cables except the power cable, and reconnect the cables one at a time (make notes as you go). If the noise level is high with all cables removed, it must come from the computer case or power cord.

Cables, Cable Shields And Ferrite Chokes

If the noise is coming from cables, verify that the source cables are shielded. If they are not, the following solutions promise little success. So get shielded cables!

Fig 4—Two shapes of snap-on ferrite cores.

Table 1
Ferrite Formulas / Effective Frequencies

Frequencies	Ferrite Material
1 MHz - 200 MHz	Radio Shack / MFJ Split Toroids
1 MHz - 40 MHz	Material 73 (Best from 1 MHz to 40 MHz)
40 MHz - 1000 MHz	Material 64 (Best from 40 MHz to 1000 MHz)
2 MHz - 1000 MHz	Material 43 (Best overall material)

Materials 73 and 64 provide better attenuation on some specific frequencies; material 43 is a good solution over a broad frequency range.

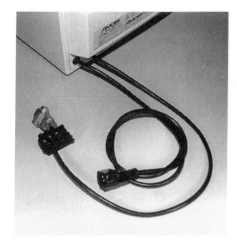

Fig 5—A noisy video card may need a ferrite core on the video cable. Here a snap-on ferrite core is installed at the computer end of a video cable.

Fig 6—An ac-line filter at the computer power supply can reduce RF leaving through the power cord.

Table 2
Sources of Supplies for EMI Problem Solving

For full addresses see the Suppliers List at the end of this book.

Ferrites / Chokes
Radio Shack "Snap-together Ferrite Data Line Filter" No. 273-105
Radio Shack "Snap-on Filter Chokes" No. 273-104
MFJ "Toroid RFI Filters" No. MFJ-701
Palomar Engineers
Amidon Associates
Fair-Rite Products

Phone Filters
K-COM
TCE TP-12

Conductive Paint
GC-Thorsen catalog no. 10-4807

Conductive Tape
3M "Conductive Copper Tape"

AC-Line Filters
Coil Craft
Corcom

To minimize cable noise, install ferrite cores on the cable (as close to the noise source as possible, see Fig 3). Good snap-together ferrite cores are available from Radio Shack and MFJ (see Fig 4). These cores are easy to use and very effective.

For problems at specific frequencies, choose a ferrite material that is most effective at the problem frequency. Ferrite material comes in several different mixes with varying permeability and frequency range. Some common mixes and their effective frequency ranges are shown in Table 1. If noise must be attenuated over two frequency ranges, use two ferrites with different materials. Simply wrap the cable through the ferrite as many turns as possible and snap on the plastic cover (Fig 5). If one core helps, but is not fully effective, add more. One or two ferrites usually removes the noise, but more may be required; try up to six on very noisy cables.

As an alternative for very noisy cables, add another layer of shielding. Unfortunately, one connector must be removed to install the added shield. It is important to have the new shield grounded at both ends. This solution is inexpensive, if you have the time to rebuild connectors.

An internal modem is a special case. The exit cable is usually standard (unshielded) telephone cable. This is usually not a problem, however, because the FCC-approved coupler transformer protects the telephone system from RF. If a modem card has problems, try split toroids on the phone cord or one of the special telephone RF filters available from TCE or K-COM (see Table 2).

AC-Line Cords

If the chassis or power cord is the noise source, make sure the computer and affected device are powered by different branches of the house wiring. This may be a suitable permanent solution. When this solution is not acceptable or it does not help, try ferrite cores on the power cord as described earlier for other cables. AC-line filters might also help. Commercial versions are available, and home-built versions are described in other chapters of this book. Extreme cases may need an ac-line filter inside the power supply. Filtered cord connectors are available at many electronics stores (see Fig 6). The connectors are designed to replace the ac plug on most personal computers. When working in the power supply, play it safe

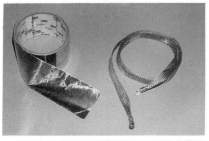

(A)

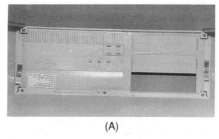

(A)

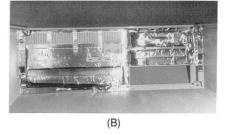

(B)

Fig 8—Before (A) and after (B) shots of a plastic computer front panel that has been shielded with copper foil.

(B)

(C)

Fig 7— (A) Copper tape can be used to "seal" in RF. A ground strap from the computer case to earth ground can also reduce the RF emissions from a computer chassis. (B) RF can leak from seams in a computer case. (C) Copper tape (or more screws) can reduce RF leaks at seams. Notice that it is also important for all Input/Output cover plates to be installed, RF can escape through these holes as well.

and apply common sense (disconnect the power and discharge all capacitors before starting work!).

Chassis Radiation

If the noise persists, it must come directly from the chassis. On metal computer cases, ensure that all screws are tightly installed (both inside and outside the computer case). The CPU (mother) board, power supply, and every accessory card must be grounded to the chassis with the proper screws. This could *solve* prob-

lems caused by missing or loose screws. It sometimes helps to ground the computer chassis (with 3/4-inch ground braid) to the same ground as the radio station. Another simple solution is to move the computer farther from the station coax and radio. This solution may not be practical, but it is worth a try.

Next look for case gaps where RF could leak out. Small holes are not a problem, but long slots are. Are the I/O cover plates tightly installed? Are the side seams tight? If you find questionable areas, place some 3M conductive copper tape over the seams. When the tape helps, either leave it as a permanent solution or devise some other method to seal the hole.

Plastic Cases

A plastic computer case may need shielding. If you elect this option, first verify that the computer does not depend on the insulating qualities of the plastic case. Will conductive materials on the inside of the case be an electrical hazard? To be sure that it is safe, examine the computer boards and the case carefully to see where they may make contact. Do not put conductive materials in these areas because they might create short circuits on the computer boards.

Use copper tape (see Figs 7 and 8), aluminum foil or conductive paint as shielding. To apply them, remove the computer boards and power supply from the plastic case. Keep copper tape and aluminum foil clear of openings, such as air vents. (The computer still needs to breathe!) Solder any connections to copper. Aluminum foil can be bonded to the plastic case with an appropriate adhesive (check to be sure the adhesive won't damage the plastic). Use screws and lockwashers for connections to aluminum. Prepare for paint application by masking any areas that must be free of paint.

Care is needed with conductive paints.

Some plastics may be damaged by the paint; others may require special primers to prevent flaking. Poor adhesion may yield a conductive "snowstorm" to play havoc with circuitry.

Other peripheral devices may cause noise problems. Do not overlook monitors, printers, Terminal Node Controllers (TNCs), keyboards and computer mice. The above techniques can be applied to all of these.

PRACTICAL SOLUTIONS (INTERFERENCE *TO* COMPUTERS)

Although interference *to* computers is not as common as interference *from* them, it does happen. Many of the principles from the "Interference From Computers" section apply here as well. In short, if RF can get out, it can also get in. The FCC and VDE do not have any RF susceptibility requirements or tests. Nonetheless, level-B computers are less susceptible to RF interference by virtue of their stricter RF-emission standards.

RF can enter the computer through any of its cables. If the hard-disk drive or controller is affected, back up the hard-disk data and make a floppy "boot" disk before working on the problem. If data is corrupted during troubleshooting, it may be impossible to boot from the hard disk!

Troubleshoot the cables as described earlier. Make notes, and apply ferrite cores and shielding as described for interference from a computer. That should solve the problem.

Keyboards And Modems

Two especially troublesome devices are keyboards and internal modems. A keyboard has its own microprocessor and logic circuits. It conveys data and receives power through the keyboard cable, which is where RF enters. Two solutions have proved effective: Apply ferrite cores on the keyboard cable (as

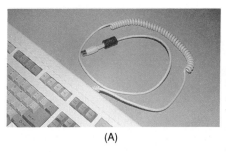

(A)

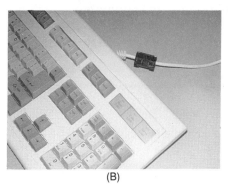

(B)

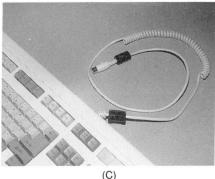

(C)

Fig 9—A ferrite core may be needed at the computer end of a keyboard cable if the main computer board is noisy. (Place ferrites close to the noise source.) At B, a noisy keyboard calls for a ferrite core close to the keyboard. (C) Rare cases need cores at both ends! This can also happen with monitors, printers and so on. Do not assume that something is okay; check it!

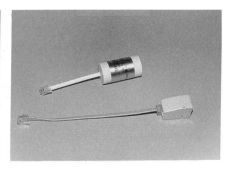

Fig 10—Phone-line filters can reduce RF escaping through an internal modem. The upper model is from TCE Labs. The lower filter was made by AT&T; it may not be widely available.

close to the keyboard as possible, Fig 9), and add bypass capacitors to the keyboard power leads. Place the capacitors from the +V lead to the ground lead, where the cable connects inside the keyboard case. Use 22-µF Tantalum capacitors (ensure that the polarity is correct during installation). More common 0.01-µF disc capacitors may also work.

The other interesting problem appears in the modem speaker. RF enters the computer through the phone line and is detected (rectified) in the modem audio circuits. A ferrite on the phone cord may help, but a phone-line filter (TCE TP-12, Fig 10) is better. If this is not entirely effective, try a low-pass filter at the speaker (as described for transistor amplifiers in the Stereos chapter of this book).

AC-Line Cords

If none of these solutions has worked, RF may be entering through the power cord. Try grounding the system as dis-

cussed earlier. Increase RF filtering in the power supply by adding a power-supply filter. A good filter has a sealed case with the schematic printed on it. Make sure the filter can handle the voltage and current that the computer requires.

CONCLUSION (IS THAT ALL?)

Computer EMI problems are no different than any others. Investigate, simplify the system, and apply shielding or filters. Make sure that all computers and accessories pass FCC or VDE Class B requirements before purchasing them, and problems will be minimized. Ferrite cores, shielded cables and a shielded computer enclosure are all important for radios and computers to coexist.

Hints & Kinks

---

from October 1988 *QST*, pp 41-42:
RFI FROM A MULTI-OUTLET BOX

After installing computerized RTTY gear, I experienced RFI in the form of "steel wool" on my monitor screen when my beam was aimed over the roof of the shack. One day, when the interference was especially intense, I decided to take another "whack" at plugging the RF leak.

I unplugged cables, one at a time, from the RTTY modem until only the ac line remained. Keying the transmitter between each cable disconnect had so far demonstrated no reduction in the interference. Obviously, the RF was coming in on the ac line or through the modem cabinet. Because grounding the cabinet did not reduce the interference, I reckoned that the ac line was the source of the leak.

As I studied the Drake LF-6 multi-

outlet box that served as the ac connection point for the station gear, I discovered that the three-sided box cover was floating above ground—even at dc. I filed the finish off the mating surfaces of the box and reassembled the box with a star washer under each screw head. I reconnected all the gear exactly as before and presto—no RFI! Three years of frustration were over.—*Sid Kitrell, W0LYM*

Chapter 13

Automobiles

By Stuart Sanders, KAØFPX
Automotive EMC Engineer
21422 Ontanga
Farmington Hills, MI 48336

The impact of automotive EMI on Amateur Radio operators is increasing. Until the 1970s, automotive systems were purely electromechanical and very robust as far as EMI was concerned. They had little or no radiated emissions (except for spark noise) and no susceptibility to on-board transmitters. This situation has changed rapidly with the application of integrated circuits and microprocessors to vehicle entertainment, communication and control systems. For instance, a typical upscale car can have 30 electronic modules. Most such modules contain at least one microprocessor and a number of low- and high-level digital and analog components. These components and microprocessors can emit copious quantities of RF noise, and they can be vulnerable to external electromagnetic fields if those factors are not considered during their design.

Most of us have heard some horror story about a ham who couldn't operate mobile because of mysterious noise emanating from the dashboard of his vehicle—or because the engine stalled during every transmission. These and other manifestations of EMI can occur in all vehicles to some degree. EMI severity varies with automobile make and model.

Automobile manufacturers have electromagnetic-compatibility departments that resolve EMI problems in the design phases. Unfortunately, their effectiveness and support in combating EMI field and customer complaints can vary. For that reason, this chapter describes some standard automotive systems, the EMI problems they can cause and steps you can take to isolate and correct them. All while

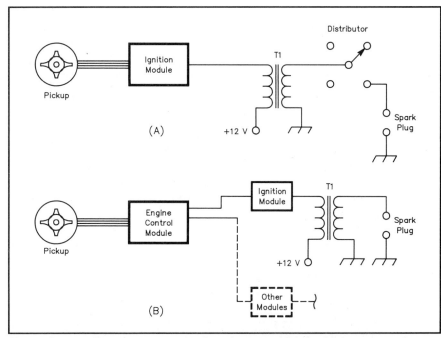

Fig 1—System diagrams representative of most vehicle ignition systems. A shows a distributor ignition system, which is common on older vehicles and some large contemporary engines. B shows an electronic-ignition system, as found on most automobiles.

maintaining the normal and safe operation of the vehicle!

SOURCES OF INTERFERENCE
Ignition Systems

The automobile ignition system has been around a very long time. Numerous articles have been written, over the years, about suppressing noise from spark plug wires, spark plugs and distributors. Most solutions involve either shielding the wires and distributor or installing some kind of suppressor device in the plug leads. Care must be taken in implement-

ing these and other fixes so that the problem is not compounded and the proper operating performance of the vehicle is maintained.

Let's look at a typical ignition system and how it operates to get an idea of how the various suppression schemes work. Fig 1 depicts systems that represent most vehicles. The first system is found mostly on older vehicles and a few new ones with large-displacement engines. During operation, a magnetic pickup in the distributor sends a trigger signal to the ignition module (essentially a high-power dc amplifier), which then drives a high-current pulse through the coil, T1. The 30- to 50-kV secondary voltage travels through the distributor rotor and its gap to the plug wire.

Newer electronic ignition systems, as shown in Fig 1B, omit the distributor in favor of one or two coils per cylinder. The *fast*-rise-time pulses from the coil current discharge across air gaps (distributor and spark plug) to create broadband ignition noise. The theoretical models (zero rise time) of such pulses are called impulse functions, in the time domain. When viewed in the frequency domain, the yield is a constant spectral energy level starting at 0 Hz and extending up in frequency. In practice, real ignition pulses have a finite rise time; the spectral-energy envelope decreases above some cutoff frequency as shown in Fig 2.

Now, if we could somehow slow down the rise time of the ignition pulse, we could make the roll-off point occur at a suitably low frequency, preferably well below the frequencies of interest. This can be done by adding damping elements (inductance and/or resistance) to the ignition circuit. Resistance is usually built into suppressor spark plugs and wires, while there is some inductance with resistance in wires, rotors and connectors. The elements can be either distributed or lumped, depending on the brand, and each technique has its own merit.

Generally speaking, the components with the most series resistance provide the most attenuation to EMI. This resistance can be measured easily with an ohmmeter. Wire with 1000 Ω/ft is typical. Typical rotors are 6000 Ω, and connectors that screw onto solid copper wire are 1000 Ω.

Measuring the inductance of a wire is trickier; it requires an impedance bridge or a network analyzer (not many hams have access to those). Visual inspection of the center conductor gives an idea of the wire turns per inch. Different values of L yield varying amounts of suppression at a given frequency, depending on the total lead length.

Grounding and Shielding

If resistor wires and plugs don't provide enough suppression, some form of improved grounding and extra shielding may help. As a preliminary step, bond the following points together with 1/2-inch (or wider) copper or stainless-steel strap: tailpipe to chassis, body to chassis (not

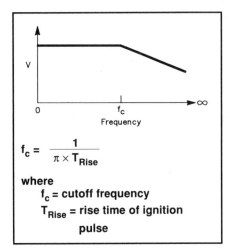

$$f_c = \frac{1}{\pi \times T_{Rise}}$$

where
    f_c = cutoff frequency
    T_{Rise} = rise time of ignition
        pulse

Fig 2—The spectral energy of ignition pulses rolls off above a certain frequency determined by the formula.

needed in unibody designs), engine block to chassis and body, trunk and hood lids to body and bumpers to chassis. This ensures that none of the metal parts act as floating antennas. This process often reduces the noise to a tolerable level. If not, go on to shielding.

Experiment by temporarily adding aluminum-foil covers to the plug-wire harnesses, coil and distributor. Ground the foil to the engine block at several points. To be effective, the ground leads must be as short as possible. Long leads have more inductance, which keeps more of the noise from reaching ground. This fast and cheap method indicates whether shielding will help, without the hassle of making up special metal parts. If the foil method is successful, buy (or make) a set of shielded plug wires. Some cases may need covers for the coil and distributor too. Expect 15 to 30 dB of noise reduction from shielding *if* it is done correctly *and* if the foil shields helped. A caution on materials: Don't use braid with less than 95% coverage to fabricate wire shields, and don't clamp dissimilar metals together to form a ground connection. The underhood heat and moisture corrodes dissimilar metals in a jiffy.

Is It Really Spark Ignition Noise?

Ignition noise manifests itself as a regular, periodic ticking in the receiver audio output (it varies with engine RPM). It sounds somewhat musical, like alternator whine, at higher speeds, but has a harsher note. There is one distinguishing feature of ignition noise: it increases in

amplitude under heavy acceleration. This results from the change in firing voltage with higher cylinder pressure. Also, ignition spark noise *usually* disappears when the antenna is disconnected (because it is usually radiated noise). By comparison, the high-current pulse in the ignition coil primarily generates a noise that is conducted on the wiring harness.

Poorly designed receivers let conducted harness noise enter the RF, IF or audio sections (usually through the power leads), interfering with the desired signals.

Alternators And Generators

Another potentially noisy automotive system is the alternator/generator. AC alternators have been used almost exclusively on vehicles since the mid 60s, because of their superior low-speed charging ability, reliability and small size. DC generators appear on older vehicles and on some off-road equipment. The noise associated with these two devices is usually broadband, sometimes extending into the UHF region. The sources of the noise are the armature brushes, the field controller (regulator) and the three-phase rectifier diodes.

Generator brushes interrupt the large dc output current (50 A). The resulting spark is primarily responsible for the "hash" noise associated with these devices.

An alternator also has brushes, but they do not interrupt current. They ride on slip rings and supply a modest current (typically 4 A) to the field winding. Hence they are relatively quiet. Most alternator regulators use a solid-state pulse-width-modulated (PWM) field-control system. This controller contributes to the RF "whine" that is characteristic of alternator noise. If alternator noise is caused by the PWM, then the noise intensity and pitch changes with current load at a constant engine speed.

Generators use a relay regulator to control field current, and thus output voltage. Its continuous sparking creates broadband noise.

Alternator or generator noise is conducted through the power wiring to the audio sections of mobile receivers and transmitters. A listening test can verify alternator noise, but if an oscilloscope is available, monitor the 12-V line feeding the affected radio. Alternator whine appears as full-wave rectified ac, coupled with commutating spikes, superimposed on the dc level (see Fig 3). Suppressing this noise requires a little trial-and-error work (in order to save money and use as few parts as possible).

Alternators rely on the low impedance of the battery for filtering. So check the wiring from the alternator output to the battery for corroded contacts and loose connectors when alternator noise is a problem. Alternator manufacturers sometimes include a capacitor from the output to ground; another 0.1- to 50-μF capacitor in parallel with the first gives some extra noise reduction. In principle, this fix helps shunt any ac component of the output to ground, just like the filter on a power supply.

Filters are available commercially for installation at a convenient point in the power leads, between the radio and the battery. It may not be a good idea to install a filter that contains series inductance between the alternator and the battery. The impedance may cause the regulator circuit to oscillate—destructively (at around 100 kHz). The number of filter elements and the total noise suppression varies with the filter price and brand. A home-built filter performs just as well for a fraction of the cost.

Add-on alternator noise suppressors act as low-pass filters whose cutoff frequency and stop-band attenuation are given by the formulas in Fig 4. Select a cut-off frequency that is 1/10 of the lowest alternator whine frequency observed. If you decide to make your own filter, look at the standard-value-capacitor Butterworth filters in The ARRL Handbook. Those tables yield values for common electrolytic capacitors and reasonable-size inductors.

The filter inductors are usually wound on ferrite toroids or pot cores. Complete inductors can be purchased from several manufacturers such as J. W. Miller or Amidon. These firms also sell empty cores so that intrepid hams can wind their own. Whatever form the inductor takes, make sure that the core material doesn't saturate from the dc current required by the load. Equations for calculating the inductance and maximum current for different cores appear in the Handbook.

DC Motors

Next to light bulbs, dc motors are probably the most common electrical devices in automobiles. A typical vehicle may have 50, ranging from small servos and stepper motors to a huge 200-A starter. Most of these motors use brushes and commutators, but some are brushless. Electronically commutated units can be found, particularly on luxury models.

Each style of motor produces a different noise spectrum. The armature current in brushless motors is controlled by switching transistors, as opposed to mechanical contacts for brush-commutator motors. Brushless motors are fairly quiet, but they can generate noise similar to that produced by an alternator.

The spikes of current drawn by a brush-commutator motor generate broadband EMI that is similar to ignition noise. (It sounds like bacon frying when heard in a receiver.) Such broadband noise generally has a more pronounced effect on

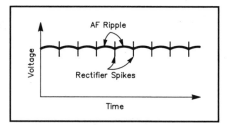

Fig 3—A time-domain plot of alternator whine contains full-wave rectified ac, along with rectifier spikes.

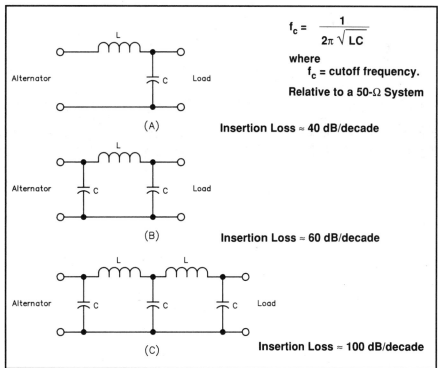

$$f_c = \frac{1}{2\pi \sqrt{LC}}$$

where
f_c = cutoff frequency.

Relative to a 50-Ω System

(A) Insertion Loss ≈ 40 dB/decade

(B) Insertion Loss ≈ 60 dB/decade

(C) Insertion Loss ≈ 100 dB/decade

Fig 4—Two, three and five-element low-pass filters to filter alternator noise.

AM receivers than FM. (An FM demodulator rejects the AM component of the noise.) Unfortunately, sparks do create some FM noise components. This can make diagnosis difficult because interference that can't be heard may still reduce the receiver sensitivity and S/N ratio.

If you suspect motor noise is the cause of the problem, obtain an AM receiver to check the frequency or band of interest. Switch on the receiver, and then activate the electric motors one at a time. When a noisy motor is switched on, the background noise increases.

Obviously, this test requires the ability to switch motors on and off. To switch fuel pumps, cooling fans and such, either pull the appropriate fuse or remove/jumper the relay that feeds the particular motor.

A note concerning fuel pumps: virtually every vehicle made since 1982 has an electric fuel pump, fed by long wires. It may be located inside the fuel tank. This motor is responsible for many EMI complaints, so keep it in mind as you look for broadband noise sources.

Motor-noise suppression is fairly straightforward. Many OEM parts suppliers make optional EMI filters for customers with land-mobile radios. The filter is usually a balanced L or PI network that mounts in the power leads, close to the device. The filters carry a regular part number and instructions. They can be ordered and/or installed by an auto dealer.

If the manufacturer doesn't list a filter for your application, try the circuit in Fig 5. The inductors may be wound on air or ferrite cores, depending on size constraints and the dc current requirement. This filter provides good suppression up to 144 MHz, if the capacitors have short leads and are the correct value. If the motor is accessible, bypass the brushes with 0.01-μF capacitors. Larger capacitors may increase brush-commutator arcing and decrease motor life.

Switches

Noisy switches can be found throughout vehicles. Untreated switch contacts can generate fast, high-frequency voltage transients that travel through the vehicle wiring harness and radiate noise. These noise bursts can break receiver squelch and cause AGC-related dropouts during a contact. Once a problem switch is located (an easy task, most are readily accessible from the driver and passenger seats), apply one of the "snubber" networks shown in Fig 6 across the terminals.

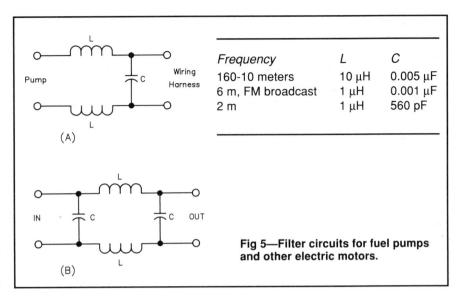

Frequency	L	C
160-10 meters	10 μH	0.005 μF
6 m, FM broadcast	1 μH	0.001 μF
2 m	1 μH	560 pF

Fig 5—Filter circuits for fuel pumps and other electric motors.

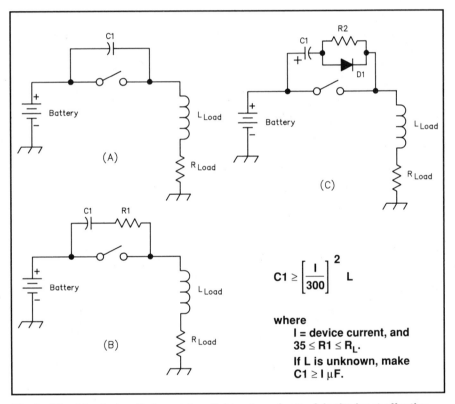

$$C1 \geq \left[\frac{I}{300}\right]^2 L$$

where
I = device current, and
$35 \leq R1 \leq R_L$.
If L is unknown, make
$C1 \geq I\ \mu F$.

Fig 6—"Snubber" circuits to quiet switching transients. A is the least effective and least costly. B is somewhat better. C is the most effective and costly.

The Ubiquitous Microprocessor

Most electronic modules in a modern vehicle contain some form of microprocessor. Microprocessors use a crystal or resonator controlled square-wave oscillator for timing. The timing pulses and harmonics are routed both inside and outside the microprocessor on interconnecting data and address buses. They can also show up as noise on the wiring harness that connects the module to the rest of the electrical system.

Clock signals cause the comb-like emissions characteristic of digital systems. (See Fig 7.) A square-wave clock signal is composed of the fundamental and all odd harmonics. There may also be spikes below the fundamental if the oscillator signal is divided inside the module to obtain a system clock. This is the case in Fig 7, where the crystal frequency is 1 MHz, and the separation is 500 kHz. If you have the equipment to look at the noise spectrum, such knowledge helps pinpoint the problem processor or module.

Before jumping into diagnostic work, let's examine how noise is generated within a module and how it can escape to show up in the amateur bands. As mentioned before, a microprocessor contains a master clock that generates square waves, which propagate through the logic circuits. As these circuits switch on and off, they draw a transient current from the power supply. Because microprocessors and other LSI chips contain many gates, the current pulses can reach rather large magnitudes and cause the V_{cc} power-supply bus to fluctuate. The fluctuations have a considerable number of HF components and can travel around the PC board, coupling to other traces and modulating any device that shares the same power bus.

The V_{cc} line is usually the worst offender, but other lines such as data, clock, and the I/O strobes can also emit significant RF. The physical dimensions, layout and filtering of the PC traces play a large part in determining how much EMI is generated because energy can only escape from the module as radiation, or as conduction onto the wiring harness.

As a rule of thumb, a conductor longer than 0.1 λ can be a good antenna. Traces on larger boards, like an instrument cluster, can reach 25 inches. By the rule, the cluster could radiate noise above 50 MHz. This kind of board is a worst-case situation from a direct-radiation standpoint. Most modules are much smaller, with correspondingly shorter conductors (2 to 6 inches). If this is true, why do small modules cause EMI at frequencies well below those that could be radiated by their PC boards? The answer is conduction.

To understand the coupling mechanism, visualize a noisy module as one or more noise-voltage sources that are magnetically and electrostatically coupled to the wiring harness via the PCB traces (see Fig 8). The coefficient of coupling, M, depends on the area of the loops formed by the two paths. The capacitive coupling depends on the length and separation of

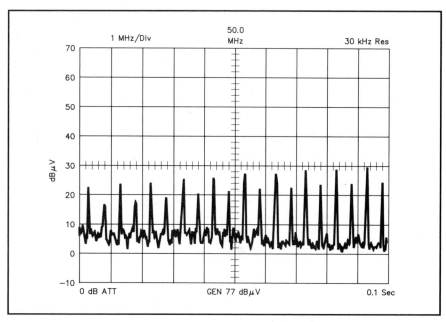

Fig 7—Spectral plot of a vehicle electronic module containing a microprocessor with a 1-MHz crystal oscillator.

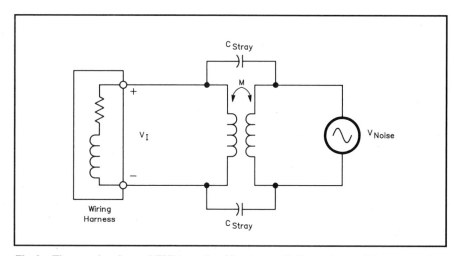

Fig 8—The mechanism of EMI transfer. V_I represents the noise voltage present at the connector of a module. Electrical noise is transferred to the vehicle wiring harness in proportion to the magnetic coupling, M, the electrostatic coupling, C_{Stray} and the frequency of the noise.

the traces. By minimizing the coupling parameters, the radiated noise from a particular board can be reduced. There are also external modifications that can be added to the module after it is installed in the car.

EMI suppression on a complex, multilayer board takes quite a bit of skill and specialized knowledge of the module functions. Only the manufacturer can ensure that the module is not impaired by any EMI modifications. Some on-board systems control such critical functions as braking and fuel control. Unauthorized

modifications that affect these systems could create performance and liability problems.

If you experience EMI from your vehicle electronics, first call or write the dealer, zone representative or the factory to see if they can help. Other mobile radio users may have the same problem; there could be a factory-approved fix for your vehicle. It makes sense to get help from the manufacturer first. Unauthorized modifications can void the vehicle warranty and affect vehicle safety.

What if the manufacturer is unable

GENERAL MOTORS CORPORATION RADIO TELEPHONE/MOBILE RADIO INSTALLATION GUIDELINES

Editor's Note: The following is an adaptation of "Radio Telephone/Mobile Radio Installation Guidelines" (August 1991 issue) from the General Motors Electrical Engineering Center. *GM makes these recommendations for their products only.*

RADIO TELEPHONE/MOBILE RADIO INSTALLATION GUIDELINES

Certain radio telephones, land-mobile radios or the way they are installed may adversely affect vehicle operations such as the performance of the engine and driver information, entertainment and electrical charging systems. Expenses incurred to protect the vehicle systems from any adverse effect of any such installation are not the responsibility of General Motors Corporation. The following are general guidelines for installing a radio telephone or land mobile radio in General Motors vehicles. These guidelines are intended to supplement, but not to be used in place of, detailed instructions for such installations, which are the sole responsibility of the manufacturer of the involved radio telephone or land mobile radio.

INSTALLATION GUIDELINE

(refer to the figures during installation)

Transmitter Location

A. Locate the RF portion of remote radios on the driver's side of the trunk as near to the vehicle body side as possible.
B. One-piece transceivers should be mounted under the dash or on the transmission hump, where they do not interfere with vehicle controls or passenger movement.
C. Great care should be taken not to mount any transceivers, microphones, speakers or any other item in the deployment path of a Supplemental Inflatable Restraint or "Air Bag."

Antenna Installation

A. The antenna should be a permanent-mount type located in the center of the roof or center of the rear deck lid. Glass mounted antennas should be kept as high as possible in the center of the rear window or windshield. If a magnet-mount antenna must be used, care should be taken to mount the antenna in the same location as a permanent mount type. If a disguise-mount antenna is used, great care should be taken to shield the tuning network from vehicle electronics and wiring, or to mount the tuning network in an area completely clear of vehicle electronics and wiring.
B. Standard metal-mount antennas may be mounted on vehicles with nonmetallic body panels by two methods. Most nonmetallic skinned vehicles have metal frames underneath. Mounting the antenna near a metal frame section and bonding the antenna mount to the frame with a short metal strap provides the ground-plane connection. Some antenna manufacturers offer "ground-plane kits" that consist of adhesive metal foil, which may be attached to the body panel to provide the ground-plane for the antenna.
C. Some vehicles use glass that contains a thin metallic coating for defrosting or to control solar gain. Glass

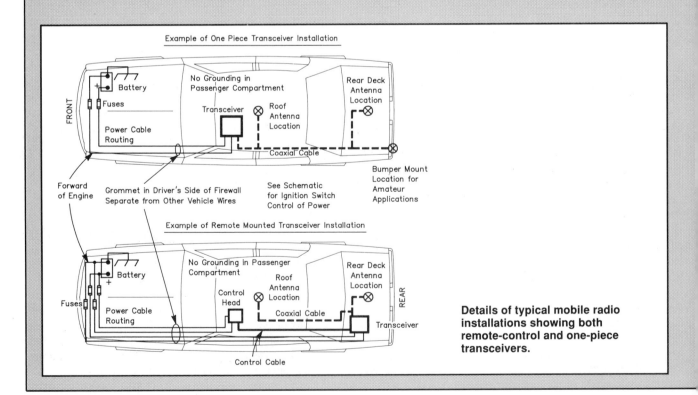

Details of typical mobile radio installations showing both remote-control and one-piece transceivers.

mount antennas will NOT function when mounted on this type of glass. Consult your GM dealer or owner's manual to determine if this glass is installed on your vehicle.

D. Each vehicle model and body style reacts to radio frequency energy differently. When dealing with an unfamiliar vehicle, it is suggested that a magnetic-mount antenna be used to check the proposed antenna location for unwanted effects on the vehicle. Antenna location is a major factor in these effects.

Antenna-Cable Routing

A. Always use high-quality coax (at least 95% shield coverage), routed away from the Engine Control Module (ECM) and other electronic modules.

B. Care should be taken to avoid routing feed lines with any vehicle wiring.

Antenna Tuning

A. It is important that the antenna is tuned properly and reflected power kept to a minimum (VSWR < 2:1).

Radio Wiring and Connection Locations

A. Transceiver power leads:

These connections, including the ground, should be made directly to the battery itself, or to the jump-start block on vehicles so equipped. GM approved methods of connecting auxiliary wiring to a side-terminal battery include the AC-Delco Side Terminal Adapter Package no. 1846855, NAPA Belden replacement battery bolts part no. 728198, or drilling (3/8-inch deep) and tapping (no. 10-32) the hex end of the original battery bolts.

NOTE:
It is recommended that a fuse be placed in the transceiver ground lead to prevent possible damage to the transceiver, in the event the battery to engine-block ground is inadvertently disconnected.

For ONE-PIECE TRANSCEIVERS where ignition-switch control is desired, a 12-V power contactor must be installed in the transceiver positive lead. The contactor should be located at the vehicle battery with the coil of the contactor driven through an appropriate in-line fuse from an available accessory circuit or ignition circuit that is not powered during cranking. The coil of the contactor must return to the negative battery terminal.

B. Handset or Control-Unit Battery and Ground:

Any ground lead from a handset or control unit must return to the negative battery terminal. It is preferable that the positive lead for a handset or control unit be connected directly to the positive battery terminal. It is recommended that the handset or control unit positive and ground leads have appropriate in-line fuses separate from those in the transceiver positive and ground leads. If ignition-switch

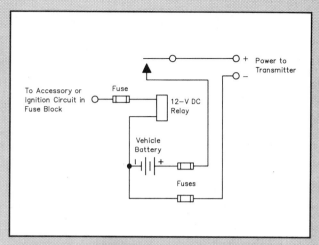

A schematic of contactor wiring for ignition-switch control of a transceiver.

control is desired, the handset or control-unit positive lead may be connected through an appropriate in-line fuse to an available accessory or ignition circuit that is not powered during cranking.

C. Connections for Multiple Transceivers and Receivers:

If multiple transceivers or receivers are to be installed in the vehicle, power leads to the trunk or under-dash area should be terminated in covered, insulated bus bars. All transceivers or receivers may then have their power leads connected to the bus bars. This makes a neater installation and reduces the number of wires running to the vehicle under-hood area.

Wire Routing

A. The power leads should be brought through a grommet in the driver's side of the firewall. For trunk-mounted transceivers, the cables should continue along the driver's-side door sill(s), under the rear seat and into the trunk through the rear bulkhead. If the battery is located on the passenger side, battery leads should cross the vehicle in front of the engine. All attempts should be made to maintain as much distance as possible between radio power leads and vehicle electronic modules and wiring.

Troubleshooting

A. Should vehicle problems develop following installation, the source of the problem should be determined prior to further vehicle operation.

B. Possible causes of vehicle problems include:
1. Power feeds connected to points other than the battery.
2. Antenna location.
3. Transceiver wiring located too close to vehicle electronic modules or wiring.
4. Poor shielding or poor connectors on antenna feed line.

or unwilling to help? With a general-coverage receiver or a spectrum analyzer, a fuse puller and a shop manual, the offending module can be identified and treated using a few basic, low-risk techniques. First, identify the frequency or frequencies of the interference. Then monitor the frequencies with a mobile rig, HT, scanner, or any other receiver with good stability and an accurate readout. A signal-strength meter is also handy to observe the effectiveness of various modifications.

As you hunt for noise, keep in mind the different frequency-drift rates for crystals and resonators. Harmonics from crystal oscillators exhibit good long-term stability even at 144 MHz. Those from ceramic resonators move quite a bit, especially with changes in ambient temperature. The inside and under-hood temperatures in a vehicle can change drastically within a few minutes and can cause perplexing intermittent resonator-related EMI. In these cases, follow the noise with a receiver until it settles down and then start the hunt.

Once the noise source is tuned in and has stabilized, find the vehicle fuse panels and pull one fuse at a time until the noise disappears. Some modules may have a "keep-alive" memory that is not disabled by pulling the main fuse. Consult the shop manual for their fuse location and any information concerning special procedures for disconnecting power. If more than one module is fed by one fuse, then locate each module and unplug it separately. This again requires a shop manual (unless you enjoy tracing wires in cramped quarters).

CURES
Ferrites And Common-Mode Currents

Wiring harnesses transport most of the emissions from vehicle electronics. Therefore, it is logical to cover wiring-harness suppression methods first. The simplest and most widespread technique is the use of ferrite in the form of common-mode chokes. The term "common mode" refers to current that flows in phase (in the same direction) along all wires in a harness. Differential-mode currents flow 180° out of phase (in opposite directions) in some wires of a harness.

Common-mode RF current is driven along the wiring by minute voltage differences between the module ground and the vehicle structure. In doing so, it radiates energy, just like an antenna. To reduce this current, an impedance is inserted in series with the harness. Ferrite works well in this application because its impedance is composed of resistive and reactive components that combine to yield a significant impedance over a wide bandwidth. The result is broadband suppression.

Many companies make split ferrite cylinders for installation over entire wire bundles without cutting any conductors. Any number of cylinders can be added for increased attenuation. Bundles can also be wound around large toroids for the same effect. Fig 9 shows several ferrite suppressors and their application to wiring harnesses. Different ferrite mixes yield a wide range of permeabilities and cover frequencies in the HF and VHF spectrum. The choice of material depends on the range of frequencies that the choke must suppress. Curves showing effective permeability v frequency are used to make the decision.

Shields

Use shields to treat modules that radiate significant energy directly from their PC boards. A shield is simply a conductive box that encloses the circuitry and is connected to a reference point (ground), usually the vehicle chassis. Ideally, shields intercept radiated noise and return it to its source through a low impedance. For a shield to be effective, it should: (1) be constructed of a good conductor such as aluminum or copper, (2) connect to "ground" at a single point to prevent ground loops, and (3) not have any openings or slots with dimensions greater than one twentieth of a wavelength at the EMI frequency.

Seams in the shield enclosure should be bonded with conductive gaskets or with fasteners spaced one inch apart, to prevent radiation. Some modules have openings, such as vents or displays, that cannot be covered with solid metal. Use aluminum or bronze mesh, and bond it to the module all around the opening perimeter. For displays, paint the mesh flat black where it may obscure numerals or pointers.

Choosing a ground point for the shield poses a problem; the presence of metal can aggravate common-mode radiation. It is best to experiment with the ground point, and see which configuration gives the best results. Keep in mind

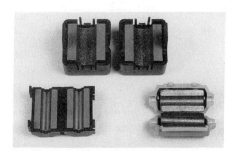

Fig 9—Examples of clamp-on ferrite cores for EMI suppression.

that the shield must not connect two separate grounds together. That creates a "ground loop" and can cause undesirable large currents to flow in the harness, possibly leading to component failure.

MOBILE TRANSMITTERS AND SUSCEPTIBILITY

Although electronic vehicle systems can interfere with mobile receivers, it is equally possible for mobile transmitters to affect the vehicle electronics. This problem (called "susceptibility") occurs less often than vehicle EMI emissions, but it can have greater consequences. Electronic modules that radiate energy can, by the principle of reciprocity, absorb energy. This energy can come from on-board transmitters, nearby radio and TV stations, lightning or any other device or event that generates an electric or magnetic field.

Symptoms of susceptibility include display scrambling, erratic engine operation, false triggering of auxiliary systems and improper operation of entertainment radios. If you experience a susceptibility problem, first determine the interference source.

Problems associated with mobile transmitters come and go as the transmitter is keyed, but off-vehicle sources may switch on and off randomly. This makes them difficult to identify. Look for an EMI-trouble pattern with respect to location; there might be an antenna tower or a microwave dish nearby. If the problem is definitely a mobile transmitter, check the following items to make sure the installation conforms to "good engineering practice":

1) The power leads should be twisted together (or run side by side) from the back of the rig all the way to the battery. *Do not use the vehicle chassis as a power return!* The leads should be routed along the body structure, away

from vehicle wiring harnesses and electronics. Each line should be fused and sufficient to carry the required current.

2) The coaxial feed line should have at least 90% braid coverage. The cable shield should be grounded to the vehicle chassis at both ends, all around the cable circumference.

3) Antenna(s) should be mounted on a good ground plane, as far from the engine and the vehicle electronics as practical. On the rear deck lid or roof, for example.

4) Match the antenna for a low SWR. That way the RF energy is radiated away from the car without a couple of trips back and forth within the coax as reflections.

The reason for 2 and 3 is simple: keep RF energy inside the coax (away from vehicle electronics). The idea behind 1 is more subtle, yet it is fundamental to EMI suppression. Minimizing the area enclosed by the conductors in a given circuit also

minimizes the energy they can absorb or radiate. Think of the current path as a loop antenna; many cable-routing schemes form excellent HF and VHF antennas.

If the installation is good and there are still problems, contact the vehicle dealer, zone representative or manufacturer for help. The rationale for vehicle emissions applies here as well (refer to

the section on microprocessor noise for a disclaimer). Also reread the earlier sections about Ferrites and Shields (in case you can't get help and want to fix the problem yourself). Since the reciprocity theory works for EMI, the suppression techniques apply; if you can keep RF energy in, you can keep it out.

Hints & Kinks

from June 1987 *QST*, p 40:

MORE ON RFI TO MICRO-PROCESSORS IN AUTOMOBILES

After installing my 25-W, 2-meter radio in a 1981 Oldsmobile Cutlass, a "valve knock" became evident when climbing hills. This would not be unusual except that it occurred only while transmitting! I suspected a voltage drop and rewired the power cable directly to the battery, but there was no improvement— not even at a 1-W input level.

It then occurred to me that my problem was RFI! At the same time, the Jan 1983 *QST* Hint of W4MJB encouraged me to investigate an RFI cure to the central processor unit. The CPU was in a metal case that was not grounded to the metal of the car, but "floated" in a plastic cover in front of the right front door, below the glove compartment. The CPU has bypass capacitors and ferrite beads on every edge connection, but only two of many wires are shielded. Simply grounding the metal case to the firewall with 1/4-inch braid

solved the problem. (The braid length is only about 5 inches.)

This problem and cure is important because future vehicles will no doubt have more and more microprocessors. Vehicle designers must be alert to RFI problems at all frequencies.—*Dave Porter, K2BPP*

from June 1979 *QST*, p 44:

RFI TO AUTOMOBILE CRUISE CONTROL

I just returned from West Virginia via the interstate. This was my first experience with 2-meter mobile. Prior to the trip, I had a Sears "add-on" cruise control installed in my 1976 Chevy pickup. The wiring for the mobile unit was rather hastily installed, and some of the radio wiring came through the same hole in the firewall as did the cruise-control wiring.

On the road with the cruise control engaged, keying the microphone in the high-power (15-W) position caused the speed to "lug-down." It returned to normal when the rig was not transmitting.

The antenna is a 1/4-wave whip on top of the cab. In the low-power (1-W) position, the problem did not occur.

On the return trip, I overheard a QSO in which one of the operators made a remark about speed lug-down while transmitting. He did not mention any particulars as to the car model or rig specifications. I wonder how many amateurs experience this problem and if someone may have found a cure.—*G. L. Baker, W5QPX*

from August 1981 *QST*, p 43:

RFI TO AUTOMOBILE CRUISE CONTROL, PART 2

After reading the item by Baker, W5QPX, I felt a great deal of empathy for him, because this is the same problem that I had with the Sears control I put into my Ford. Since I am running essentially the same type of equipment, I thought my experience would help him and others. I put a Pace 25-W rig in my Mustang, with a Hustler colinear antenna. I noted that when I was transmitting with an SWR greater than 1.5:1, I had the lag or bogging down of the cruise control with a speed loss until I unkeyed. When I readjusted the antenna to a 1:1 SWR, the bogging disappeared. I also found that when I got close to CB operators using too much power and modulation, I experienced the bogging down of the cruise control. Feeling that this interference was from harmonics in the UHF region, I watched carefully for UHF repeaters while mobile, and I did experience the same lag or bogging when I crossed the path of a UHF system. The effect was a loss of power to the control, necessitating a resetting adjustment.

A simple cure was foremost in my mind; hence I wrapped the control unit in aluminum baking foil to provide some shielding of the previously unshielded solid-state unit from RF. I have not experienced the problem since. The local repeater group still chuckles over the RFI problem another operator had in his car. RF would get into his electric broadcast antenna system, and the antenna would glide up and down during his amateur transmissions.—*Bill Richards, II, WB5ZAM*

from September 1964 *QST*, p 58:
BLACK-MAGIC INTERFERENCE REDUCER

I was having a great deal of trouble using my mobile station because of a tremendous amount of ignition noise from my car. After trying several cures, I accidentally found that the noise was almost completely eliminated by connecting a lead from the bumper to the chrome ring that surrounds the driver's side tail light. There is, of course, nothing mysterious about this scheme. It is mentioned here to point out how one must try almost everything when it comes to eliminating mobile interference. Grounding pieces of metal that could be acting as antennas is only one method. Other things that can be tried during a noise-reduction session are: (1) grounding the exhaust pipe at both ends and even at several spots along its length, (2) grounding the engine hood to the body at several spots, and (3) grounding the engine block to the body at several spots. Conventional methods of noise suppression using suppressor resistors and coaxial feedthrough capacitors should also be used along with bonding and grounding operations.—*J. G. Michaud, K2UBE*

Chapter 14

EMI Regulations and Standards

By Dennis Bodson, W4PWF
ARRL Technical Advisor
233 N Columbus St
Arlington, VA 22203

David Peach
US Dept of Commerce
NTIA/1TS.N1
325 Broadway
Boulder, CO 80303

Robert Cass, KBØEVX
785 39th St
Boulder, CO 80303

John C. Hennessee, KJ4KB
ARRL Regulatory
Information Specialist

RFI, radio frequency interference, is considered a byproduct of "wireless" transmission systems. RFI is frequently perceived as a transmit problem, but it can be a receive-mode problem as well.

THE EARLY DAYS OF RFI

Radio frequency interference has been around since the advent of radio. Amateur Radio operation began around the turn of the century, and RFI has always been a problem. In the "good old days," no rules and regulations governed Amateur Radio. Bedlam reigned. There were no standards, and amateurs essentially did as they pleased. More than 75 years ago, most interference complaints came from government and commercial radio operators. There were few consumer electronic devices. Hams used high-power spark-gap transmitters, and the transmitted signals were several hundred kilohertz wide. All transmissions were, in effect, controlled interference. The inevitable was about to happen...

Amateurs first got into trouble with the government over interference complaints in 1909. Many amateurs had better and more powerful stations than those used by the Navy and commercial services. Amateur indifference to the pleas of these operators, to cease operating when there was murderous interference, was sublime. Amateurs of the day knew they had superior equipment, and they outnumbered commercial operators. (In 1910, there were 4,000 hams and fewer than 1,000 government and commercial stations.) Hams weren't about to cease operation because their properly operating gear was causing interference to the

---

KING SPARK!

Grown now to full maturity, developed and perfected by years of prewar and World War I experience, it reached its highest peak in the succeeding eighteen months. Glorious old sparks! Night after night they boomed and echoed down the air lanes. Night after night the mighty chorus swelled, by ones, by twos, by dozens, until the crescendo thunder of their Stentor bellowings shook and jarred the very universe! A thousand voices clamored for attention. Five-hundred-cycle's high metallic ring. The resonant organ basso of the sixty-cycle "sync." The harsh resounding snarl of the straight rotary.

Character: Nervous, impatient sparks, hurrying petulantly. Clean-cut business-like sparks batting steadily along at a thirty-word clip. Good-natured sparks that drawled lazily and ended in a throaty chuckle as the gap coasted down-hill for the sign-off.

Survival of the fittest. Higher and higher powers were the order of the day.
The race was on, and devil take the hindmost.
Interference.
Lord, what interference!
Bedlam!
Well, it could not be Utopia!

by A.L. Budlong, W1BUD

(This is taken from page 60 of *200 Meters & Down*, by Clinton B. DeSoto, published by ARRL. It originally appeared in an early ARRL publication, *The Story of the American Radio Relay League*, circa 1925.)

"outdated" equipment of commercial stations.[1] The battle lines were drawn—it was the amateurs against the government and the commercial stations. The issue was RFI.

Between 1902 and 1912, 28 bills pertaining to the regulation of radio were introduced. The government, meaning the Navy, thought that their interference problem would disappear if they could have amateurs declared lawbreakers

[1]C. B. DeSoto, *200 Meters & Down* (Newington: ARRL, 1981), p 28.

when they interfered with registered government or commercial stations. It was obvious that amateurs would not be "king of the air" for much longer (DeSoto, pp 28-30).

The Radio Act of 1912 was passed: amateurs were required to obtain a license and limited to 1,000 W. They were also limited to the "useless" wavelengths below 200 meters. Willful and malicious interference was made punishable by law (DeSoto, pp 32-33). Amateur Radio was, at last, given the right to exist. Under the banner of the newly formed ARRL, amateurs worked hard to ensure that Amateur

Radio would continue. They often wondered if they were amateurs—or full time lobbyists! (DeSoto, p 64.)

Amateurs Control RFI

The second interference crisis came in 1921. The Department of Commerce delivered an informal but flat ultimatum to Amateur Radio, as personified by ARRL, saying something like this: "We can't control you, but you must be controlled if you are to exist; therefore, control yourselves, or cease to exist." (DeSoto, p 68.) Commercial broadcast stations began springing up everywhere. Thousands—indeed, hundreds of thousands—of nontechnical broadcast listeners with crude equipment experienced RFI from hams. People grew resentful—their pleasure was being disrupted. They sought redress through politicians—local, state and national—in an effort to get rid of "those damned amateurs." "Quiet hours" were also introduced. The potential danger to Amateur Radio was great. Needless to say, amateurs began to deal constructively with their interference problems in the early 1920s.

ARRL faced this new challenge "head on" by undertaking a massive publicity campaign to educate the public about RFI. The ARRL created a Field Organization of amateurs, to help resolve RFI problems. Cooperation between amateurs and broadcast listeners was the watchword (DeSoto, p 76). Based largely on their good-faith efforts to resolve interference complaints, amateurs won the gratitude of government officials and even Presidents (DeSoto, p 83). Ham stations became more advanced with the new vacuum-tube technology, and amateurs made significant advances on nearly every front. By 1936, there were 46,850 amateurs (DeSoto, p 133). During the late '40s, the FCC received RFI complaints at a rate of 7000 to 8000 per year, a "manageable" amount.

Evolution Of Regulation

In the early days of radio regulations, jurisdiction over wire and radio communications was handled at various times by the Department of Commerce, Post Office Department, Interstate Commerce Commission and the Federal Radio Commission. The advance of technology and increasing interference problems necessitated the coordination of these functions in a single agency. The Communications

Act, signed June 19, 1934, created the Federal Communications Commission (FCC) for that purpose.

Prior to the Communications Act of 1934, the Radio Act of 1912 and the Radio Act of 1927 had provided some regulation for Amateur Radio. The word "amateur" was used for the first time in the Radio Act of 1927. This law had created the Federal Radio Commission (FRC) and gave the FRC authority to classify radio stations, prescribe the nature of the service to be rendered by each class, assign frequencies and allot power to the various classes and stations, determine their location, regulate their apparatus, make regulations either for the prevention of interference or to carry out provisions of the Act, and require logs and records of transmission. This Act allowed for the beginning of RFI regulation.

The TVI Devil

During the 1950s, the use of televisions increased dramatically, as did RFI complaints. During 1953, there were 21,000 complaints. TVI brought a new challenge for amateurs. Almost all TV viewers lived in "fringe areas" because there weren't many television stations. Nearly every issue of *QST* carried at least one TVI article. The FCC knew that RFI was caused by design deficiencies, but they did not have the power to do anything.

As the need for more efficient communications increased, better and more affordable radios and other electronic devices evolved. The appeal of radio includes the ability to communicate without the need for cumbersome wires and cables. The appeal became so great, in fact, that these electronic devices manifested in just about every nook and cranny of the modern world. Therein lies the problem of RFI and the need for regulation.

Along with the proliferation of radio services, within the physical confines of the radio spectrum, comes the need for regulatory controls. Controls have developed over the years as the need arose, and they have been implemented at both the national and international levels. Why international? Because radio waves know no geographical bounds—they do not stop for customs inspections. Although RFI energy radiates at a somewhat lower level, the interaction and upset is still present.

As a result, the regulation and control of RFI matured along with the development of regulations for the whole of Amateur Radio.

The Plot Thickens

Through the '60s and '70s, the growing problem of RFI continued to seriously threaten the well being of the Amateur Radio Service. Amateurs were faced with more than just radio and television interference: CATV systems, power lines and a host of new consumer devices, to name a few. The Citizens Band (CB) radio boom increased the probability that a home-electronic device would be located near a transmitter. The ARRL tried to explain to disgruntled consumers that typical RFI situations involving radio amateurs result from design deficiencies in the affected device. Consumers, however, found it difficult to believe that a "passive" device, such as a TV or stereo, could be the "source" of interference. After all, when the ham was not transmitting, there was no problem! Of course the perception was that "the ham down the street" caused the RFI.[2]

Could the RFI situation become even worse? Yes, it could and would. During the '70s, some local governments began adopting ordinances making it "illegal" to interfere with television or radio reception. These laws were based on a "causing a public nuisance" concept. No one wanted to hear the ham explain. The situation became intolerable.

The Coming Of The Law

Amateur Radio was being blamed for the inability of electronic devices to reject unwanted radio signals, and a solution was needed. At the time, the FCC did not have the authority to set minimum rejection standards for consumer electronic devices, and ARRL leaders knew the situation would only get worse. Amateur Radio needed a law that would amend the Communications Act of 1934 giving the FCC exclusive jurisdiction over RFI matters. This law would preempt regulation by state or local governments. During the "old days," amateurs did not want RFI law, but it became evident in the '70s that a law was needed—if the Amateur Radio

[2]W. Clift, "`RFI Bill' Becomes Law; Amateur Radio Benefits!," November 1982 *QST*, pp 11-13.

Service was to survive. By 1981, the FCC was receiving over 62,000 RFI complaints annually. It was estimated that for every reported case, over 100 were unreported. Ninety-seven percent of these complaints involve a service *other* than the Amateur Radio Service.

The ARRL took a leading role by lobbying Congress for RFI legislation. Early attempts were unsuccessful; many bills were introduced, but they all died. It became clear that not everyone shared the ARRL's enthusiasm for giving the FCC authority to set RFI-immunity standards.

The industry was "dead set" against standards because they might be forced to add a $5 filter to the cost of a product. The Electronic Industries Association (EIA) lobbied fiercely against the bills.

Barry Goldwater, K7UGA, the only licensed ham in the Senate, introduced the "Goldwater Bill" and Rep Timothy Wirth introduced HR 5008. These bills, called the Communications Amendments Act of 1982, were passed and signed into law by the President on September 13, 1982. Finally, after 10 years of lobbying by the ARRL, the FCC was given exclusive jurisdiction over RF susceptibility of home entertainment equipment (Clift, p 13). That's the good news. The bad news is that Public Law 97-259 is not yet implemented because the FCC has opted for voluntary standards. We'll discuss Public Law 97-259 again later in this chapter.

Public Law 97-259 is not a cure all, however. On March 30, 1989, the FCC amended its rules to permit Part 15 (low-power unlicensed) devices, with higher power than would otherwise be permitted, in four amateur bands. The battle over RFI continues. The Part 15 section discusses this in detail.

FCC Structure and Regulatory Process

We tend to think of "radio" strictly within the context of "Amateur Radio," but it is just a fraction of the radio world. It is, however, necessary to consider the overall structure of the FCC and of its rules in order to understand how these many rules affect Amateur Radio, either directly or indirectly.

The regulation as it applies to radio-frequency interference (RFI) is a part of the total law and regulations of the Amateur Radio Service. The Amateur Radio Service is just one of 34 Private Radio Services currently administered by the Federal Communications Commission Private Radio Bureau. FCC regulation and the FCC treatment of RFI is discussed in greater detail later in this chapter.

The FCC does not function under any other Government department. It is an independent federal agency created by Congress and, as such, reports directly to Congress. The FCC staff is organized on a functional basis. There are four operating bureaus (Mass Media, Common Carrier, Field Operations, and Private Radio) and seven staff offices: Managing Director, Science and Technology, Public Affairs, Plans and Policy, General Counsel, Administrative Law Judges and the Review Board. Its headquarters are in Washington, DC.

The FCC Field Operations staff is located in field offices throughout the country. These staff personnel locate sources of interference and suggest remedial measures. The control of RFI includes the closing of unauthorized transmitters.

RFI and Policy in the United States

There are five determiners of Regulatory Policy in the United States: the FCC, industry, citizen groups, the Courts and the White House. In addition, the Congress frequently interacts with all five of these groups. The result is a political bias of the regulation. RFI regulation is, at times, affected by the political actions of all of these groups.

Although the FCC retains the overall regulation authority, it does not regulate Federal Government radio operations; this is done by the National Telecommunications and Information Administration (NTIA), under delegated authority from the President. In theory, the FCC and NTIA jointly develop telecommunication objectives, with input from the above-named groups. The responsible bureaucracy moves very slowly because of numerous stalls initiated by political groups.

RFI And International Law

The regulations of the United States Amateur Radio Service, as they relate to RFI, apply as long as the operator is within the territorial confines of this country or in international waters. One who wishes to operate within the confines of another country's territory, may do so only with a guest (or reciprocal-operating) permit from that country.

The primary driving force for international RFI regulation outside the United States is the European Economic Community (EEC). The EEC is committed to establishing a joint European market in 1992. Within the Community at present, differences in national regulations among the 10 member states prevent the sale of a common item across Europe. The national regulations refer to many aspects of product design, and EMC/RFI is just one area that has been recognized as requiring harmony. The aim is to establish an internal market among the member states, much like that among the states within the USA.

WHAT IF...YOUR TOWN HAS AN ORDINANCE REGULATING RFI?

Some local municipalities do have ordinances that try to regulate RFI—even though the FCC clearly has exclusive jurisdiction. ARRL Counsel Chris Imlay, N3AKD, received a letter from FCC General Counsel Robert L. Pettit reaffirming the fact that RFI matters have been preempted by the FCC. Imlay had requested an opinion on a Pierre, South Dakota, ordinance regulating RFI.

The letter from the FCC attorney stated that Congress has preempted any concurrent state or local regulation of RFI pursuant to the provisions of the Communications Act. The Act provides that the FCC "may, consistent with the public interest, convenience and necessity, make reasonable regulations (1) governing the [RFI] potential of devices which...are capable of emitting radio frequency energy...in sufficient degree to cause harmful interference to radio communications..." The FCC said that the legislation provides explicitly that the Commission shall have exclusive authority to regulate RFI. *Congress declared "Such matters shall not be regulated by state or local law, nor shall radio transmitting be subject to state or local regulation as a part of any effort to resolve an RFI complaint."*

A copy of the "Pettit letter" is available from the Regulatory Information Branch at ARRL HQ for an SASE.—*John C. Hennessee, KJ4KB, ARRL Regulatory Information Branch*

The EEC's Electromagnetic Compatibility (EMC) Directive defines two principal objectives, which all electrical and electronic products must satisfy before they can be marketed: (1) the product should not be a source of interference to others, and (2) it should be immune to interference.

To demonstrate compliance with the EMC Directive, a manufacturer can use a number of routes. First, he can show compliance with the relevant technical standards, which are agreed upon throughout Europe. They are known as European Norms (ENs). If no ENs apply, he can use any member state's national standards that have been deemed to meet the requirements of the Directive (by a committee known as the Article 5 Committee). Where technical standards do not exist, he must keep a technical file that records the EMC performance of the product. The file can be inspected by the regulatory body should the need arise.

FCC REGULATIONS AND RFI

There are not many regulations that directly affect RFI caused by Amateur Radio operation. The Communications Amendments Act of 1982, Public Law 97-259, made a significant impact on the Amateur Radio Service. This law (among other things) gives the FCC authority to regulate the susceptibility of electronic equipment to RFI. State and local entities sometimes try to regulate RFI. See the accompanying sidebar, "What If...Your Town Has an Ordinance Regulating RFI?" Public Law 97-259 gives the FCC exclusive jurisdiction in such cases.

The rules and regulations governing Amateur Radio are found in Part 97 of the Code of Federal Regulations, Title 47. Rules and regulations governing RFI are found throughout Title 47, which is composed of five volumes. Part 97 is one of over 200 sections of Title 47 (Telecommunications). Amateurs should be aware of these other regulations if they become involved in solving an RFI problem. Part 97 shows only one side of the problem. The rules concerning the affected consumer device show the other side; that may make it easier to find a solution.

The related parts of Title 47 of the Code of Federal Regulations include: Part 15 covering Radio Frequency Devices (unintentional radiators such as personal computers and CB receivers); Part 18 for Industrial, Scientific, and Medical Equip-

President Reagan, shown here with Senator Goldwater, K7UGA, promptly signed the Goldwater-Wirth Amateur Radio legislation into law. *(Official White House photo by Karl H. Schumacher.)*

ment (including domestic microwave ovens); and Part 76 for the Cable Television Service. Every ham should have a current copy of Part 97, but it is also useful to obtain copies of the other Parts listed here. Copies of Title 47 can be ordered from the government.[3] The complete and up-to-date Part 97 is contained in *The FCC Rule Book*, which includes a great deal of interpretation material. The book is available from ARRL HQ.[4]

Public Law 97-259

Several major changes in the regulations were made on September 14, 1982, when President Reagan signed into law a measure that affected RFI as it relates to the amateur service. The changes affect Amateur Radio and RFI in the following areas:

[3]Superintendent of Documents, US Government Printing Office, Washington, DC 20402; 202-783-3238.

[4]Contact ARRL HQ, 225 Main St, Newington, CT 06111; 203-666-1541.

1. The FCC was given the authority to regulate susceptibility of electronic equipment to RFI. In essence, this is an attempt to stem the flow of electronic devices that cannot function normally in the presence of RF energy.
2. The FCC was authorized to use the services of volunteers in monitoring for rules violations. When assisting the FCC, the volunteers can issue advisory notices to apparent violators, but volunteers are not authorized to take enforcement actions.

The legislative history of the bill gave the FCC jurisdiction over RF susceptibility of home entertainment equipment. It states, in part:

This law clarifies the reservation of exclusive jurisdiction over RFI matters to the Federal Communications Commission. Such matters shall not be regulated by local or state law, nor shall radio transmitting apparatus be subject to local or state regulation as a part of any effort to resolve an RFI complaint. The FCC believes that radio operators should not be subject to fines, forfeitures or other liability imposed by any local or state authority as a result of interference appearing in home electronic equipment or systems. Rather, the Commission's intent is that regulation of RFI phenomena shall be imposed only by the Commission.

It is important to note that the FCC has opted for voluntary standards, rather than formal rules and regulations. The FCC and the ARRL are both active in an industry group (in cooperation with the American National Standards Institute, ANSI) that is studying "RF immunity issues." This group, called the Ad Hoc Committee on Public Law 97-259 of the Accredited Stan-

WHO IS ULTIMATELY RESPONSIBLE FOR RESOLVING RFI PROBLEMS?

ARRL HQ is asked this question many times a day, and the answer is: both parties. RFI may result from: (1) harmonics, (2) spurious emissions, and (3) fundamental overload. Fundamental overload may produce RFI when the affected device is inadequately shielded, has unfiltered leads or is located very close to the RF source. If, however, the transmitter is being operated in a completely legal manner, and the radiated RF energy is within the limits of FCC regulations, the interference is probably caused by design deficiencies in the affected device. Typical solutions include adding filters and/or shields.

If the interference is caused by a licensed transmitter, its owner is responsible for transmitter compliance with the FCC rules and regulations. While it is useful to determine in each particular case whether the problem is in the amateur gear or the neighbor's home-entertainment equipment, RFI problems are best resolved in a "no fault" cooperative environment. *This is crucial*. If blame is the main concern, satisfactory solution of RFI problems is unlikely. In rare cases, problems remain even when the transmitter is 100% "clean" and the affected device is not at fault. Compromise may be the ultimate solution.— *John C. Hennessee, KJ4KB, ARRL Regulatory Information Branch*

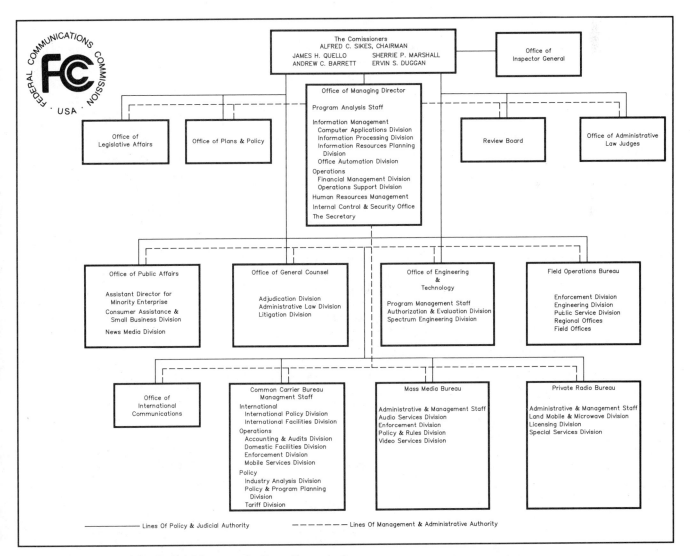

Organization chart of the Federal Communications Commission.

dards Committee C-63, has set an RFI-rejection standard for certain home electronic equipment. The voluntary standard provides for TVs and VCRs to reject unwanted-signal field strengths of not less than one volt per meter (1 V/m). This standard is not sufficient for full-power amateur stations under "worst case" conditions, but it covers the majority of cases.

Part 97

To summarize the Part 97 regulations concerning RFI: (1) do not occupy more bandwidth then necessary for the information rate and type of emission; (2) use good engineering practice to reduce spurious emissions; (3) use the minimum power necessary to carry out the desired communications; and (4) the FCC may impose restricted hours of operation if the amateur station is found to be causing interference with receivers of good engineering design.

The specific sections of Part 97 that relate or directly apply to RFI are: §97.15 Station antenna structures, §97.121 Restricted operation, §97.307 Emission standards, and §97.313 Transmitter power standards. Excerpts from these sections of Part 97 appear in this chapter.

While §97.15 is not directly related to RFI, antenna location and height can contribute to RFI problems. ARRL HQ has conducted studies that show there is less potential for interference when an amateur antenna is raised far above the level of existing antennas. Raising the antenna decreases the potential for fundamental overload. The study, entitled "Antenna Height and Communications Effectiveness," is available from the Regulatory Information Branch at ARRL HQ for an SASE.

Section 97.15(e) is of particular importance because it codifies PRB-1, the Federal preemption of overly restrictive state and local ordinances. Obviously, the higher the antenna, the less chance of causing RFI. Section 97.15(e) states:

§97.15 Station antenna structures.

⋮

(e) Except as otherwise provided herein, a station antenna structure may be erected at heights and dimensions sufficient to accommodate amateur service communications. [State and local regulations of a station antenna structure must not preclude amateur service communications. Rather, it must reasonably accommodate such communications and must constitute a minimum practicable regulation to accomplish the state or local authority's legitimate purpose. See PRB-1, 101 FCC 2d, 952 (1985) for details.]

Section 97.121 offers some of the most direct wording the FCC has regarding RFI and Amateur Radio operation. It details restrictions the FCC may impose on the operation of an amateur station causing RFI while the problem is being resolved.

"Uncle Charlie" can assign you "quiet hours" until an RFI investigation is completed. It is in the amateur's best

---

"PRB-1 KIT" AVAILABLE TO AMATEURS

Of particular interest to amateurs facing RFI problems is §97.15(e). This section codifies PRB-1, the Federal preemption of state and local antenna restrictions. Local governments can still regulate such things as safety and aesthetics, but amateur antennas must be reasonably accommodated. The Regulatory Information Branch at ARRL HQ has available a large packet of information that includes cases and sample ordinances as well as the text of PRB-1. HQ can also send you a list of ARRL Volunteer Counsels—amateurs who are also lawyers—who can give you further legal guidance. PRB-1 is not a cure all, and it cannot help if you are faced with a covenant or other deed restriction. The "PRB-1 kit" is available for a 9- × 12-inch SASE with 10 units of postage (either First- or Third-Class delivery).—*John C. Hennessee, KJ4KB, ARRL Regulatory Information Branch*

WHAT IF...YOUR LANDLORD THREATENS TO EVICT YOU OVER RFI?

If a landlord threatens eviction over RFI, call ARRL HQ and obtain the name of a Volunteer Counsel immediately. Lease agreements are private contractual agreements between the owner and the person leasing the building. Certain restrictions may prohibit antennas and RFI. The courts generally rule that the tenant was free to go where there are no such restrictions. Even if the agreement doesn't specifically state that antennas are prohibited, the landlord can do whatever he deems necessary to protect his property. Since such agreements "run with the land," PRB-1 offers no protection. A Volunteer Counsel or other lawyer may be able to give further guidance.—*John C. Hennessee, KJ4KB, ARRL Regulatory Information Branch*

WHAT IF...YOUR NEIGHBOR TAKES YOU TO COURT OVER RFI?

In rare instances, neighbors take amateurs to court over RFI matters. If this happens to you, *call ARRL HQ immediately and obtain the name of an ARRL Volunteer Counsel (VC) in your area.* A VC is a lawyer who is familiar with Amateur Radio. Since law is their livelihood, they are not expected to always give free advice. VCs have indicated that the initial consultation is always without charge, and they often give reduced rates to fellow amateurs.

Never try to defuse the situation without the aid of a lawyer, even when the law is on the side of Amateur Radio. While the courts clearly do not have jurisdiction over RFI matters, preparing to appear before a judge is potentially expensive. A VC can often resolve the matter before the amateur is taken to court. A neighbor may call the police and try to have an amateur declared a "public nuisance" for causing interference. *Only the FCC has jurisdiction over RFI cases!* ARRL HQ can send you sample RFI court cases.—*John C. Hennessee, KJ4KB, ARRL Regulatory Information Branch*

interest to cooperate fully with the FCC. An amateur must be sure that transmissions are not causing interference to his (or her) own home electronic devices.

§97.121 Restricted operation

(a) If the operation of an amateur station causes general interference to the reception of transmissions from stations operating in the domestic broadcast service when receivers of good engineering design, including adequate selectivity characteristics, are used to receive such transmissions, and this fact is made known to the amateur station licensee, the amateur station shall not be operated during the hours from 8 p.m. to 10:30 p.m., local time, and on Sunday for the additional period from 10:30 a.m. until 1 p.m., local time, upon the frequency or frequencies used when the interference is created.

(b) In general, such steps as may be necessary to minimize interference to stations operating in other services may be required after investigation by the FCC.

Section §97.307 addresses the emission standards an amateur station must meet. Paragraphs (c), (d) and (e) relate directly to RFI; detailing amateur operator action required to keep from interfering with other services, and the allowed levels of spurious emissions from amateur equipment. In summary: (1) do not occupy more bandwidth than necessary for the information rate and type of emission; and (2) use good engineering practice to reduce spurious emissions.

§97.307 Emission standards.

(a) No amateur station transmission shall occupy more bandwidth than necessary for the information rate and emission type being transmitted, in accordance with good amateur practice.

(b) Emissions resulting from modulation must be confined to the band or segment available to the control operator. Emissions outside the necessary bandwidth must not cause splatter or key-click interference to operations on adjacent frequencies.

(c) All spurious emissions from a station transmitter must be reduced to the greatest extent practicable. If any spurious emission, including chassis or power line radiation, causes harmful interference to the reception of another radio station, the licensee of the interfering amateur station is required to take steps to eliminate the interference, in accordance with good engineering practice.

(d) The mean power of any spurious emission from a station transmitter or external RF power amplifier transmitting on a frequency below 30 MHz must not exceed 50 mW and must be at least 40 dB below the mean power of the fundamental emission. For a transmitter of mean power less than 5 W, the attenuation must be at least 30 dB. A transmitter built before April 15, 1977, or first marketed before January 1, 1978, is exempt from this requirement.

(e) The mean power of any spurious emission from a station transmitter or external RF power amplifier transmitting on a frequency between 30-225 MHz must be at least 60 dB below the mean power of the fundamental. For a transmitter having a mean power of 25 W or less, the mean power of any spurious emission supplied to the antenna transmission line must not exceed 25 μW and must be at least 40 dB below the mean power of the fundamental emission, but need not be reduced below the power of 10 μW. A transmitter built before April 15, 1977, or first marketed before January 1, 1978, is exempt from this requirement.

Section 97.313 is concerned with transmitter power specifications and limits. Again, this section does not specifically address RFI issues, however, as with antenna location and height, excessive transmitter output power can aggravate RFI problems. Therefore, one of the cardinal rules of good amateur practice, "use the minimum power necessary to carry out the desired communications," directly applies.

§97.313 Transmitter power standards.

(a) An amateur station must use the minimum transmitter power necessary to carry out the desired communications.

(b) No station may transmit with a transmitter power exceeding 1.5 kW PEP. Until June 2, 1990, a station transmitting emission A3E is exempt from this requirement provided the power input (both RF and direct current) to the final amplifying stage supplying RF power to the antenna feed line does not exceed 1 kW, exclusive of power for heating the cathodes of vacuum tubes.

(c) No station may transmit with a transmitter power exceeding 200 W PEP on:
  (1) The 3.675-3.725 MHz, 7.10-7.15 MHz, 10.10-10.15 MHz and 21.1-21.2 MHz segments;
  (2) The 28.1-28.5 MHz segment when the control operator is a Novice or Technician operator; or
  (3) The 7.050-7.075 MHz segment when the station is within ITU Regions 1 or 3.

(d) No station may transmit with a transmitter power exceeding 25 W PEP on the VHF 1.25 m band when the control operator is a Novice operator.

(e) No station may transmit with a transmitter power exceeding 5 W PEP on the UHF 70 cm band when the control operator is a Novice operator.

(f) No station may transmit with a transmitter power exceeding 50 W PEP on the UHF 70 cm band from an area specified in footnote US7 to §2.106 of the FCC Rules, unless expressly authorized by the FCC after mutual agreement, on a case-by-case basis, between the EIC of the applicable field facility and the military area frequency coordinator at the applicable military base. An Earth station or telecommand station, however, may transmit on the 435-438 MHz segment with a maximum of 611 W effective radiated power (1 kW equivalent isotropically radiated power) without the authorization otherwise required. The transmitting antenna elevation angle between the lower half-power (−3 dB relative to the peak or antenna bore sight) point and the horizon must always be greater than 10°.

(g) No station may transmit with a transmitter power exceeding 50 W PEP on the 33 cm band from within 241 km of the boundaries of the White Sands Missile Range. Its boundaries are those portions of Texas and New Mexico bounded on the south by latitude 31° 41' North, on the east by longitude 104° 11' West, on the north by 34° 30' North, and on the west by longitude 107° 30' West.

Part 15

Other regulations govern low power, unlicensed devices that could cause interference to the Amateur Radio Service. Part 15 of Title 47 looks primarily at non-communication electronic equipment that generates RF energy. The prime example is the personal computer. Because of their high-speed clocks, they generate RF energy that can contribute to RFI. Part 15 addresses the emission limits for such electronic equipment. While not directly related to RFI from amateur station operation, this part may interest amateurs. Computers and other high-speed digital circuits are becoming a part of Amateur Radio operation. These can cause interference to amateur equipment and neighboring consumer electronics.

A number of very important changes were made to Part 15 in the past few years. In March 1989, the FCC amended its rules regarding permitted leakage from receivers, VCRs and stereos. The rules were changed to the more stringent Class-B computing limits, but noncompliant devices were "grandfathered" for 10 years. At the same time, the FCC adopted seven new "consumer bands" where intentional-radiation devices, such as garage-door openers, could operate at higher power than was previously possible. Four of these bands are located in amateur bands (902 MHz, 2.4-, 5.6- and 24-GHz) on a primary or secondary basis.

The ARRL strongly opposed these changes and feels that the FCC failed to deal effectively with the issue: The easiest solution would be to require a label stating that the device is susceptible to RFI. The ARRL also voted to seek reconsideration and whatever injunctive relief is necessary to protect amateur interests. The ARRL filed an emergency motion for stay in the US Court of Appeals for the District of Columbia Circuit, but it was denied.[5]

The sections of Part 15 that are most applicable include: §15.5(a), (b) and (c) Conditions of Operation, §15.13 Incidental Radiators, §15.17 Susceptibility to Interference and sections from Subpart B-Unintentional Radiators (§15.101 Equipment authorization of unintentional radiators, §15.105 Information to the user, §15.107 Conduction limits, and §15.109 Radiated emission limits). The pertinent information from these sections follows.

Section 15.17 is interesting in that it advises designers and manufacturers of electronic devices that use RF energy to be aware of the potential for RFI from outside sources, such as Amateur Radio. It also advises manufacturers to take appropriate measures to control the susceptibility of their equipment to RFI.

§15.17 Susceptibility to interference.

(a) Parties responsible for equipment compliance are advised to consider the proximity and the high power of non-Government licensed radio stations, such as broadcast, amateur and land mobile stations, and of US Government radio stations when choosing operating frequencies during the design of their equipment so as to reduce the susceptibility for receiving harmful interference. Information on non-Government use of the spectrum can be obtained by consulting the Table of Frequency Allocations in 2.106 of this chapter.

(b) Information on US Government operation can be obtained by contacting: Director, Spectrum Plans and Policy, National Telecommunications and Information Administration, Department of Commerce, Room 4096, Washington, DC 20230.

§15.5 General conditions of operation.

(a) Persons operating intentional or unintentional radiators shall not be deemed to have any vested or recognized right to continued use

[5]J. Cain, "Happenings," February 1991 *QST*, pp 54-55.

of any given frequency by virtue of prior registration or certification of equipment, or, for power line carrier systems, on the basis of prior notification of use pursuant to Section 90.63(g) of this chapter.

(b) Operation of an intentional, unintentional or incidental radiator is subject to the conditions that no harmful interference is caused and that interference must be accepted that may be caused by the operation of an authorized radio station, by another intentional or unintentional radiator, by industrial, scientific and medical (ISM) equipment, or by an incidental radiator.

(c) The operator of the radio frequency device shall be required to cease operating the device upon notification by a Commission representative that the device is causing harmful interference. Operation shall not resume until the condition causing the harmful interference has been corrected.

(d) Intentional radiators that produce Class B emissions (damped waves) are prohibited.

Section 15.13 of the Code of Federal Regulations states:

Manufacturers of these devices shall employ good engineering practices to minimize the risk of harmful interference.

The kinds of equipment the FCC considers "unintentional radiators" and their allowed emission limits are discussed in Subpart B of Part 15. Section 15.101 of this subpart identifies the kinds of equipment that must be certified by the FCC or independently verified as meeting the emission limits spelled out in §15.107 and §15.109. These include personal computers and their associated peripheral devices along with receivers that operate between 30-960 MHz and CB receivers.

Two general classes of digital equipment are defined in §15.105: Class-A digital equipment typically used in a commercial environment, and Class-B digital equipment that is used in a residential installation. (§15.105 appears as a sidebar of the Computers chapter.) The conduction limits for Class-B digital equipment and any equipment that is designed for connection to an ac power line is listed in §15.107, where:

> …the RF voltage within the band 450 kHz to 30 MHz conducted back into the power line shall not exceed 250 microvolts.

For Class-A digital equipment, the conduction limits are relaxed somewhat to: 1000 µV for RF emissions from 450 kHz to 1.705 MHz and 3000 µV for RF emissions from 1.705 to 30 MHz. As listed in §15.109, the radiated emission limits for unintentional radiators are shown in Tables 1 and 2. For CB receivers, the field strength of the radiated emission within the 25-30 MHz frequency range shall not exceed 40 µV/m at a distance of 3 meters.

It is interesting to note the distinction between Class-A digital equipment and Class-B digital equipment. Class-B equipment, intended for use in a residential environment where the likelihood of RFI is greater, must meet much stricter RF emission limits.

Part 15, Subpart C, addresses RF emission limits from intentional radiators such as carrier-current and perimeter-protection systems. Although the systems are not generally used in a residential environment, Amateur Radio operators should be aware of such devices; they may contribute to an RFI complaint. Interested readers should refer to the FCC rules for further information about this equipment and its associated emission limits.

Part 18

The FCC rules in Part 18 of CFR 47 deal with industrial, scientific, and medical equipment (ISM) that emits electromagnetic energy in the RF spectrum. As with the equipment covered by Part 15, the ISM equipment addressed here is also a potential source of RFI. The most applicable sections of Part 18 are:
§18.107 Definitions,
§18.109 General technical requirements,
§18.111 General operating conditions,
§18.115 Elimination and investigation of harmful interference,
§18.301 Operating frequencies,
§18.303 Prohibited frequency bands,
§18.305 Field strength limits, and
§18.307 Conduction limits.

The basic FCC definitions of ISM equipment are given in five paragraphs of §18.107. We all know the most-common piece of ISM equipment: the microwave oven.

Table 1

Radiated Emission Limits for Unintentional Radiators Other than Class-A Digital Equipment

Frequency of Emission (MHz)	Field strength (µV/m, 3 m from radiator)
30-88	100
88-216	150
216-960	200
Above 960	500

Table 2

Radiated Emission Limits for Class-A Digital Equipment

Frequency of Emission (MHz)	Field strength (µV/m, 10 m from radiator)
30-88	90
88-216	150
216-960	210
Above 960	300

§18.107 Definitions.

⋮

(c) Industrial, scientific, and medical (ISM) equipment. Equipment or appliances designed to generate and use locally RF energy for industrial, scientific, medical, domestic or similar purposes, excluding applications in the field of telecommunication. Typical ISM applications are the production of physical, biological, or chemical effects such as heating, ionization of gases, mechanical vibrations, hair removal and acceleration of charged particles.

(d) Industrial heating equipment. A category of ISM equipment used for or in connection with industrial heating operations utilized in a manufacturing or production process.

(e) Medical diathermy equipment. A category of ISM equipment used for therapeutic purposes, not including surgical diathermy apparatus designed for intermittent operation with low power.

(f) Ultrasonic equipment. A category of ISM equipment in which the RF energy is used to excite or drive an electromechanical transducer for the production of sonic or ultrasonic mechanical energy for industrial, scientific, medical or other noncommunication purposes.

(g) Consumer ISM equipment. A category of ISM equipment used or intended to be used by the general public in a residential environment, notwithstanding use in other areas. Examples are domestic microwave ovens, jewelry cleaners for home use, ultrasonic humidifiers.

Section 18.109 directly addresses the interference issues of ISM equipment. Basically ISM equipment should be of "good engineering design" with adequate filtering to reduce RFI.

§18.109 General technical requirements.

ISM equipment shall be designed and constructed in accordance with good engineering practice with sufficient shielding and filtering to provide adequate suppression of emissions on frequencies outside the frequency bands specified in §18.301.

Section 18.111 outlines the basic RFI operating conditions for users of ISM equipment and §18.115 outlines procedures for eliminating and investigating harmful interference from ISM equipment. In summary, these sections state that the operator of ISM equipment causing harmful interference to any authorized radio service "shall promptly take appropriate measures to correct the problem." However, this provision "shall not apply in the case of interference to an authorized radio station or a radiocommunication device operating in an ISM frequency band." And, it "shall not apply in the case of interference to a receiver arising from direct intermediate frequency pickup by the receiver of the fundamental frequency emissions of ISM equipment operating in an ISM frequency band and otherwise complying with the requirements of this part." These conditions are very similar to the RFI operating conditions for the Amateur Radio Service. In other words, if the ISM equipment is operating properly in its assigned band, any RFI it is causing must be corrected by the entity that is receiving the interference.

Sections 18.301, 303, 305 and 307 deal with the operating frequencies and field-strength limits of the ISM bands. These bands are distributed across the spectrum in small segments from 6.78 MHz to 245 GHz. The field-strength and conduction limits depend on the band of operation and the equipment used. Interested readers should refer to these sections of the FCC rules and regulations for a complete listing of the bands and limits.

Part 76

The operation of Cable Television Service (CATV) is covered by Part 76 of Title 47. Because CATV systems can, and do, operate on frequencies assigned to other services, there exists a potential for interaction between these systems and other radio services, including Amateur Radio. The two main sections of Part 76 pertinent to the amateur operator are §76.605 Technical standards and §76.613 Interference from a cable television system.

Section 76.605 covers the technical standards for the operation of a CATV system, such as video-signal level and signal-to-noise ratio. This section also specifies the maximum radiation allowed from a CATV system (see Table 3).

Regarding harmful interference from a CATV system, §76.613 clearly places responsibility (see "What If A Leaky Cable TV System is Interfering With Your Signal?").

In essence, Amateur Radio operators

WHAT IF...A LEAKY CABLE TV SYSTEM IS INTERFERING WITH YOUR SIGNAL?

Complete instructions for dealing with CATV systems appear in the CATV section of the Televisions chapter. Look there for guidance. The following federal regulations apply:

§76.613 Interference from a cable television system.

(a) Harmful interference is any emission, radiation or induction which endangers the functioning of a radio navigation service or of other safety services or seriously degrades, obstructs or repeatedly interrupts a radiocommunication service operating in accordance with this chapter.

(b) The operator of a cable television system that causes harmful interference shall promptly take appropriate measures to eliminate the harmful interference.

(c) If harmful interference to radio communications involving the safety of life and protection of property cannot be promptly eliminated by the application of suitable techniques, operation of the offending cable television system or appropriate elements thereof shall immediately be suspended upon notification of the Engineer in Charge (EIC) of the Commission's local field office, and shall not be resumed until the interference has been eliminated to the satisfaction of the EIC. When authorized by the EIC, short test operations may be made during the period of suspended operation to check the efficacy of remedial measures.

---

who receive interference from local CATV operations have legal recourse to cause the CATV company to "clean-up their act." Nonetheless, if their signals can exit the system, amateur signals can penetrate it and interfere. As with all RFI problems, "cooperation" is a key word when solving CATV problems.

THE CREATION OF NEW REGULATIONS
RFI Committees and Working Groups

IEEE Standards Committee C63 on Electromagnetic Compatibility (EMC). The following listed working groups are engaged in development of EMC/RFI standards:

- C63.13/D8—Recommended Practice for Electromagnetic Compatibility Limits.
- C63.14/D2—C63/EMC Definitions.

Table 3
CATV Leakage Limits

Frequency (MHz)	Field Strength (µV/m)	Distance (ft)
0-54	15	100
54-216	20	10
216+	15	100

- C63.15/D2—Recommended Practice for the Method of Immunity Measurement of Electrical and Electronic Equipment.

EIA/TIA—Telecommunications Industry Association in association with the Electronic Industries Association.

Joint Telecommunications Standards Steering Group (JTSSG). The JTSSG is a Department of Defense planning committee that oversees the development of military standards. Current work deals only with standards for electronic-equipment resistance to RFI such as high-altitude electromagnetic pulse (HEMP) and other sources of interference or upset.

The following standards and handbooks are currently in development or revision under direction of the following working groups:

- MIL-STD-188-124B—Grounding, Bonding, and Shielding for Common Long Haul/Tactical Communications Systems Including Ground-based Communications Facilities and Equipment.
- MIL-STD-188-125—High Altitude Electromagnetic Pulse (HEMP) Protection for Ground-based Facilities Performing Critical Time Urgent Missions.
- MIL-STD-188-148—HF AJ Waveform
- MIL-STD-188-242-1—Interoperability and Performance Standards for VHF Frequency Hopping Tactical Radio Systems.

- MIL-STD-188-244—Interoperability and Performance Standards for UHF Frequency Hopping Tactical Radio Systems (HAVE QUICK IIA).
- MIL-HDBK-423—High-Altitude Electromagnetic Pulse (HEMP) Protection for Fixed and Transportable Ground-based Facilities, Volume I: Fixed Facilities.

Current And Future Progress On The Legal Front

The sources of RFI have gradually increased over the years. More and more electronic and electrical devices have been invented and sold to consumers. The "good news" is that these devices have made our life more efficient; the "bad news" is the devices create unwanted interference and spurious signals.

Amateur Radio is affected in two ways: the need to provide radios that are resistant to interference, and the need to reduce or control spurious and primary radiation of transmitting stations. Recent changes to the FCC regulations (that is, Public Law 97-259) have given the FCC authority to regulate equipment RFI susceptibility. Public Law 97-259 also attempts to ease the load on the FCC by allowing volunteers to assist in monitoring for rules violations. The FCC must then respond if enforcement is necessary.

Amateur Radio is on the upswing. An FCC rule change, effective February 14, 1991, established a new class of license that does not require Morse code proficiency. A consequence of this development could be an increase in the number of radio transmitters on the air (and energy radiated). Increased activity brings the possibility for more RFI and violations. It also increases the need for monitoring and possibly modification of regulations.

The development of features that allow for automated radio operation (for example, automatic linking) may affect methods of use and expand the use of Amateur Radio. Will this development affect RFI? Let's hope the results will be positive: greater understanding and increased RFI immunity. The potential for radiation of spurious signals (RFI), however, will obviously increase.

Chapter 15

The ARRL EMI/RFI Report Form

By Ed Hare, KA1CV
Senior ARRL Laboratory Engineer

Here is a valuable EMI-fighting tool: the EMI/RFI Report Form. Over the years, EMI *reporting* has declined substantially. Telephone calls and letters to ARRL HQ, however, indicate that the number of actual EMI problems has only declined slightly.

Most cases of EMI are not reported, to the FCC, to the Electronic Industries Association (EIA) or to ARRL HQ. I have done informal EMI surveys at club talks and ARRL-Convention Technical Forums. When I ask how many amateurs have been involved in some form of EMI problem, nearly every hand goes up. When I ask how many have reported that problem, nearly all hands go down.

The ARRL HQ staff is committed to helping amateurs fight EMI. However, a handful of people in Newington can't do it alone. We need details of the EMI problems that hams face all around the world. When some new device (such as the touch controlled lamp) appears in the marketplace, we need to know whether it presents an EMI problem. The ARRL EMI/RFI Report Form is the best way to send this valuable information to HQ. Please use the report form to help.

ARRL HQ wants to ensure that the EMI problem is fully understood by all involved persons and organizations. On-going ARRL programs and discussions with groups like the FCC and the EIA require accurate EMI data for success. For example, the EIA has characterized the number of EMI complaints that involve the Amateur Radio Service over the past few years as "very small." That view probably reflects decreased reporting more than reduced equipment susceptibility. Increased reporting can clarify the situation.

Past work with the FCC, Congress and industry has led to improvements in the area of TVI. Electronic-equipment manufacturers have created and complied with voluntary standards for EMI susceptibility. A growing number of homes are connected to relatively leak-free cable television systems. The number of television and VCR interference problems has declined.

On the other hand, telephone deregulation and technological changes have put susceptible electronic circuitry into such devices as cordless phones, automobiles and appliances. This has significantly increased the interference potential near radio transmitters.

Filling In The Form

We don't need every bit of information for your report to be useful. First, do you want the Technical Department EMI information package? If so check the box. The first block is for the amateur station. (A case number will be filled in by ARRL HQ when the report is received.) Have the complainant fill out the second block, and you should fill in the complaint area together. If you are completing the form at the beginning of the EMI work, don't worry about the optional area. If the problem is already cured, complete the optional area as well.

When your report arrives at HQ, it will join the ARRL EMI database. The database allows HQ staff to track the number of EMI problems, the kinds of EMI problems and the kinds of susceptible devices. Copies of all unresolved EMI reports will be sent to the appropriate ARRL Section Technical Coordinator for follow-up. Thanks for helping us keep track of EMI!

ARRL RFI REPORT

Send to: ARRL Headquarters, Box RFI
225 Main St
Newington, CT 06111

Case No. _____

Amateur

☐ Send ARRL Technical Department EMI Information Package

NAME	CALL
ADDRESS	PHONE NO. — —
CITY, STATE, ZIP	

Complainant

☐ Send ARRL Technical Department EMI Information Package

NAME	
ADDRESS	PHONE NO. — —
CITY, STATE, ZIP	

Complaint

	TYPE (VCR, ETC)	MANUFACTURER	MODEL NO.
Transmitting Equipment			
Susceptible Equipment			

Description of problem (nature of interference, severity, fixes tried)

OPTIONAL

AGENCIES NOTIFIED (FCC, EIA, NCTA, MANUFACTURERS?)

WAS THE PROBLEM RESOLVED? IF SO, BY WHOM AND HOW?

ARRL Form TD1/92

Fig 1—The ARRL EMI reporting form. Please fill out a copy for every EMI complaint and mail it to ARRL HQ. From an official viewpoint, unreported EMI problems do not exist.

Chapter 16

Filter Performance

By Zack Lau, KH6CP/1
ARRL Laboratory Engineer

Many cases of EMI that result from Citizens Band (CB) or Amateur Radio transmissions can be eliminated or substantially reduced by use of filters. There are low-pass, high-pass and power-line filters, each type named for its function. The information presented here is intended to help you choose a filter (or filters) to solve an EMI problem.

FILTER FUNDAMENTALS

You must have an idea of the frequencies that are involved to understand how filters work and determine which is right for your application. Amateur Radio operators (hams) use a wide range of frequencies between 1.8 and 29.7 MHz. Hams use higher frequencies too, but those operations are not covered by this report. CB operators use a narrow band of frequencies at about 27 MHz. Television and FM radio broadcasting takes place on frequencies above 54 MHz. In a perfect world there would be no interference to TV and FM radio receivers from ham and CB transmissions; unfortunately, this is not the case. Transmitting stations can emit unwanted signals that cause interference. To some extent, receivers respond to signals that they shouldn't receive at all. Filters can help.

A *low-pass filter* is used between a CB or Amateur Radio transmitter and the station antenna to suppress unwanted signals above the operating frequency, yet freely pass signals at the transmitting frequency. The unwanted signals usually occur at multiples of the transmitter operating frequency. They can be received by TVs as interference. For example, an unwanted signal from an amateur transmitter operating in the 21-MHz band could cause interference to TV channel 3 (60 to 66 MHz). An ideal low-pass filter

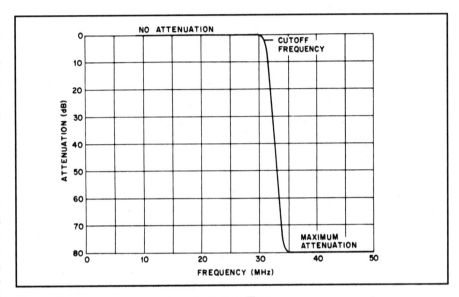

Fig 1—Frequency response for a low-pass filter.

for this application would pass all signals below 29.7 MHz with no attenuation, but pass nothing above that frequency.

In real life, filters cannot switch abruptly from passing everything to passing nothing at a single frequency. In addition, low-pass filters cannot remove unwanted signals completely. They can, however, greatly reduce (attenuate) the strength of unwanted signals so that they are no longer bothersome. The frequency at which a filter attenuates signals by a substantial amount is called its *cutoff frequency*, or cutoff. Most low-pass filters cutoff somewhere above 30 MHz; maximum attenuation usually occurs around 54 MHz, the beginning of the TV channels. Fig 1 shows how a very good low-pass filter responds to signals of various frequencies.

High-pass filters are installed in the line between a TV set (or FM receiver)

and its antenna to prevent unwanted signals from entering the tuner. An ideal high-pass filter would pass all frequencies above 54 MHz and nothing below that frequency. EMI is often generated within a TV when a strong signal enters the set. The signal need not be at the frequencies of television channels. For example, a CB transmitter at 27 MHz could cause interference to TV channels well above 54 MHz, even if the CB station used a low-pass filter and generated no unwanted signals at TV frequencies. A high-pass filter reduces the unwanted signal, and the interference is never created. TV high-pass filters cutoff at about 40 to 50 MHz. Fig 2 shows how a very good high-pass filter responds to signals of various frequencies.

Power-line filters are used in the ac cords of electronic equipment. (A TV set, radio or audio amplifier may respond to

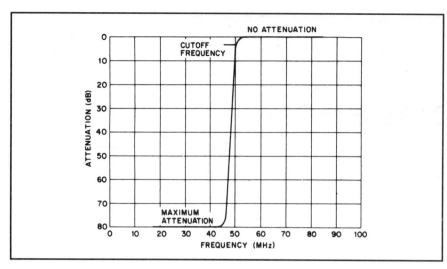

Fig 2—Frequency response for a high-pass filter.

conducted RF energy.) A line filter is usually installed between a 120-V ac-line cord and the ac outlet, but some filters are used in 12-V dc automobile power lines. Power-line filters prevent radio frequency (RF) energy from entering equipment through the power lines. An ideal power-line filter would pass energy at the 60-Hz line frequency and would pass nothing above that frequency. Fig 3 shows how a very good power-line filter responds to signals of various frequencies.

Other filters are used in special cases. Nearby FM broadcast stations may cause TV interference. FM traps are designed to prevent unwanted FM broadcast energy from entering a TV. FM traps usually attenuate signals between approximately 80 and 130 MHz, and pass all other frequencies. Some FM traps are adjustable so that maximum attenuation occurs at the frequency of a specific FM station. Still other filters are designed to eliminate interference to audio equipment. These filters keep RF energy from entering audio equipment through interconnecting cables or speaker leads.

FILTERS TESTED

The Federal Communications Commission (FCC) tested 43 interference filters in their laboratory. FCC compiled the test results in an Internal Technical Note, and the American Radio Relay League (ARRL) obtained a copy of the report under the Freedom of Information Act. ARRL also tested a number of filters in their Newington, Connecticut, Laboratory. FCC and ARRL test results are presented in the pages that follow. Although

the FCC lab tested three samples of each filter, data from only one sample is presented here. Of the three filters tested,

one was selected that seemed to best represent the typical filter performance. Only one sample of each filter was tested in the ARRL Lab.

New filters appear in the marketplace each year, and others are discontinued. Although the FCC and ARRL tried to make these tests as complete as possible, all available filters may not be represented. Some of the filters in the test are no longer available.

Filter Test Procedure

Filters were tested in the ARRL Lab using a spectrum analyzer and tracking signal generator. The results were recorded and plotted. The test setup for all but power-line filters is shown in Fig 4. The FCC laboratory setup, shown in Fig 5, is similar except that a computer was used to process and plot the results.

For low-pass filter tests, the output of the tracking signal generator was con-

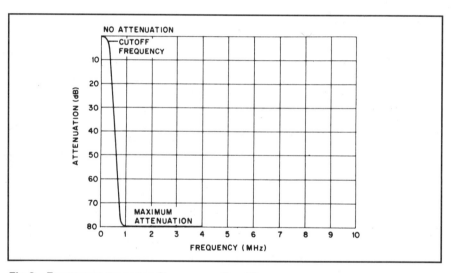

Fig 3—Frequency response for a power-line filter.

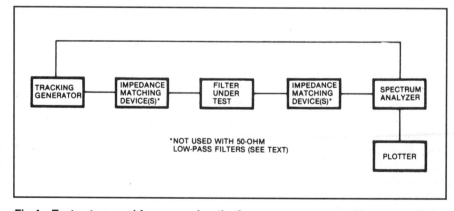

Fig 4—Test setup used for measuring the frequency response of low-pass, high-pass and FM-trap filters in the ARRL Lab.

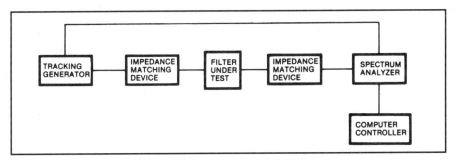

Fig 5—Test setup used for measuring the frequency response of low-pass, high-pass and FM-trap filters in the FCC Lab.

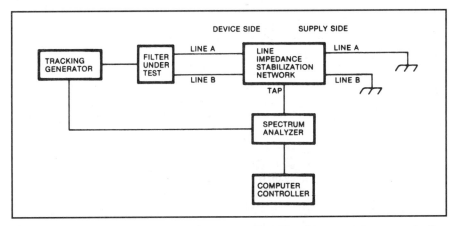

Fig 6—Test setup used for measuring power-line filter frequency response in the ARRL Lab.

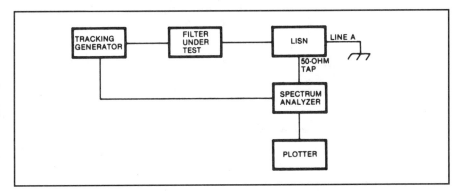

Fig 7—Test setup used for measuring power-line filter frequency response in the FCC Lab.

nected directly to the filter input. Filter output was then connected directly to the input of the spectrum analyzer. Spectrum analyzer measurements were made as the frequency of the tracking generator swept through the frequency range indicated on the graphs.

High-pass filters and FM traps were tested using a similar setup. The test equipment has a 50-Ω impedance, and the impedance of the filters is 75 Ω or 300 Ω. Impedance-matching devices were used to ensure that the filter under test, spectrum analyzer and tracking generator were properly terminated. In the ARRL Lab setup, a resistive matching pad was use to match between 50 Ω and 75 Ω. For 300-Ω filters, a two-step matching system was used. 50 Ω to 75 Ω matching pads were followed by a 4-to-1 balun transformer to match between 75 Ω and 300 Ω.

Power-line filters were tested with the setup shown in Figs 6 and 7. The line impedance stabilization network (LISN) is a passive network that provides a 50-Ω tap for the spectrum analyzer so that accurate measurements may be made. See Fig 8. ARRL and FCC used identical LISN devices for the tests. Power-line filters are normally symmetrical, so normally only one side of each filter is shown. In some cases, there were differences between the lines, so both are shown.

The results of ARRL and FCC filter tests are reproduced in the following pages. The response of each filter is plotted. The make, model, and purpose of each filter is given in the caption. This report was prepared by the American Radio Relay League, 225 Main Street, Newington, CT 06111, 203-666-1541.

CHOOSING A LOW-PASS FILTER

Given a choice, pick one that has a money-back guarantee, so you can take it back if it doesn't work. To the "dangerously" informed, simply comparing loss curves might seem to be the best method. Loss curves, however, only give part of the story. To the truly informed, a physical inspection of the filter is more significant than loss curves.

Look at the quality of the chassis and components. Pass up filters with plastic cases. Sheet-metal cases are better than plastic. Die-cast metal is best. How close together are the screws that close the case? They should be no more than two inches apart; more screws are better. What kind of connectors are used? UHF (SO-239) connectors are used on most HF filters. BNC connectors are better; N connectors are best. If the case and connectors are of high quality, chances are the design and performance are good as well.

Do you know how most low-pass filters actually work? Most reflect the unwanted signal back to the source. They convert the load impedance (normally assumed to be 50 Ω) into something quite different. Consider a filter as the opposite of a Transmatch; it's designed to *cause* mismatches (at unwanted frequencies). Unfortunately, the load and source are rarely 50 Ω at the unwanted frequencies. In fact, since transmitters usually have a low-pass filter at the output, it is almost certain that they do not present matched sources to the filter at unwanted frequencies.

We Can't Know Everything

Someone interested in a display of

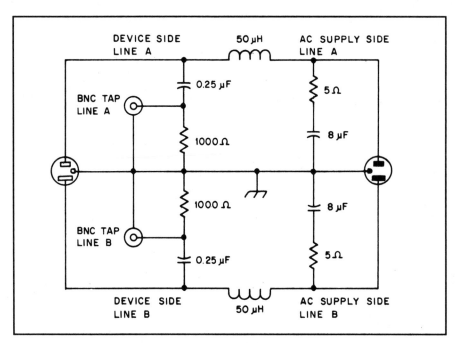

Fig 8—Diagram of the line impedance stabilization network (LISN) used by the ARRL and FCC for power-line filter testing.

technical supremacy, with the right equipment and software ($100,000 circa 1991), along with lots of time, could systematically evaluate filters. First, measure the complex impedance of the antenna and feed line across the VHF/UHF spectrum with a network analyzer. Then the effective load presented to the transmitter at all frequencies of interest can be calculated.

The key step (perhaps the most difficult) is modeling the transmitter. What harmonic content does it supply to the load (consisting of the low-pass filter, feed line, and antenna)? Harmonic-balance software can supposedly accomplish this task—but it's slow and expensive, even on very fast computer systems.

We could just measure the low-pass filter, feed line and antenna in one step, but the measurement would only be significant at one feed-line length. In a mismatched system, the electrical length of the feed line is important. Here is the problem: There is no easy way to calculate the needed length—unless you can compute the harmonic content of a transmitter feeding an arbitrary load. Still, we can attack the problem intelligently without knowing everything.

Optimizing Feed-Line Lengths

First of all, concentrate on a frequency of interest, say a TV channel with interference. (If the interference affects all channels, the problem is likely front-

end overload. Try a high-pass filter at the affected device. If that does not help, suspect harmonic interference at the TV IF, 42.25 MHz.) Once you know the frequency, calculate the electrical wavelength for the cable you are using. The factor used to convert the wavelength in free space to one in a media such as a transmission line is known as the velocity factor.

$$\lambda_f = \lambda \times VF$$

where

λ_f = wavelength in feed line
λ = wavelength in free space
VF = velocity factor.

There are two "magic" lengths—multiples of 1/2 λ and odd multiples of 1/4 λ. Varying the feed-line length by multiples of 1/2 λ (say 3 half wavelengths) normally doesn't change anything at the frequency of interest. However, adding 1/4 λ (or an odd multiple thereof) changes the nature of mismatch considerably. It may be possible to markedly change the performance of a filter by intelligently changing the cable lengths.

Of course, proper filter performance often depends on a good ground for the filter. Remember that you need a ground for VHF—54 MHz and up. A seemingly short length of 4.3 feet is terrible—it's 1/4 λ at 54 MHz. Since ground wires should be a maximum of a few inches long, it may make more sense to put a "ground" at the transmitter. Do this by placing a large

metal sheet under the transmitter. Given that the troublesome harmonics are at VHF/UHF, 20 square feet is often adequate. The best ground performance can be obtained by placing the sheet on top of concrete or some other lossy dielectric. "Eliminate TVI with Common-Mode Current Controls," by Richard Buchan, WØTJF (May 1984 *QST*, pp 22-25) describes this grounding technique in depth. Of course, it is sometimes possible to locate a filter at the earth ground; this allows very short connections.

What Makes A Good Low-Pass Filter?

One big filter performance factor (which you can determine from visual inspection) is shielding. Are there shields between the filter sections? A box with no intermediate shields doesn't work well at VHF or UHF (TV frequencies). How good are the connectors? The better female UHF connectors have silver plated contacts to maximize conductivity. Cheaper chassis jacks often have a socket contact consisting of a tube split down one side only. Construction techniques deserve consideration. Choose a case held together by screws rather than rivets. By disassembling the filter, you can discern the network used. Screw closures also allow for repairs or modifications.

A Rough Comparison of Networks

To some extent, one can guess at the impedance presented to the transmitter by noting what is connected to the filter input terminal:

- A shunt connected capacitor tends to short harmonics to ground.
- A series connected inductor tends to present an open circuit to harmonics.
- A shunt connected series-resonant LC network presents a very good short at the resonant frequency. The inductive reactance dominates at much higher frequencies, however, and presents a relatively high input impedance.
- A series connected parallel-resonant LC network presents a good open circuit at the resonant frequency. The rest of the network has a great effect at higher frequencies.

By the judicious choice of shunt and series networks, it is possible to get a variety of impedances at different frequencies, hopefully optimized to reduce the amount of harmonic energy. Of course, a 1/4-λ of coax acts as an impedance transformer, further complicating the issue.

Often, a filter that compliments the existing transmit filter(s) works best. If the transmitter has a pi output filter (one with a shunt capacitor across the output) for instance, filters with series inductors work better than those with shunt capacitors.

How to Evaluate Loss Curves

A close examination of the curves can yield some useful information. For instance, curve steepness tends to indicate the number of filter sections used: Filters with steeper curves have more sections. To some extent, filters with poor shields show steady degradation as the frequency increases.

Discount the presence of sharp nulls. Manufacturing tolerances may easily shift the null frequencies by 10% or more.

Attenuation values greater than 70 dB are outside the measurement range of the spectrum analyzer. Therefore, a reading of 77 dB could well be 71 or 86 dB. Also, the equipment used corrects for generator inaccuracies and cable losses at a single point—this explains why some of the FCC data shows passive filters with gain at some frequencies.

The attenuation of the amateur signal in a low-pass filter depends on two factors: the resistive losses and mismatch losses. The meaning of the first is pretty obvious: nonideal components lose some power as heat.

Mismatch losses result when the filter presents the wrong impedance. If amateur transmitters presented a resistive 50-Ω impedance, it would be pretty easy to figure out the power loss. With actual broadband transmitters, mismatches can actually result in more power output!

With a tuned output stage, such as that found in many vacuum-tube linear amplifiers, the mismatch loss is reduced to zero if the amplifier is tuned properly into the resultant load.

However, if the system requires a Transmatch for proper operation, the low-pass filter should still be placed between the rig and the Transmatch, not between the Transmatch and the antenna. While either location allows the mismatch loss to be tuned, placing a filter where the SWR is high reduces the power that the system can handle.

Even if you know everything, it's obvious that you can't simultaneously achieve maximum harmonic suppression and minimum passband loss with real passive filters. Equipment designers commonly choose the minimum acceptable harmonic suppression, which typically leads to filters with the lowest possible insertion loss.

Choosing A High-Pass Filter

You may have thought the low-pass filter situation was confusing; high-pass filters are worse! It is safe to say that both the source and load impedances are unknown in most cases. It is also a good assumption that the TV set is not grounded for RF. While it is possible to build VHF/UHF ground by placing a large conductive sheet on a lossy dielectric, it is impractical (in many cases) to actually use such a ground. As a result, it is possible for the TV set to be affected by common-mode currents (RF that flows on the coax shield and chassis). Fortunately, common-mode filters are available (often touted as filters for CATV). These filters, while quite effective when applied cor-

rectly, may have little effect on strong signals picked up by an ordinary TV antenna. Common-mode filters often work best when placed at the TV set, while ordinary high-pass filters may work best when located to allow good grounding. Graph 42 shows the improvement possible with good grounding.

Choosing DC Power-Lead Filters

There are at least two basic kinds of dc power filters. One eliminates AF interference, such as that generated by an alternator. The other eliminates RF interference that could enter through the power leads. It's not likely that a single filter would do both.

Choosing Telephone Filters

It may be good to ask what RFI the filter is designed to remove. Many filters are optimized for a specific band of frequencies. They often provide uncertain performance elsewhere. Keep in mind that some phones respond directly to RF in the environment—no external filter can help. Substitute a known good phone to detect these problem phones.

Filter Saturation—A Problem?

While looking at the filter charts, you might note the power handling test of the TCE Labs telephone filter. When subjected to high power levels (watts of RF), this filter lost a few dB of attenuation. Conceivably, an amplifier combined with a low antenna might produce a field strong enough to cause such problems. For normal amateur installations, filter saturation is more of a theoretical curiosity than cause for worry.

Table 1
Filter Illustration and Performance Summary

Graph No.	Manufacturer	Model No.	1.8-29.7 MHz (Ham) Less = Better	Attenuation 26.9-27.5 MHz (CB) Less = Better	54-216 MHz (TV) More = Better
Low-pass Filters					
1	Barker & Williamson	425	0.3	0.3	56
2	Cornell Dubilier	CBTVI-1	1.0	0.8	26
3	Heath	HDP-3700	0.4	0.4	59
4	J. W. Miller	C-514-T	0.5	0.5	77
5	Microwave Filter Co.	Interfilter	0.2	0.2	70
6	R. L. Drake	TV-3300-LP	0.5	0.5	68
7	R. L. Drake	TV-3300-LP	0.4	0.4	76
8	R. L. Drake	TV-42-LP	0.3	0.2	62
9	R. L. Drake	TV-42-LP	0.5	0.5	73
10	Unknown	Unknown	0.2	0.2	70
11	Wm. M. Nye Co.	250-0020-001	0.3	0.3	59
12	Ten-Tec	5061	0.2	0.2	46

Graph No.	Manufacturer	Model No.	More = Better	More = Better	Less = Better
High-pass Filters					
13[†]	Archer	15-580	65	67	0.5
14	Archer	15-580	73	74	2.5
15	Blonder Tongue	FR-CB-75	17	30	1.0
16	Channel Master	7203	46	48	0.5
17	Channel Master	7203	40	47	1.1
18	J. W. Miller	C-513-2	62	63	0.5
19	Marine Technology	EMI-TV75	68	68	0.6
20	R. L. Drake	TV-75-HP	64	65	1.0
21	R. L. Drake	TV-75-HP	70	74	2.2
22	RMS Electronics	CA-2700	32	33	0.6
23	Winegard	HP 2700	35	37	0.4
24	Ten-Tec	5060	37	52	1.4
25	AMECO	HP-75T	65	68	2.5
26[††]	Archer	15-851	39	44	2.0
27	Archer	15-851	42	44	2.4
28	Antsco	MT-11CB	22	24	3.1[*]
29	Caltronic	CB/TV Noise Filter	0.9	0.9	0.7
30	Channel Master	0211	16	18	0.6[**]
31	Blonder Tongue	FR-CB-300	21	35	2.0
32	J. W. Miller	C-513-T1	69	69	3.9[***]
33	J. W. Miller	C-513-T3	62	62	2.3

(continued at top of next page)

Graph No.	Manufacturer	Model No.	1.8-29.7 MHz (Ham)	Attenuation 26.9-27.5 MHz (CB)	54-216 MHz (TV)
			More = Better	More = Better	Less = Better

High-pass Filters

Graph No.	Manufacturer	Model No.	1.8-29.7 MHz (Ham)	26.9-27.5 MHz (CB)	54-216 MHz (TV)
34	J. W. Miller	C-510-T	22	23	1.4
35	Marine Technology	EMI-TV300	35	37	8.0
36	R. L. Drake	TV-300-HP	38	40	1.3
37	R. L. Drake	TV-300-HP	39	38	1.3
38	Unknown	Unknown	42	45	2.8
39	Vanco	TV-1	12	14	1.0
40	Vecor	922	24	26	1.3
41	Workman	TV-5	14	16	0.6
42	AMECO	HP-300T	63	>70	3.8

† nos. 13-25 for coax; †† nos. 26-42 for twin lead

Power-line Filters

Graph No.	Manufacturer	Model No.	More = Better	More = Better
43	Archer	15-1110	32	38
44	Archer	15-1110	8	—
45	Cornell Dubilier	CBAPF	14	29
46	Cornell Dubilier	NF 10286-1	20	21
47	Cornell Dubilier	NF 1A 364-3	45	45
48	J. W. Miller	7811	8	8
49	J. W. Miller	C-508-L	44	44
50	J. W. Miller	C-508-L	33	—
51	J. W. Miller	C-509-L	37	—
52	J. W. Miller	C-509-L	39	39
53	J. W. Miller	C-517-L1	27	—
54	Unknown	Unknown	7	7
55	Marine Technology	EMI-120 V	8	32

Miscellaneous Filters

Graph No.	Manufacturer	Model No.
56	J. W. Miller	C-506-R
57	J. W. Miller	C-505-R
58	Winegard	T-FM7
59	Winegard	T-FM3
60	Winegard	TR3-2FM
61	Winegard	FT-760
62	Radio Shack	273-104
63	TCE Labs	BX #2-33
64	TCE Labs	TP-12
65	Samas Telecom	SNI/CB-HAM
66	Samas Telecom	SNI/CB-HAM
67	Samas Telecom	000-143-826
68	Radio Shack	272-1085
69	Radio Shack	270-055
70	Radio Shack	270-030A

NOTES: All attenuation figures are worst case for the indicated frequency ranges.

*Includes losses in test setup. This filter rejects 6-m ham frequencies.

**This filter rejects FM broadcast and 2-m ham frequencies.

***Includes losses in test setup.

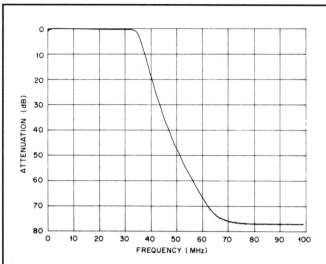

Graph 1—ARRL swept frequency-response plot of the Barker & Williamson model no. 425 low-pass filter with a 35-MHz cutoff, designed for use in 50-Ω coaxial transmission lines.

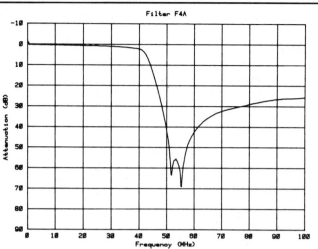

Graph 2—FCC swept frequency-response plot of the Cornell Dubilier model no. CBTVI-1 low-pass filter with a 40-MHz cutoff, designed for use in 50-Ω coaxial transmission lines.

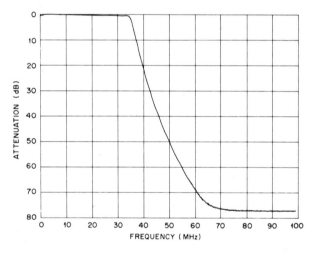

Graph 3—ARRL swept frequency-response plot of the Heath model HDP-3700 low-pass filter with a 34-MHz cutoff, designed for use in 50-Ω coaxial transmission lines.

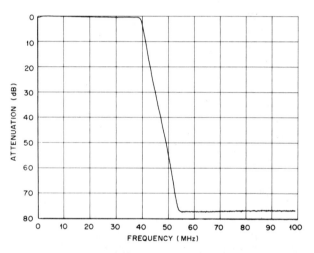

Graph 4—ARRL swept frequency-response plot of the J. W. Miller model no. C-514-T low-pass filter with a 40-MHz cutoff, designed for use in 50-Ω coaxial transmission lines.

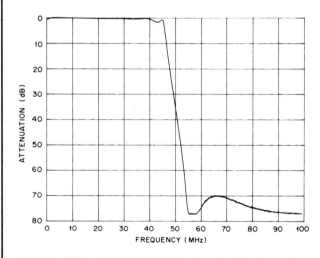

Graph 5—ARRL swept frequency-response plot of the Microwave Filter Co "Interfilter" model low-pass filter with a 45-MHz cutoff, designed for use in 50-Ω coaxial transmission lines.

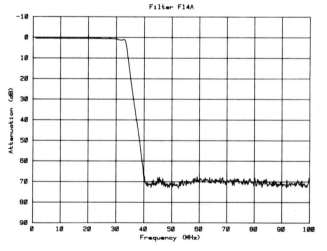

Graph 6—FCC swept frequency-response plot of the R. L. Drake model no. TV-3300-LP low-pass filter with a 30-MHz cutoff, designed for use in 50-Ω coaxial transmission lines.

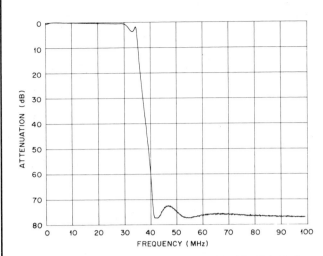

Graph 7—ARRL swept frequency-response plot of the R. L. Drake model no. TV-3300-LP low-pass filter with a 30-MHz cutoff, designed for use in 50-Ω coaxial transmission lines.

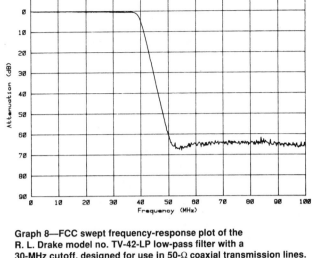

Graph 8—FCC swept frequency-response plot of the R. L. Drake model no. TV-42-LP low-pass filter with a 30-MHz cutoff, designed for use in 50-Ω coaxial transmission lines.

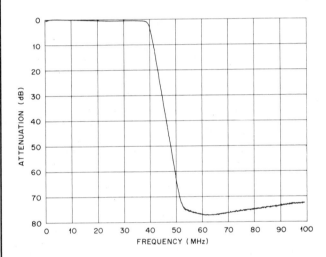

Graph 9—ARRL swept frequency-response plot of the R. L. Drake model no. TV-42-LP low-pass filter with a 30-MHz cutoff, designed for use in 50-Ω coaxial transmission lines.

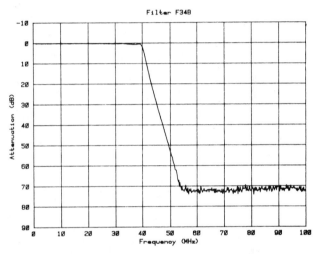

Graph 10—FCC swept frequency-response plot of a generic (not marked, manufacturer unknown) low-pass filter with a 40-MHz cutoff, designed for use in 50-Ω coaxial transmission lines.

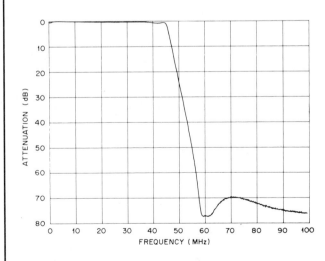

Graph 11—ARRL swept frequency-response plot of the Wm. M. Nye Co. model no. 250-0020-001 low-pass filter with a 46-MHz cutoff, designed for use in 50-Ω coaxial transmission lines.

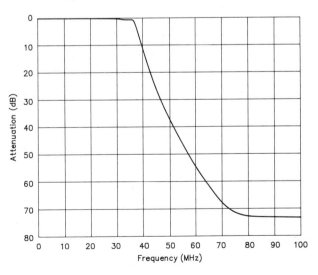

Graph 12—ARRL swept frequency-response plot of the Ten-Tec model 5061 low-pass filter with a 30-MHz cutoff, designed for use in 50-Ω coaxial transmission lines.

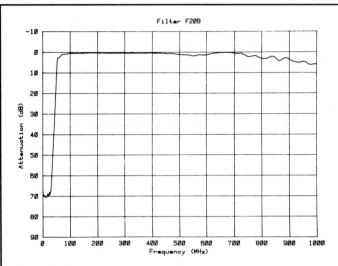

Graph 13—FCC swept frequency-response plot of the Archer model no. 15-580 high-pass filter with a 50-MHz cutoff, designed for use in 75-Ω coaxial transmission lines.

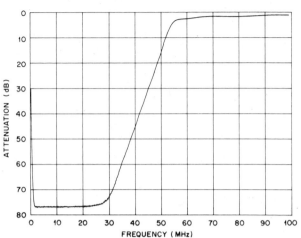

Graph 14—ARRL swept frequency-response plot of the Archer model no. 15-580 high-pass filter with a 50-MHz cutoff, designed for use in 75-Ω coaxial transmission lines.

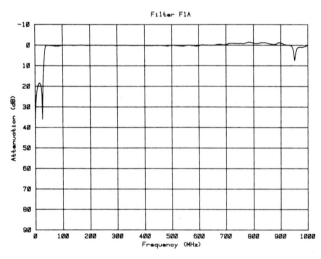

Graph 15—FCC swept frequency-response plot of the Blonder Tongue model no. FR-CB-75 high-pass filter with a 40-MHz cutoff, designed for use in 75-Ω coaxial transmission lines.

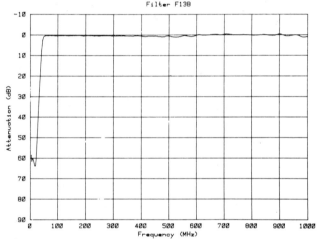

Graph 16—FCC swept frequency-response plot of the Channel Master model no. 7203 high-pass filter with a 50-MHz cutoff, designed for use in 75-Ω coaxial transmission lines.

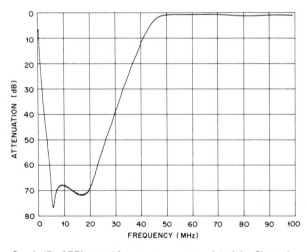

Graph 17—ARRL swept frequency-response plot of the Channel Master model no. 7203 high-pass filter with a 50-MHz cutoff, designed for use in 75-Ω coaxial transmission lines.

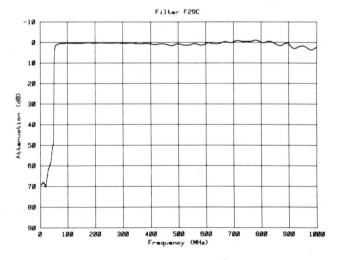

Graph 18—FCC swept frequency-response plot of the J. W. Miller model no. C-513-2 high-pass filter with a 50-MHz cutoff, designed for use in 75-Ω coaxial transmission lines.

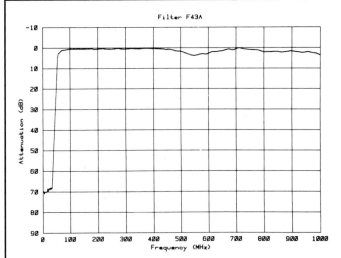

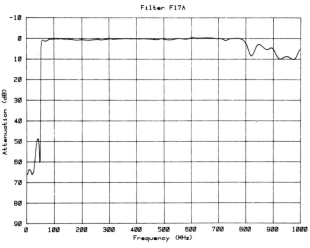

Graph 19—FCC swept frequency-response plot of the Marine Technology model no. EMI-TV75 high-pass filter with a 54-MHz cutoff, designed for use in 75-Ω coaxial transmission lines.

Graph 20—FCC swept frequency-response plot of the R. L. Drake model no. TV-75-HP high-pass filter with a 50-MHz cutoff, designed for use in 75-Ω coaxial transmission lines.

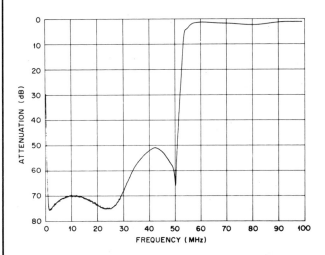

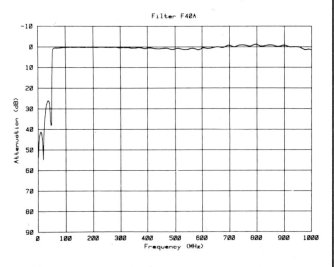

Graph 21—ARRL swept frequency-response plot of the R. L. Drake model no. TV-75-HP high-pass filter with a 50-MHz cutoff, designed for use in 75-Ω coaxial transmission lines.

Graph 22—FCC swept frequency-response plot of the RMS Electronics model no. CA-2700 high-pass filter with a 54-MHz cutoff, designed for use in 75-Ω coaxial transmission lines.

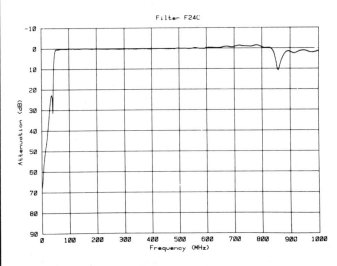

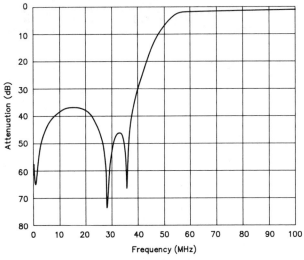

Graph 23—FCC swept frequency-response plot of the Winegard model no. HP 2700 high-pass filter with a 50-MHz cutoff, designed for use in 75-Ω coaxial transmission lines.

Graph 24—ARRL swept frequency-response plot of the Ten-Tec model no. 5060 high-pass filter with a 40-MHz cutoff, designed for use in 75-Ω coaxial transmission lines.

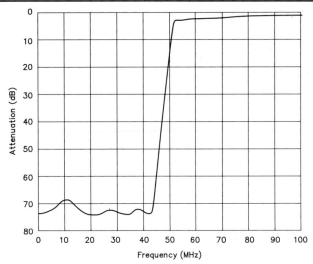

Graph 25—ARRL swept frequency-response plot of the AMECO model no. HP-75T high-pass filter designed to attenuate 0-52 MHz, for use in 75-Ω coaxial transmission lines.

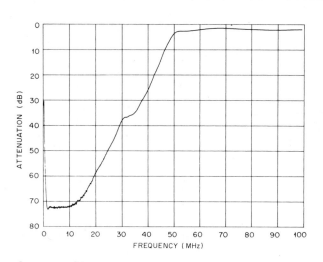

Graph 26—ARRL swept frequency-response plot of the Archer model no. 15-851 high-pass filter with a 50-MHz cutoff, designed for use in 300-Ω balanced transmission lines.

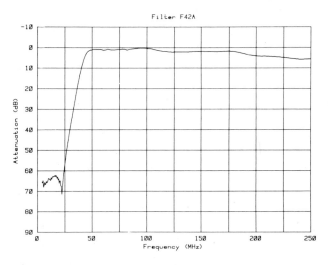

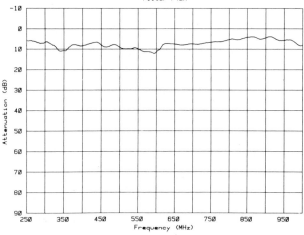

Graph 27—FCC swept frequency-response plot of the Archer model no. 15-851 high-pass filter with a 50-MHz cutoff, designed for use in 300-Ω balanced transmission lines.

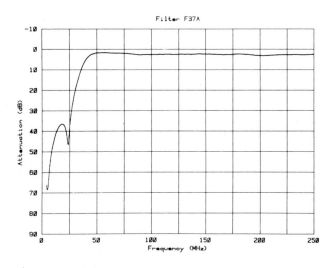

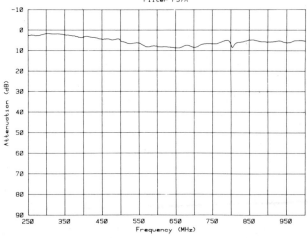

Graph 28—FCC swept frequency-response plot of the Antsco model no. MT-11CB combination balun (300-Ω balanced to 75-Ω unbalanced) and high-pass filter with a 50-MHz cutoff. Since the impedance-matching devices used in testing this filter were different, their insertion loss could not be factored out of the plotted data. Therefore, the insertion loss shown is the loss of the filter, plus the loss of a 300-Ω to 50-Ω balun, plus the loss of a 75-Ω to 50-Ω transformer.

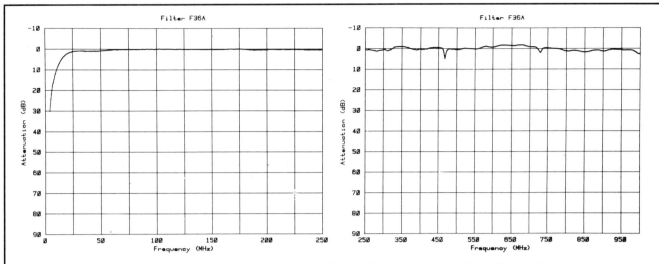

Graph 29—FCC swept frequency-response plot of the Caltronics CB/TV Noise Filter model high-pass filter with a 50-MHz cutoff, designed for use in 300-Ω balanced transmission lines.

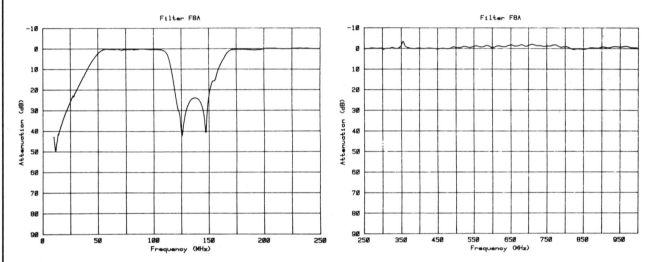

Graph 30—FCC swept frequency-response plot of the Channel Master model no. 0211 filter for use in 300-Ω transmission lines. Called a sub-mid band filter by the manufacturer, it is designed to reject signals below 50 MHz and in the band from 110 to 170 MHz.

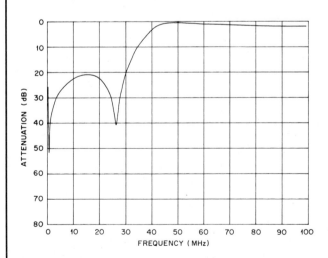

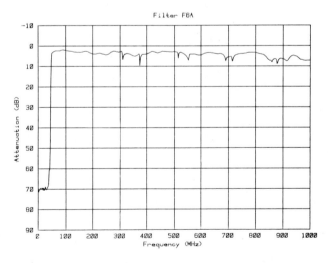

Graph 31—ARRL swept frequency-response plot of the Blonder Tongue model no. FR-CB-300 high-pass filter with a 45-MHz cutoff and a notch at 26.5 MHz, designed for use in 300-Ω balanced transmission lines. The notch is designed to provide additional attenuation in the Citizens Band.

Graph 32—FCC swept frequency-response plot of the J. W. Miller model no. C-513-T1 combination balun (300-Ω balanced to 75-Ω unbalanced) and high-pass filter with a 50-MHz cutoff. Since the impedance-matching devices used in testing this filter were different, their insertion loss could not be factored out of the plotted data. Therefore, the insertion loss shown is the loss of the filter, plus the loss of a 300-Ω to 50-Ω balun, plus the loss of a 75-Ω to 50-Ω transformer.

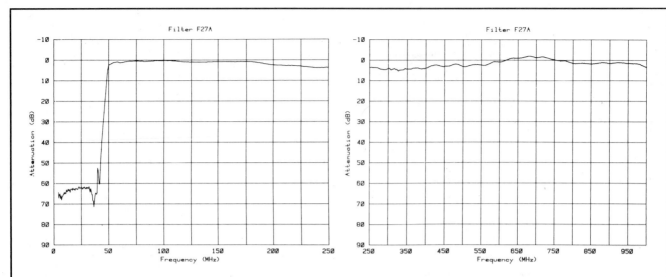

Graph 33—FCC swept frequency-response plot of the J. W. Miller model no. C-513-T3 high-pass filter with a 50-MHz cutoff, designed for use in 300-Ω balanced transmission lines.

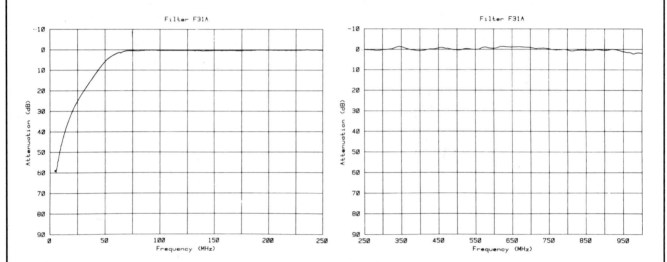

Graph 34—FCC swept frequency-response plot of the J. W. Miller model no. C-510-T high-pass filter with a 50-MHz cutoff, designed for use in 300-Ω balanced transmission lines.

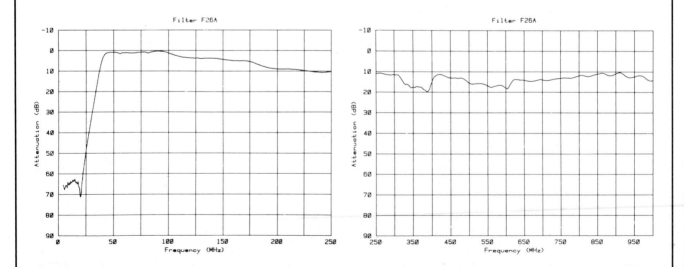

Graph 35—FCC swept frequency-response plot of the Marine Technology model no. EMI-TV300 high-pass filter with a 40-MHz cutoff, designed for use in 300-Ω balanced transmission lines.

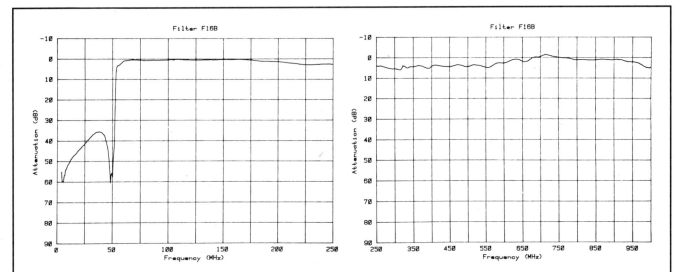

Graph 36—FCC swept frequency-response plot of the R. L. Drake model no. TV-300-HP high-pass filter with a 50-MHz cutoff, designed for use in 300-Ω balanced transmission lines.

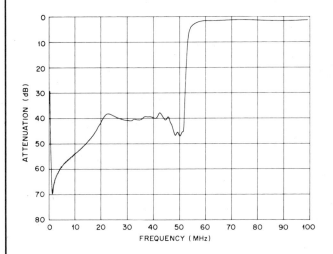

Graph 37—ARRL swept frequency-response plot of the R. L. Drake model no. TV-300-HP high-pass filter with a 50-MHz cutoff, designed for use in 300-Ω balanced transmission lines.

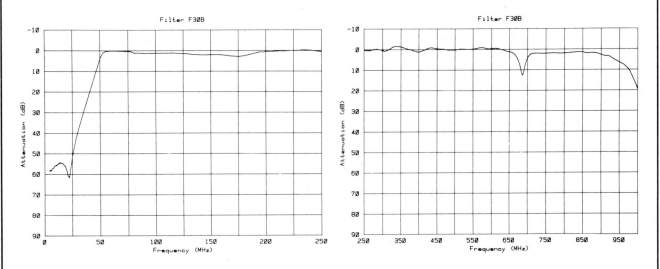

Graph 38—FCC swept frequency-response plot of a generic (not marked, manufacturer unknown) high-pass filter with a 50-MHz cutoff, designed for use in 300-Ω balanced transmission lines.

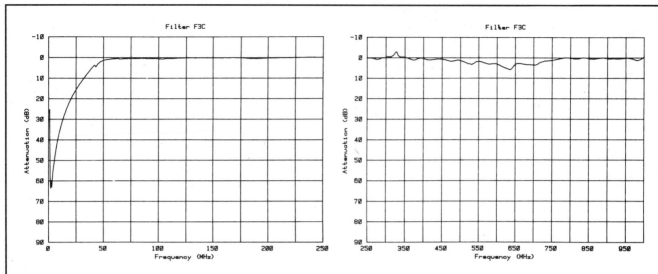

Graph 39—FCC swept frequency-response plot of the Vanco model no. TV-1 high-pass filter with a 50-MHz cutoff, designed for use in 300-Ω balanced transmission lines.

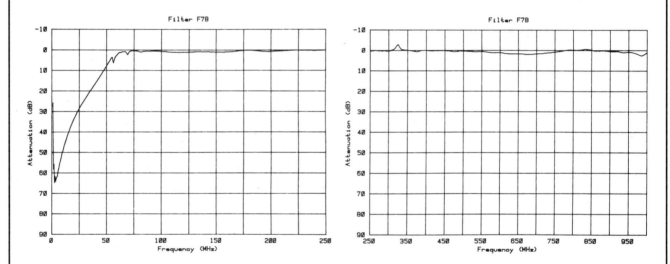

Graph 40—FCC swept frequency-response plot of the Vecor model no. 922 high-pass filter with a 50-MHz cutoff, designed for use in 300-Ω balanced transmission lines.

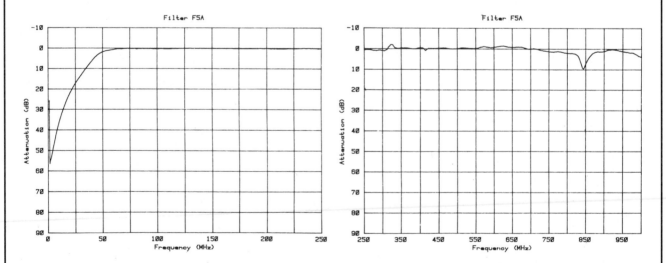

Graph 41—FCC swept frequency-response plot of the Workman model no. TV-5 high-pass filter with a 50-MHz cutoff, designed for use in 300-Ω balanced transmission lines.

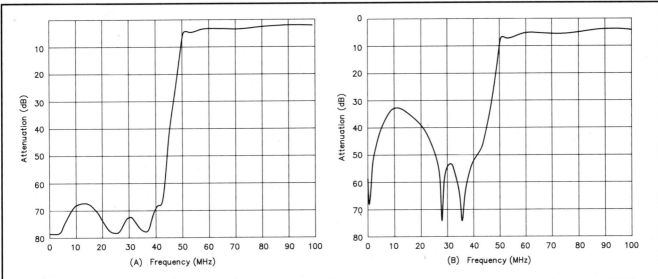

Graph 42—(A) ARRL swept frequency-response plot of the AMECO HP-300T high-pass filter designed to attenuate 0-52 MHz, for use in 300-Ω balanced transmission lines. B shows a frequency-response plot of the same HP-300T filter without the case grounded. The lack of grounding degrades performance.

Power-Line Filters

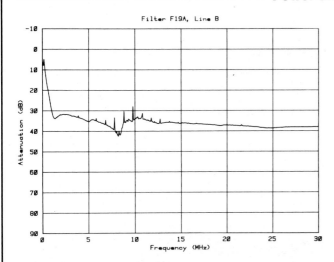

Graph 43—FCC swept frequency-response plot of the Archer model no. 15-1110 power-line filter designed for use in 120-V ac systems. This is a two-wire, no-ground, device. Line A of the LISN was grounded, and line B only was tested.

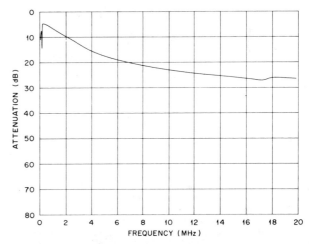

Graph 44—ARRL swept frequency-response plot of the Archer model no. 15-1110 power-line filter designed for use in 120-V ac systems. This is a two-wire, no-ground, device. Line A of the LISN was grounded, and line B only was tested.

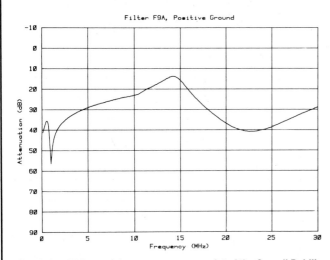

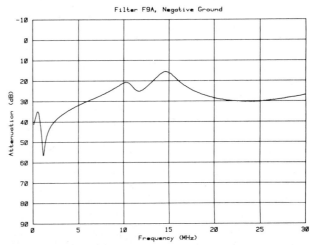

Graph 45—FCC swept frequency-response plot of the Cornell Dubilier model no. CBAPF power-line filter, designed for use in vehicles with either a negative or positive ground.

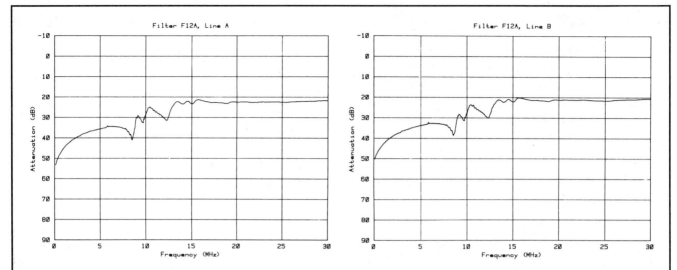

Graph 46—FCC swept frequency-response plot of the Cornell Dubilier model no. NF 10286-1 power-line filter, designed for use in 220-V ac systems.

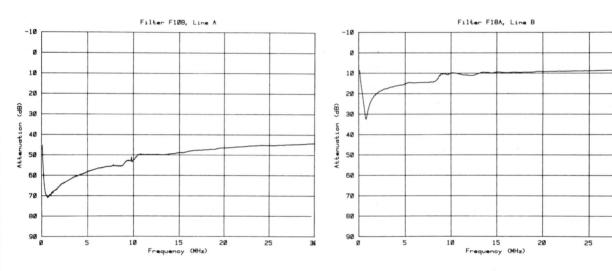

Graph 47—FCC swept frequency-response plot of the Cornell Dubilier model no. NF 1A 364-3 power-line filter designed for use in 120-V ac systems. One filter is inserted in each side of the ac line. Since the A and B lines of the LISN are identical, this filter was tested in one line only.

Graph 48—FCC swept frequency-response plot of the J. W. Miller model no. 7811 power-line filter designed for use in 120-V ac systems. This is a two-wire, no-ground, device. Line A of the LISN was grounded, and line B only was tested.

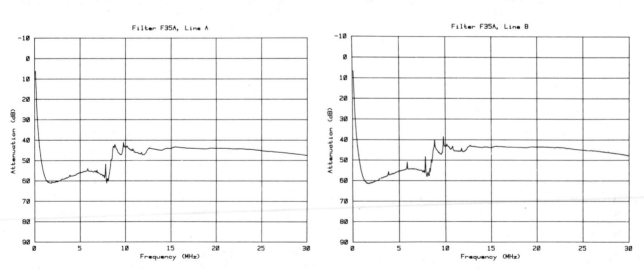

Graph 49—FCC swept frequency-response plot of the J. W. Miller model no. C-508-L power-line filter, designed for use in 120-V ac systems.

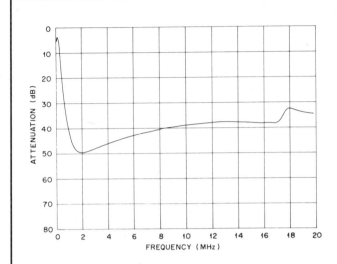

Graph 50—ARRL swept frequency-response plot of the J. W. Miller model no. C-508-L power-line filter, designed for use in 120-V ac systems. This plot is of line A.

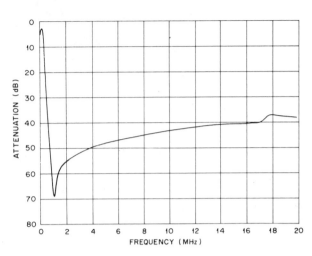

Graph 51—ARRL swept frequency-response plot of the J. W. Miller model no. C-509-L power-line filter, designed for use in 120-V ac systems. This plot is of line A.

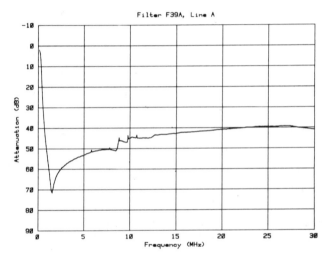

Filter F39A, Line A

Filter F39A, Line B

Graph 52—FCC swept frequency-response plot of the J. W. Miller model no. C-509-L power-line filter, designed for use in 120-V ac systems.

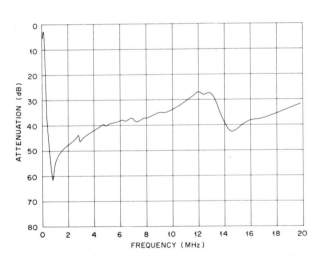

Graph 53—ARRL swept frequency-response plot of the J. W. Miller model no. C-517-L1 power-line filter, designed for use in 120-V ac systems. This plot is of line A.

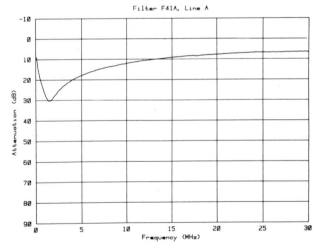

Filter F41A, Line A

Graph 54—FCC swept frequency-response plot of a generic (unmarked, manufacturer unknown) power-line filter designed for use in 120-V ac systems. This is a two-wire, no-ground, device. Line B of the LISN was grounded, and line A only was tested.

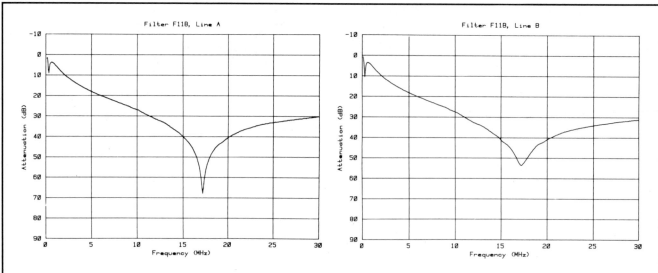

Graph 55—FCC swept frequency-response plot of the Marine Technology model no. EMI-120V power-line filter, designed for use in 120-V ac systems.

Miscellaneous Filters

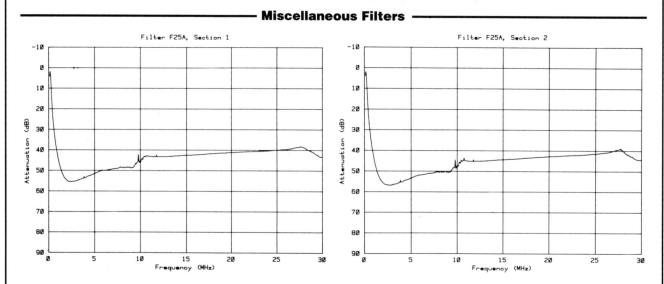

Graph 56—FCC swept frequency-response plot of the J. W. Miller model no. C-506-R RF choke, designed to be installed in the speaker leads of an audio amplifier. It is intended to isolate the amplifier from RF currents that might be picked up by the speaker leads.

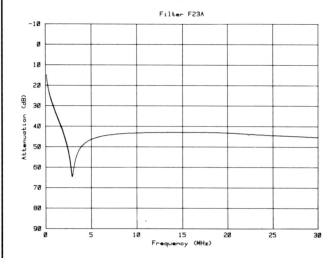

Graph 57—FCC swept frequency-response plot of the J. W. Miller model no. C-505-R coaxial RF choke, designed to be installed at the input of an audio amplifier. It is intended to isolate the amplifier from RF currents that might be picked up by the input leads.

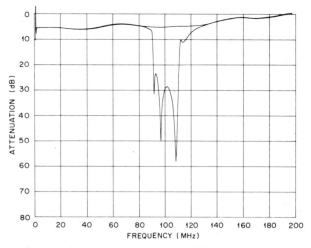

Graph 58—ARRL swept frequency-response plot of the Winegard model no. T-FM7 FM trap, designed for use in 75-Ω coaxial transmission lines. This filter is designed to suppress interference from FM broadcast stations operating between 85 and 130 MHz.

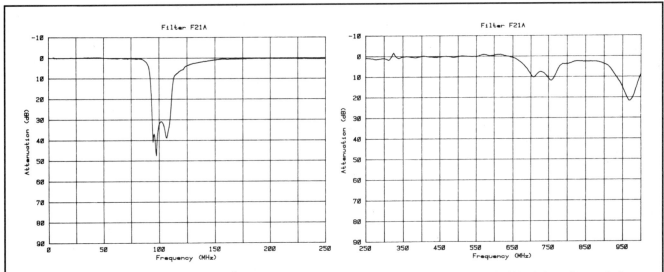

Graph 59—FCC swept frequency-response plot of the Winegard model no. T-FM3 FM trap, designed for use in 300-Ω balanced transmission lines. This filter provides a notch in the FM broadcast band, approximately 88 to 108 MHz.

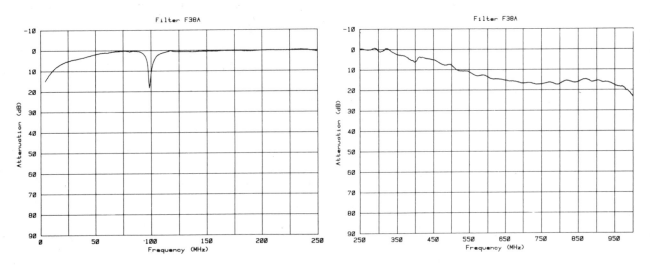

Graph 60—FCC swept frequency-response plot of the Winegard model no. TR3-2FM VHF-FM trap, designed for use in 300-Ω balanced transmission lines. This trap is designed to notch out undesired signals in the 54- to 216-MHz range.

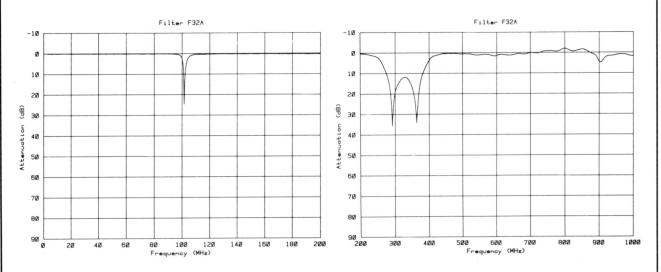

Graph 61—FCC swept frequency-response plot of the Winegard model no. FT-760 tunable FM trap, designed for use in 75-Ω coaxial transmission lines. This trap is designed to notch out one or two FM broadcast stations.

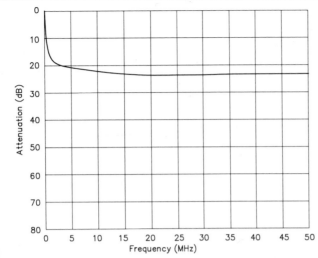

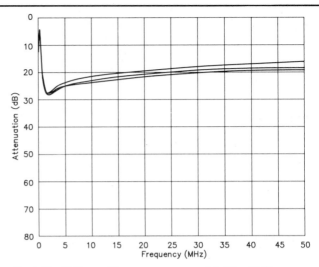

Graph 62—ARRL swept frequency-response plot of the Radio Shack 273-104 snap-on choke using two cores and 8 turns of no. 22 hookup wire. The plot shows the insertion loss in a 50-Ω system. The filter was swept with 5 W and showed no sign of saturation. Unfortunately, a load-insensitive high-power amplifier was not available for further testing.

Graph 63—ARRL swept frequency-response plot of the shield conductor of the TCE labs model BX #2-33 (serial no. 3163). The plot shows the insertion loss in a 50-Ω system. Multiple traces show the effects of varying the filter position (relative to ground). This is a common-mode filter for CATV, not a high-pass filter. The filter was swept with 5 W and showed no sign of saturation. Unfortunately, a load-insensitive high power amplifier was not available for further testing.

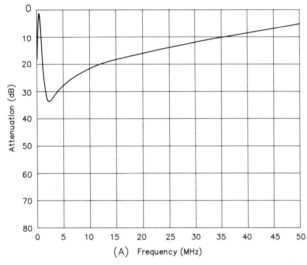

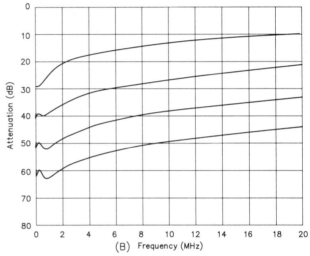

Graph 64—(A) ARRL swept frequency-plot of the red wire of the TCE Labs model TP-12 (serial no. 3125) telephone RF filter. The plot shows the insertion loss in a 50-Ω system. The green wire has a similar plot, while the black and yellow wires appear to be an open circuit for dc. Phone systems that use all four wires will not work with this filter. (B) swept frequency-plot of the red wire of the TCE Labs TP-12 at different power levels. The plot shows the insertion loss in a 50-Ω system. The top trace shows the response at +38 dBm (6 W). Lower traces show the response as power is reduced (in 10-dB increments). Ideally, the lines would be spaced evenly by 10 dB. The TCE TVI filter did not show performance loss at high power levels.

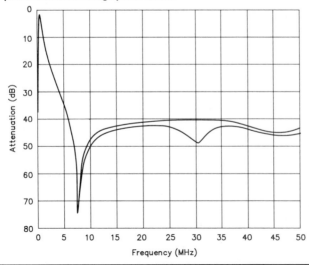

Graph 65—ARRL swept frequency-response plot of the Samas Telecom SNI/CB-HAM RFI, Noise Suppressor no. 000-218-339. Plot shows the insertion loss in a balanced 300-Ω system. A second trace shows the degraded performance of the filter when grounded, which might be considered unusual.

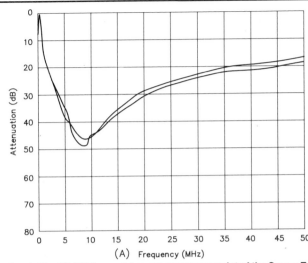

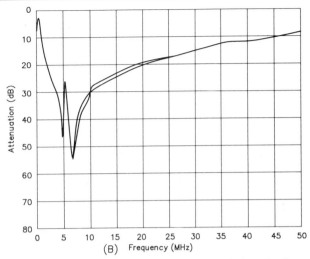

Graph 66—(A) ARRL swept frequency-response plot of the Samas Telecom SNI/CB-HAM RFI Noise Suppressor. Plot shows the insertion loss in a 50-Ω system with the unit grounded. Each side of the line is measured separately with the other floating. (B) frequency-response plot of the Samas Telecom SNI/CB-HAM RFI Noise Suppressor used without a ground in the same 50-Ω system.

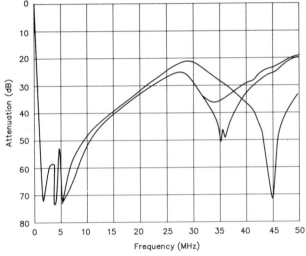

Graph 67—ARRL swept frequency response plot of the Samas Telecom RFI Noise Suppressor no. 000-143-826. Plot shows the insertion loss in a 300-Ω balanced system. Different traces show the effect of lead dress on the VHF performance (as one might expect) of a filter featuring band-selector switches for optimizing AM broadcast rejection.

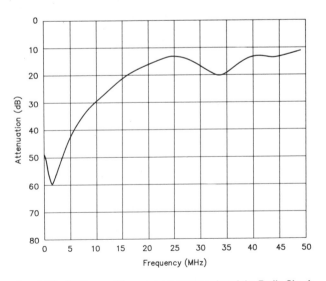

Graph 68—ARRL swept frequency-response plot of the Radio Shack 272-1085 noise filter capacitor (for automobile systems). The plot shows the insertion loss in a 50-Ω system.

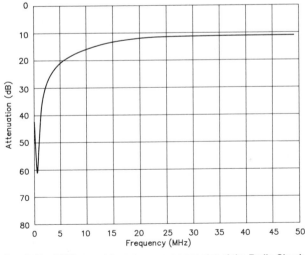

Graph 69—ARRL swept frequency-response plot of the Radio Shack 270-055 20-A, heavy-duty automotive noise filter. The plot shows insertion loss in a 50-Ω system. The filtering is primarily at audio rather than radio frequencies.

Graph 70—ARRL swept frequency-response plot of the Radio Shack 270-030A automotive 4-A noise filter (choke and capacitor). The plot shows the insertion loss in a 50-Ω system.

Suppliers List

This list gives address details for suppliers mentioned in this book. The list includes both retailers and manufacturers. Some of the companies listed may not be willing to sell small quantities to individuals. Use these addresses to make contact for product information and to inquire about retailers in your area.

3M Electrical Specialties
6801 Riverplace Blvd
Austin, TX 78726
800-233-3636

Amidon Associates
PO Box 956
Torrance, CA 90508
213-763-5770
fax 213-763-2250

Bell Industries
J. W. Miller Division
306 E. Alondra Blvd
PO Box 2859
Gardena, CA 90247-1059
213-515-1720
fax 213-515-1962

CoilCraft
1102 Silver Lake Rd
Cary, IL 60013
708-639-2361

Corcom, Inc
1600 Winchester Rd
Libertyville, IL 60048
708-680-7400
fax 708-680-8169

Cornell-Dubilier Electronics (CDE)
1605 Rodney French Blvd
New Bedford, MA 02744
508-996-8561

Digi-key Corp
701 Brooks Ave S
PO Box 677
Thief River Falls, MN 56701
800-344-4539

Fair-Rite Products Corp.
PO Box J
Wallkill, NY 12589
914-895-2055

GC-Thorsen
1801 Morgan St
PO Box 1209
Rockford, IL 61102-1209
815-968-9661
fax 815-968-9731

Instrument Specialties Co
PO Box A
Delaware Water Gap, PA 18327-0136
717-424-8510 east
714-579-7100 west

K-COM
Box 82
Randolph, OH 44265
216-325-2110

MFJ Enterprises
PO Box 494
Mississippi State, MS 39762
601-323-5869

Mouser Electronics
2401 Hwy 287 N
Mansfield, TX 76063
800-346-6873

Palomar Engineers
PO Box 455
Escondido, CA 92033
619-747-3343

RADIOKIT
PO Box 973
Pelham, NH 03076
603-635-2235
Telex: 887 697

Radio Shack
1100 One Tandy Center
Fort Worth, TX 76102

Richardson Electronics, Ltd
40W267 Keslinger Rd
La Fox, IL 60147
800-323-1770
fax 708-208-2550

Samas Telecom, Inc
3425-F Pomona Blvd
Pomona, CA 91768
714-598-0250
fax 714-594-6212

Schaffner EMC, Inc
9-B Fadem Rd
Springfield, NJ 07081
201-379-7778
fax 201-379-1151

Sprague Electric Co
Distribution Division
41 Hampden Rd
PO Box 9102
Mansfield, MA 02048-9102
508-339-8900

TCE Laboratories
RR 9, Box 243D
New Braunfels, TX 78133
800-545-5884

Tech Spray
PO Box 949
Amarillo, TX 79105
806-372-8523

Index

Notes

Notes

RADIO FREQUENCY
INTERFERENCE

PROOF OF
PURCHASE

FEEDBACK

Please use this form to give us your comments on this book and what you'd like to see in future editions.

Where did you purchase this book? □ From ARRL directly □ From an ARRL dealer

Is there a dealer who carries
ARRL publications within: □ 5 miles □ 15 miles □ 30 miles of your location? □ Not sure.

License class:

□ Novice □ Technician □ Technician with HF privileges □ General □ Advanced □ Extra

Name

_____ Call sign _____

Address _____

City, State/Province, ZIP/Postal Code _____

Daytime Phone (   ) _____ Age _____

If licensed, how long? _____ ARRL member? □ Yes □ No

Other hobbies _____

_____

Occupation _____

From _____

_____

_____

EDITOR, RADIO FREQUENCY INTERFERENCE
AMERICAN RADIO RELAY LEAGUE
225 MAIN ST
NEWINGTON CT 06111-1494

····· please fold and tape ·····